Marx, Engels, and Marxisms

The Marx renaissance is underway on a global scale. Wherever the critique of capitalism re-emerges, there is an intellectual and political demand for new, critical engagements with Marxism. The peer-reviewed series Marx, Engels and Marxisms (edited by Marcello Musto & Terrell Carver, with Babak Amini, Francesca Antonini, Paula Rauhala & Kohei Saito as Assistant Editors) publishes monographs, edited volumes, critical editions, reprints of old texts, as well as translations of books already published in other languages. Our volumes come from a wide range of political perspectives, subject matters, academic disciplines and geographical areas, producing an eclectic and informative collection that appeals to a diverse and international audience. Our main areas of focus include: the oeuvre of Marx and Engels, Marxist authors and traditions of the 19th and 20th centuries, labour and social movements, Marxist analyses of contemporary issues, and reception of Marxism in the world.

Salvatore Engel-Di Mauro

Ecosocialism

An Introduction

palgrave
macmillan

Salvatore Engel-Di Mauro
Geography and Environmental
Studies
State University of New York
New Paltz, NY, USA

ISSN 2524-7123 ISSN 2524-7131 (electronic)
Marx, Engels, and Marxisms
ISBN 978-3-031-74287-3 ISBN 978-3-031-74288-0 (eBook)
https://doi.org/10.1007/978-3-031-74288-0

Cover credit: Liz Couldwell & Susan Doyle/Getty Images

This Palgrave Macmillan imprint is published by the registered company Springer Nature Switzerland AG
The registered company address is: Gewerbestrasse 11, 6330 Cham, Switzerland

Contents

CHAPTER 1

Introduction

Global Environmental Impacts

Deleterious environmental transformations have turned planetary for now several centuries. There can still in part be a recovery from some of the destructive impacts, like deforestation, air pollution, and acid rain. Some, though, are irreversible (at least within our time scales), like biodiversity decline (species extinctions) and radiative forcing leading to global climate change.

The number of species going extinct is unprecedented in the planet's history in terms of speed and causal mechanisms, amply meriting the label of Sixth Mass Extinction (Ceballos et al. 2017; Meena et al. 2020; Sánchez-Bayo and Wyckhuys 2019; Şekercioğlu et al. 2012; WWF 2020). The main direct process leading to the high extinction rates is habitat loss, which results from many kinds of impacts, such as wetland drainage, permanent deforestation, pollution with organic (e.g., dioxins, herbicides) and inorganic (e.g., radionuclides, heavy metals) substances, and the diverse regional repercussions of global climate change. Clear-cutting can be tantamount to extinguishing tree species whose spatial extent is limited, such as in the often highly biodiverse Tropics. The loss of forests also means a potential acceleration of soil erosion and a net transfer of carbon into the atmosphere, instead of it being stored in soils and trees. More soil erosion has led to choking of rivers, lakes, and coastal waters with often nutrient-rich sediment (eutrophication), frequently resulting in

S. Engel-Di Mauro, *Ecosocialism*, Marx, Engels, and Marxisms,
https://doi.org/10.1007/978-3-031-74288-0_1

dead zones (mass death of most aquatic organisms) from oxygen depletion due to a sharp rise in mainly algae. Eutrophication also increases greenhouse gas emissions, a main culprit of global climate change (Li et al. 2021).

Concern over global climate change has since the early 1990s overshadowed all other environmental issues once mainstream media finally gave it prominence. Within the atmospheric and meteorological sciences, the globally climate-altering effects of greenhouse gas output into the atmosphere was studied and debated since at least the 1950s. By the end of the 1970s, the connection was verified independently by scientific experts in multiple countries (Weart 2008, 22–37; Zillman 2009). Since the 1850s, with the increasing shift to fossil-fuel energy at first in a few imperialist and settler colonial countries, anthropogenic greenhouse gas emissions have risen to unprecedented levels and have kept increasing, with CO_2 emissions in particular spiralling upwards from 5.6 to 36.3 Gt from 1950 to 2021 (IEA 2022). The consequences of such profligacy have been more frequent extreme weather events, the melting and disappearance of glaciers, rising sea-levels, and submersion of coastlines and entire islands, destroying the lives and livelihoods of millions of people especially in Tropical zones, where greenhouse gas emissions have been minute to negligible compared to the now wealthiest countries (Herring et al. 2020). Perhaps a better term would be meteorological volatility to describe the ever more rapid and extreme weather swings associated with increasing global average surface temperature (global warming). The combined effects of global climate change are also directly deleterious to human health (Haines and Patz 2004; Jacob and Winner 2009) and food production (Arora 2019; Malhi et al. 2021) in highly unequal ways because of structural social inequalities.

Ecologically nefarious effects from human activity are nothing new, but they used to be infrequent, low magnitude, relatively constrained in terms of ecosystem ramifications, and mainly circumscribed to small areas. Even the height of Roman Empire lead emissions, which raised atmospheric levels tenfold within parts of Europe, is a fiftieth of those that occurred just within the last century (McConnell et al. 2018; Preunkert et al 2019). It is the past half millennium, that is, the mainly Western European genocidal and slavery-based invasions of the Americas that marks the beginning of what has come to be known as the ecological crisis, intensified by a steep general spike in environmental damage since the 1950s in many parts of the world.

Global deforestation offers one way to show how recent the environmental crisis really is. Up to a third or two thousand million hectares of forested areas is estimated to have been lost worldwide, from roughly six to four thousand million hectares, equal to a change from 57 to 38% of land surface cover (Ritchie 2002; Williams 2003). But this masks tremendous differences over time and space and prevents any clarity about causes. In the Mayan city states between 400 and 1000 CE, for example, people mainly conserved forests while exerting much environmental pressure through urban expansion (McNeil et al. 2010). In the European region, it took about 2000 years to arrive at substantive deforestation, between 1000 BCE and 1000 CE (Kaplan et al. 2009). Compare this to 90% old growth forest loss within a handful of centuries in North America resulting from European settler colonialism. In China, there were thousands of years of development, spread, and intensification of farming that eventually led to major forest decline, but the fastest and greatest drop was over the 1750–1950 period, from 25% to 5–10% of total land-cover area (Marks 2017, 287–288). Globally, the bulk of the damage, more than 1.5 of the two thousand million hectares from 52 to 38% happened since 1750. That is, it is linked to a drastic shift in social structure in a few places, mainly in the European region, where demand for and consumption of wood and other forest resources started to skyrocket and soon violently engulfed the rest of the world.

The kinds of impact occurring especially since the late 1800s to early 1900s are also unique in Earth's history. For example, there are tens of thousands of chemical compounds that have been developed and released historically from a handful of countries and without much of any scrutiny. What is more, they have never existed on this planet before, such as stratospheric ozone-layer impairing chlorofluorocarbons. It is virtually impossible not only to foresee or keep track of the long-term effects of each new chemical, but to get a grasp of the ecological and human health effects from the thousands of possible interactions among those compounds and between those chemicals and other pre-existing ones, whether prior to the existence of humans or from past human activities. Yet plastics, persistent toxic substances (PCBs, lead, etc.), oil spills, radioactive waste, and much poisonous else keep being unleashed. These are just a few samplings of the vast differences between impacts in the recent and remote past and among different kinds of society.

The forms of environmental change wrought by recent kinds of human impact intimately linked to capitalist ways are interrelated and mutually transforming. For example, combined deforestation, wetland drainage, and high-magnitude fires, which can also be regionalised effects of global climate change, have climate-altering, soil erosive, air pollution, and habitat-destructive implications that, in turn, can shape global climate, dampening or magnifying trends in global average temperatures. More recently, it has been discovered that high-magnitude wildfires, whose greater potential frequency is related to climate change (Di Virgilio et al. 2019), are undermining the stratospheric ozone layer's still very gradual recovery. The rebuilding of stratospheric ozone is an outcome of the Montréal Protocol's implementation. It led to phasing out emissions of ozone formation disrupting compounds except for the highly impacting and simultaneously global climate-altering nitrous oxides (mainly from agrochemicals-intensive farming and, to a significant degree, nylon production; see Tian et al. 2020), whose reactivity with ozone is ironically moderated by carbon dioxide emissions (Stolarski et al. 2015). Stratosphere-reaching organic substances released through wildfires absorb chloride-bearing compounds, such as hydrogen chloride, and catalyse the formation of highly reactive chlorine molecules that break up ozone (Solomon et al. 2023). These are but a couple of ways to demonstrate that even those environmental phenomena (including destructive ones) whose causes are ascertained cannot be grasped in terms of direct input–output correspondence or linear cause-effect relations. There are multiple nonhuman interconnections and mutually shaping interactions triggered by environmental impacts from human activities (see, for example, Wallace et al. 2018). As this is not a book on the complexity of environmental change in relation to human activities and social systems, nor a compendium of environmental impacts through human activities, the few above-described examples must suffice and without any deep exploration of intricacies and uncertainties. They, in any case, take nothing away from the effective causes of the numerous, cumulative, and planetary-scale environmental transformations deleterious to ever more people's lives.

Socio-Ecological Crisis

The unprecedented and novel nature of this recent form of impact from within a single species has motivated many to describe the predicament with words like environmental or ecological crisis and, more recently, the Anthropocene. Naming a process is certainly important to making it easier to talk about it, but the identifier used can hide actual causes or paper over fundamental issues. A small fraction of the human species has in fact had the political-economic power and social influence to induce environmental practices that have eventually wreaked planetary havoc, which has been offloaded to health-impairing and deadly effect mainly on the least powerful. Terms like the Anthropocene and ecological crisis miss the mark quite a bit on this score (Foster and Clark 2021; Malm and Hornborg 2014).

Using a term that identifies the main causes of current world environmental degradation, including what some have started to call climate chaos (Isla 2018; Schwartzman and Schwartzman 2019), clarifies immediately what is at stake and is certainly preferable to whitewashing words like the Anthropocene, on which more is stated below (see also Barry and Maslin 2016; Crutzen et al 2013; Finney and Edwards 2016; Foster and Clark 2021; Luciano 2022). But in this work, I opt to describe the issue as a socio-ecological crisis. This wording is intended to articulate what others have already pointed out. Wallis (2018), for instance, understands the current crisis as simultaneously capitalist and environmental. There is a slight difference in what socio-ecological denotes here, however. Instances of degradation and devastation from localised to planetary scale are caused socially (generating destructive human impacts) and expressed ecologically, in that the repercussions and processes involve many organisms, including people, and physical processes. The ecological consequences, in other words, are simultaneously social but also have social, specifically capitalist causes. For example, the inherently war-prone feature of capitalism and its ever-greater intensification and duration in socially and environmentally devastating effects reveal an aspect of the socio-ecological nature of the crisis.

To elaborate on this further, the crisis is social in that the causes of specifically human impacts leading to negative environmental and ecological changes are generated by social relations. It is in the differences of how humans treat each other that one can identify the sources of human impacts. For example, many especially in countries with high per capita Gross Domestic Product (GDP) rely on household cleaning products that are often toxic and involve production systems that generate pollution even prior to the household cleaning products being used. Those scrubbing sinks or toilets in their homes, oftentimes women, have no say over what products are made available and how they are made, much less how the containers and discarded substances are treated after use. They are also often expected to clean the home (sometimes by means of threats of or actual patriarchal violence) and are bombarded with advertisements trying to convince them of the effectiveness or appropriateness of such chemicals. The most widely available products are not ecologically sustainable, either in content or in terms of the product's life cycle.

Alternatives exist of substances that are much less harmful if not ecologically sounder, like vinegar, lemon juice, borax, and rhubarb, but how those alternatives are produced or distributed is not up to the mostly racially minoritised women who would be using them. Even awareness of their existence and then finding the time to learn how to use them effectively is hindered by working-time constraints, lack of access to information, and much else that increases the effort necessary to shift to ecologically more sustainable cleaning products, which also tend to be made more expensive when they are sold in readily usable form. There are also many kinds of substances in many household items (e.g., oils, tars, battery acids) that require often toxic industrial grade chemicals for neutralisation or dilution to safe concentrations.

Another example is individualised car travel. The reasons why many people drive cars are hardly for enjoyment. People are often compelled to drive motorised vehicles because of how the society in which they live is structured, where public transport is made unavailable, or workers are forced to live far from workplaces because they cannot afford nearby housing, or frequent car driving is tied to toxic forms of masculinity.

In both cases of cleaning products and car driving, primacy is not people's health or impacts on ecosystems. It is efficiency in mainly women's housework to provide a clean enough environment for workers. The result is exposure to toxic substances for mostly unpaid houseworkers and all the communities near the industries extracting, processing, and putting together the raw materials to make the household cleaning equipment. This is not to mention all the polluting activities and resource consumption involved in storing, transporting, and selling such cleaning products. Relative to individualised car use, it is the rapidity of travel mainly to get to work, to shop, to facilitate if not expedite the completion of work for employers and sales of commodities. The outcome is diffuse air and water pollution from constant car use and the uneven exposure to pollutants from fossil-fuel extraction, refining, processing, and much else, as well as from all the raw material extraction, processing, and what is not needed to make, store, distribute, sell, use, and fix cars.

In other words, the primary objective is profitability, the accumulation of capital. Even if cars and cleaning products were made through ecologically sustainable activities and raw materials, the pressure to constantly raise the number of sales will make for ecologically unsustainable endless resource use and waste generation. If one is interested in reducing harmful human impacts on the environment, then it is imperative to change society so that all people are incentivised to use less materials and energy and have access to and can use ecologically sustainable products and transport for their everyday activities. This implies increasing social equality, which is not possible in a system of private property, where a few people influence, if not dictate how the rest of society treats environments and ecosystems.

The social is, then, one main component and effective cause of the socio-ecological crisis, in that human impacts have social causes (Kovel 2007; O'Connor 1998). The ecological part entails much more than human processes are involved. Though it is not trees that make demands on people to engage in logging operations, it is still forests being cut and trees possibly regrowing or not. Forest dynamics are not just dependent on whether logging will happen. The crisis is, then, more widely speaking, ecological. Many people are facing or dying from physical environments that are being or have been degraded relative to what we need as humans to live healthy lives or just to live at all, but it is also nonhuman organisms, often those on which we depend directly, who are being harmed or wiped out and many people suffer the consequences of that as well.

The ecological aspect implied in the concept of socio-ecological crisis is especially important if one is to avoid the delusion that changing society will be enough to solve the problem. Soil lead contamination and pollution will not go away once capitalism is overcome. Wetlands and forests that have been annihilated will not reconstitute themselves just because people are treating each other differently. So, using the term socio-ecological crisis is to call attention to and underline the fact that it is a crisis in how we relate to and are affected by the rest of an ecosystem or the planet we inhabit.[1] We are always in and active participants of

[1] Another way to describe it is in terms of mutually transforming processes at all scales that can be disastrous or constructive relative to human lives in particular; this is part of what is otherwise known as a dialectical materialist framework, about which more is written below.

ecosystems, that is, a web of interactions between many forms of life and physical processes (Engels 1876 [2010]).

Renaming this geological epoch as the Capitalocene (Moore 2017), for example, under-represents the ecological part of our lives and of the problems at hand. It implicitly reduces vast ecological and physical environmental changes to the activities of a single species, and just its capitalist fractions at that. The concept of Capitalocene is an important way of calling attention to what is whitewashed by advocates of the term Anthropocene, but the scope of Capitalocene is too narrow. Identifying the main or fundamental social cause of a socio-ecological problem is important in devising political strategies to transform society and suppress ecologically harmful human impacts, but the pinpointing of social causes cannot exhaust the range of analyses of and necessary politics to confront all the socio-ecological processes and disasters set in motion, especially those that are beyond human influence and those that have been bequeathed from decades or centuries of past impacts. This is the reason for preferring the concept of socio-ecological crisis, even while recognising the political merits of using the term Capitalocene.

Those who pay attention to what other species do and to how physical environments change, with or without humans, realise that environmental or ecological change is not just the result of what a single organism does, and certainly not reducible to the impacts of a population within a single species. How climate changes, for instance, is not just related to greenhouse gas emissions and land surface alterations (like deforestation), but also with nonhuman processes, like the tilt of the Earth's axis and solar activity, and nonhuman actors, like the level of methane emitted by microbial respiration in different soils, such as those that developed in permafrost over thousands of years and are currently melting. It is the conjuncture of nonhuman with human processes that determines what sort of environmental change is unfolding.

Climate change, for example, is a regular feature of this planet, with or without humans in it. This may at first seem a platitude (perhaps not to those who intimate, through slogans, that climate change can be stopped), but the implication is that the kind and direction of climate change or environmental change broadly is not a given. It is not predetermined by human action alone. There is at the very least a range of possible turns and it so happens, for now, that what is known about that range is that even the most benign will bring rough living to mass death to many if not most of us. The point is that there is also the possibility, however remote for now, that environmental changes, even those that are negative and due to human or rather capitalist activities, could bring about the opposite of what is currently happening or expected in future. A series of catastrophic volcanic eruptions, for instance, could revert the global warming trend and bring about disastrous cooling. One should not count on it, as the overwhelming evidence is to the contrary, that is, most of humanity is forced to face a grim future already even when they have done little to nothing to cause the devastation. But the mitigation or exacerbation of environmental degradation is not just by means of human action, but of human action in concert with the action of thousands of other organisms and with the dynamics of physical processes. The above are reasons to regard the matter as a socio-ecological crisis.

Capitalist Causes

The relationship between the recent and unprecedented destructiveness of environmental impacts and capitalist social relations should be considered self-evident to all. Consider the relationship between Gross Domestic Product (GDP) and CO_2 emissions at the global level. GDP is an estimate of sales of goods and services (monetary value) within a country over some specified duration. That is, the higher the GDP, the more energy and material resources are consumed. The intensity of resources used for each good or service made and sold may vary considerably, but an increase in selling whatever item or activity implies that more energy and matter have been used. Most of the energy consumption, about 84%, propelling the production of goods and services that manage to get sold is derived from fossil-fuel combustion, which releases greenhouse gases, mainly CO_2 (Ritchie et al. 2022; Schwartzman and Schwartzman 2019). This is the main reason that, as has long been shown, GDP and CO_2 emissions tend to rise together (Tucker 1995), as shown in Fig. 1.1 with available data ranging from 1900 to 2018. The close relationship is confirmed by a high Pearson correlation coefficient ($r = 0.866$) between the two variables.

The curve resulting from plotting the two variables could suggest that beyond 51 trillion International 1990 USD further increases in GDP lead to less steep CO_2 emission rates. But even if so, the overall trend is more, rather than less emissions over time (Fig. 1.2), making a veritable mockery of all high-level talks on climate change and any initiatives or practices attempting to curb emissions or capture and store carbon. What is more, CO_2, once airborne, takes decades to hundreds of years to

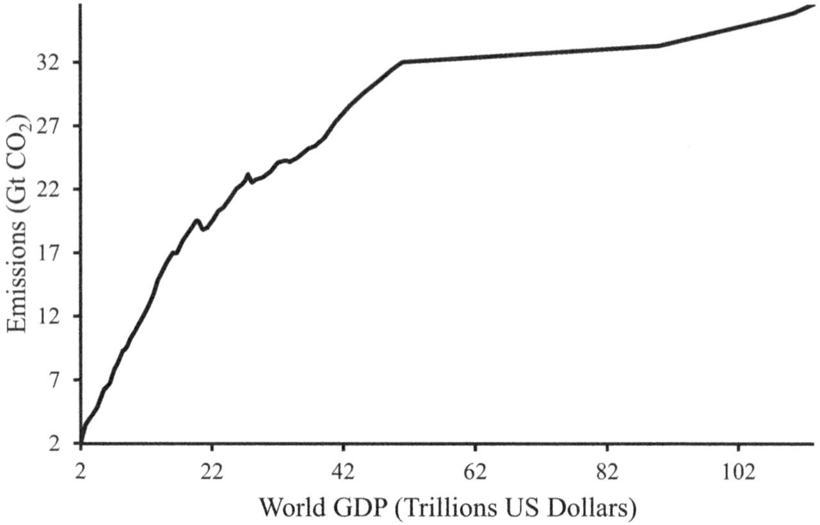

Fig. 1.1 Relationship between World GDP (1990 International US dollars) and CO_2 emissions (Gigatonnes), 1900–2018. (*Source* Ritchie et al. [2020])

be cycled out of the atmosphere. It builds up over time, and so lowering emission rates still adds to the total CO_2 in the atmosphere, intensifying radiative forcing. Cumulative emissions have risen exponentially since the 1800s (Fig. 1.3). To put it another way, atmospheric CO_2 concentrations have increased from about 283 ppm in 1800 to 419 ppm in 2022. This is the highest known level in three million years (see also Bereiter et al. 2015). As our species has only existed for about 300,000 years (Gibbons 2017), the sharp and fast rise in atmospheric CO_2 sets unprecedented challenges for this and future generations. Even with the cessation of all anthropogenic CO_2 emissions, it would take several centuries for most of emitted anthropogenic CO_2 to be cycled out of the atmosphere. This does not include all other greenhouse gases, like methane and nitrous oxide, which have tens to hundreds of times the radiative forcing effect of CO_2 (Ciais et al. 2013; NOAA 2022). Lasting damage has already been done and is being intensified.

Many nevertheless argue that switching to renewable energy sources, coupled with carbon capture technologies, will eventually turn the tide in

Fig. 1.2 CO_2 emissions (Gigatonnes), 1850–2022. (*Source* Ritchie et al. [2024])

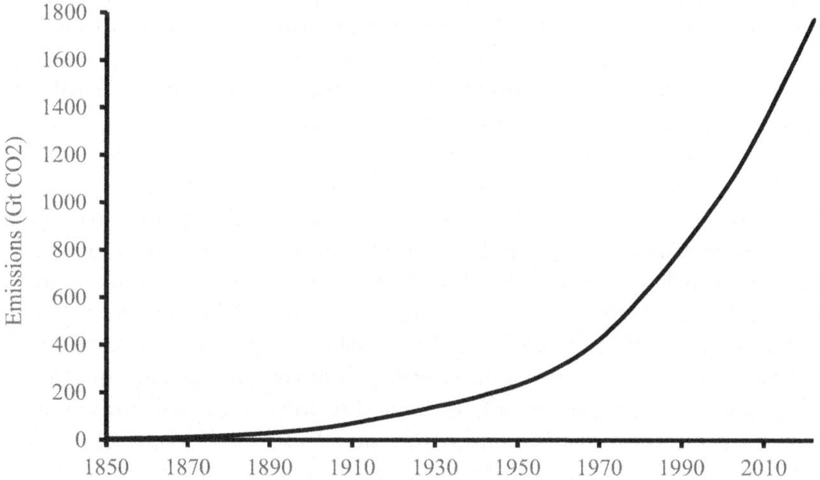

Fig. 1.3 Cumulative CO_2 emissions (Gigatonnes), 1800–2022. (*Source* Ritchie et al. [2024])

favour of reducing emissions and solve the problem. To hold any credibility at all, such a viewpoint must ignore, among other inconvenient facts, decades of net increases in environmental destruction in general, the continuing overwhelming reliance on fossil-fuel energy sources, and the manifest consistent failure to reduce greenhouse gas emissions. The fundamental problem with such sanguine reasoning (and related policies) is the failure to notice that the issue and its causes are social and involve political struggle.

To simplify, one is dealing with a kind of society, capitalism, predicated on compelling the working majority to produce or help produce (by means of unrecognised or unpaid work) ever-increasing amounts of goods and services for ever higher sales for business owners to raise or maintain profit rates (or just to remain viable) and for other owners to get profits from rent. It is a system where competition and individualism take pride of place and impede the cooperation and resource sharing needed to address problems effectively (rather than displacing them to others elsewhere or in future), especially when confronting global environmental problems.

A case in point is the chronic prevarication of businesses about the social and environmental repercussions of the products they sell. A recent instance is the knowledge that the largest fossil fuels corporations have had by the 1970s, if not earlier, regarding the global climate change implications of fossil-fuel combustion (Franta 2018). This is hardly the first time that businesses have concealed from the public crucial information about the deadliness of their products. Tobacco and asbestos companies are among the more notorious examples. All the corporations that stood to benefit from leaded petrol and leaded paint production and sale similarly denied any harmful effects to their products or feigned lack of knowledge. Such denial and active suppression of information is standard business practice, as it must be, if businesses are to thrive or survive competition. Moralising about malfeasance and grasping for individual culprits, as still in vogue in much of environmentalism, might lead to a few important changes and local solutions (e.g., emissions regulation to prevent smog) but ultimately derail efforts to comprehend and act on the effective causes. At the same time, individualising blame, or moralising (such as about population growth), is exactly the way to avoid confronting the social relations of power that promote or even compel such destructive behaviour (for a fuller discussion of this, see Kovel 2007, 26–29).

The devastating effects of doing regular business follow from the predominant ways that societies are organised and structured in a capitalist framework. The working majority make all the goods and services that enable businesses to exist but the profits from sales and the derived wealth go to the business-owning minority to do with such wealth as they see fit. The working majority (whether paid or not) may at some point become sufficiently organised and militant to succeed in getting a greater share of the wealth they produce through higher wages and state provisions (e.g., healthcare), but this tends to be temporary historically and very uneven and circumscribed geographically in favour of sections of a handful of settler colonial or imperialistic countries. The reason is generally because state institutions (often including military or paramilitary outfits) intervene on behalf of the business-owning minority to restore or enhance profit rates if there is a dent in profits (also known as "economic growth" or GDP) that undermines competitiveness or the political power of the business-owning minority.

Such a society is made of institutions prioritising profits over people's well-being, and certainly over the functioning of ecosystems, the health of other species, or the state of liveability of this planet, and anything else. Businesses must constantly make profits (expand) just to stay afloat. This spells ever more resource use. Switching energy source or technologies will not alter this fact. Technical, market-friendly, or parliamentary solutions will not address the constantly increasing rates of raw materials being extracted and consumed and the constantly increasing amounts of often toxic waste being produced and dumped. In a system based on using and often abusing workers to produce stuff endlessly for an endless expansion of sales, any energy source or renewable materials will be consumed to ecologically unsustainable and destructive levels and raise health risks mainly for the unpaid and paid working many. The destruction includes frequent war, which is eventually what happens when there are ruling economic groups constantly vying to grab resources and extract labour from others. Business and their supporting state institutions engage in countless manoeuvring to outcompete each other and to outgun ruling economic groups somewhere else for the sake of endlessly accumulating economic power over the rest of society, if not the world. War is a main way to gain the upper hand in the endless competition required for the endless accumulation of capital. It is capitalism's regular functioning that generates environmental destruction of magnitudes, scales, rates, and complexities unparalleled in any other social system and in the history of humanity. When the foundations are rotten, they must be replaced as a matter of priority. Repairing only what rests on rotten foundations is to build a self-deceiving impression of stability.

The foundational problem is not unknown even in the mainstream, where it is at least partially recognised. It is "the structures of power and decision-making that continue to promote climate inaction", as noted in a recent BBC report (Timperley 2020). In fact, one should rather speak of a very active and vigorous state repression of climate activism, seeing the criminalisation, violent treatment, and even assassination of environmental activists in liberal democracies like Italy, UK, Germany, the US, Colombia, Honduras, and Brazil (Global Witness 2020).

Scientists as well are coming around to appreciate the enormous power chasms and capitalist causes (Ceddia 2020; Wiedemann et al. 2020). A still small minority are seeing the light and identifying capitalism as the problem, even pointing to Marx for inspiration and socialism as workable alternative (Wiedemann et al. 2020). But those relatively sparse moments of enlightened understanding slip quickly into ingenuous calls for scientific interdisciplinarity, engagements with a wider public, constructive policy formulation and implementation, and other lofty ideas disconnected from socialist struggles and from the most elementary understanding of class and multiple forms of oppressive relations shaping science and the political economy of scientific work in a capitalist society (Lewontin and Levins 2007). As Marx spotted the problem long ago,

> The weak points in the abstract materialism of natural science, a materialism that excludes history and its process, are at once evident from the abstract and ideological conceptions of its spokesmen, whenever they venture beyond the bounds of their own specialty. (Marx 1867/1992, 352, fn. 2)

More than a century later, most scientists continue to evade historical processes (like settler colonialism), including those that led to the sort of material conditions enabling the science they practise, and appeal to technical or regulatory solutions based on the same environmentally and socially disastrous capitalist logic that results in the problems to be solved (Ripple et al. 2020).

They refuse to re-educate themselves on the wealth of historical experiences and resulting ideas (including in the social sciences and humanities) and novel forms of organising for a society free of domination. It should be considered embarrassing that the scientific mainstream, in a spectacular failure to apply basic scientific methodology to the study of environmental issues and their own sciences, does not even bother looking up the copious research already done on the social causes of inequalities and environmental damage. Even more disgracefully, such scientists continue the pretence of objectivity and/or neutrality all the while politically committed to liberalism, capitalism's main ideological plank (whether expressed as conservative, reformist, radical, "illiberal", etc.). These are characteristics of scientists' work in capitalist societies that, while essential to understanding biophysical processes, are inadequate or may even become an impediment to addressing environmentally or ecologically destructive practices.

What is stubbornly unacknowledged, including among most academics, is that the unprecedented rise in and forms of environmental impacts, including greenhouse gas output, started when ruling classes in several countries, mainly in Western Europe, transformed into or were replaced by capitalist ruling classes. As they imposed capitalism as the main way to organise, structure, and govern the rest of society, such ruling classes presided over the development of the most formidable imperialist and colonising projects of the day, eventually engulfing the entire globe with unheard of levels and kinds of violence, including multiple and still ongoing genocides, chattel slavery, systematic land dispossessions, ever more mass-destructive weaponry, and civilian-targeting and urbicidal warfare (Tilly 1990). The process necessitated plenty of energy supplies and eventually fossil-fuel sources suited the push for constant profits for private individual and corporate property owners and its corollary aggregate requirement of gaining and securing control over ever-greater amounts of material resources, that is, global expansionism. By the 1850s, ruling classes were adopting and turning to fossil-fuel energy as the main energy basis of economic and military power (Malm 2016).

Yet within the same society from which the most destructiveness has been unleashed, it is most obvious that its most powerful members are ultimately responsible. Or at least it should be obvious. The majority in the most ecologically destructive societies have never had a say in what and how technologies are developed and implemented and to what ends, much less in most of the rest of the world. Most people do not consume vast quantities of resources or emit copious amounts of greenhouse gas emissions. A mere 10% of humanity is linked to about half of all human-produced CO_2 emissions, a figure roughly mirrored in half of humanity correlating with about 10% of emitted CO_2 (Gore 2020). An economy that produces enormous material inequalities and, especially since the 1970s, intensifies them further, unsurprisingly yields high inequalities in resource consumption and thereby emissions. This chronic tendency to ever wider inequalities is intrinsic to capitalism. It is what makes the market "free"—that is, free to a minority and very costly, if not deadly to the many. Under a global capitalist regime, efforts at developing markets in greenhouse gas emissions (e.g., emissions trading) and carbon capturing technologies and production systems (e.g., regenerative agriculture) are resource-wasting attempts to continue with an environmentally and socially destructive system, which includes maintaining the privileges of the rich to consume at will.

Capitalism-Compliant Explanations

Among the more popular notions, especially in conventional forms of environmentalism, is that population growth is the main if not the most fundamental culprit of environmental degradation. This is a form of determinism called populationism. Most populationists do not confine themselves to demographic matters, but will explore consumption patterns, technological change, politics, and much else. What sets them apart is fixating on population growth as a cause of the socio-ecological crisis on a par with other causal processes that they (may) consider. The flaws in and politically reactionary character of populationist arguments have been exposed in numerous works (e.g., Angus and Butler 2011). Populationists presume that having more people leads inexorably to environmental degradation, ignoring the many instances of constructive ecological relations many human communities have developed, typically in non-capitalist societies. To make their case, populationists must pretend away the vast differences in political power and extreme social inequalities typifying capitalist society. In so doing, they are capitalism apologists by default. Populationists may differ over the kinds of population control measures they advocate, depending on where they stand in the range of political manifestations of capitalism (e.g., liberal, conservative, fascist), but they all ground their ideas on baseless or tendentious arguments. Furthermore, by failing to explain why human population growth itself happens, populationists can offer no solution to the alleged problem they themselves identify.

Populationism partially overlaps but also significantly diverges from those upholding capitalist technocentrism or "green" capitalism, which is a form of fundamentalism where environmental problems are viewed as the result of mismanagement, technical incompetence, and like judgements. That is to say, the problem is presented as a failure to follow what the technical experts know is the right way to do things, which never calls into question the prevailing ways in which society is structured. Alternatively, environmental destruction is regarded as the result of not putting capitalist principles into practice in the correct way. There is a tendency for a belief that environmental problems are resolvable by means of technological and/or institutional innovations, but only within a capitalist framework. As a result, they end up promoting the very factors that cause the problems they want to solve. An example is renewable energy. In a capitalist system, wind turbines and solar panels, for example, will be over-produced relative to human needs since it is profits for the few, not the needs of the many, that are to be fulfilled, thereby increasing mining impacts and producing more waste streams, among other forms of environmental damage.

A third approach is ostensibly dissatisfied with if not outright critical of the ways things are, remains anchored to classical liberal ideology and politics, and/or espouses anti-revolutionary (especially anti-communist) positions. There is acknowledgement that there are social causes to environmental degradation, even that it involves social inequality, including racial and gender oppression. Those holding such views often stop short of finding out about, much less questioning the specifically capitalist relations of power that induce people to do what they do in the environment.

However, there also exist often self-described critical perspectives that may even expressly identify capitalism as the cause of the socio-ecological crisis but tend to bury capitalist causes under mounds of contingencies and complexities and with the pretext of avoiding "grand narratives" (a thinly veiled metonym for perspectives like Marxism). The tendency in such self-described critical frameworks is to draw up policy wish lists or ideas of social transformation without providing political strategies or programmes or by calling for strategies that conform to or are! limited to anti-revolutionary or liberal politics. This is typical of perspectives like Actor-Network Theory or what are often called post-structural approaches (see also Gare 1995).

What these diverse frameworks have in common is capitalist reformism, meaning that a vague postcapitalism or emancipatory politics is deemed preferable to actually existing socialism or as the only viable solution to accept and support a reformed social system acknowledged to be associated with or causative of the socio-ecological crisis. According to such worldviews, capitalism must be mended, even radically transformed, to usher in a future of social justice, to resolve or attenuate environmental harms, and/or to prevent environmental and wider social injustices. However, the foundations of capitalism, like the private appropriation of the wealth produced by workers and the private ownership of the means of production (like land), are to be left substantially or are not addressd at all.

These three wide-ranging dominant perspectives—populationist, "green" capitalist, and critical—are not mutually exclusive, though generally accommodating to capitalism in contrasting ways. They converge substantively even as some of them are ensconced in radical terminology. Howard Parsons is among those who long ago pithily described the foundational abstractions of these perspectives as "incomprehensible mumbo-jumbo of ancient religions", mystifying as they do the capitalist structural and power dynamics underlying environmental degradation and social inequalities (Parsons 1977, 13). What those perspectives intimately share is a concern for upholding or reinforcing one or another form of capitalist ideology, such as free-market, liberal democracy, or anti-communist social democracy (the post-1950s variety). As a result, they contribute to exacerbating the socio-ecological crisis. The critique of those worldviews has been amply and thoroughly made by others of mainly Marxist and Socialist Ecofeminist persuasion (their works are cited in Chapters 5 and 6) and will not be further countenanced here.

Ecosocialism

The socially destabilising effects of capitalism were recognised early on by stalwarts of capitalism themselves, like John Stuart Mill, who fantasised the possibility of a stationary economy that would favour overall balance in society, in Nature. Self-described communists Karl Marx and Friedrich Engels were among the first to understand the impossibility of stationarity for a capitalist economy as a result of it being predicated on endless capital accumulation (Engels 1972 [1875–1882]; Marx 1992 [1867]). Perhaps to state the obvious, capitalists of whatever economic sector are not exactly interested in renting or selling at a loss relative to maintenance or production costs. Businesses are individually compelled to grow economically, punishment being reduced competitiveness and profit margins, and possible bankruptcy. If none of this mattered in a capitalist society, then capitalists and their state supporters should be indifferent to contracting profit margins and economic downturns.

It is to defy everyday common experience in a capitalist society to claim that businesses can exist without constantly seeking profits just to stay afloat and to pretend away the ever-increasing rates of resource consumption caused by such inherent economic expansionism. As proven repeatedly with data furnished from capitalist countries' institutions themselves, "economic growth fails to reduce unemployment, poverty, and income inequality, and it is limited by finite resources" (Xue 2016a, 84). The hardly secret capitalist tenet and policy constant is unlimited economic expansion, as if the planet had unlimited resources. Since capitalist economics are about prioritising profits, belief in the intrinsic goodness of capitalism or the free market translates into a systematic subordination of most of humanity's needs and of ecosystems to business imperatives.

The contemporary socio-ecological crisis is about how centuries of a predominantly capitalist world have brought about devastation to planetary levels, to the point that radical changes must be made to environmental practices, and therefore to society as a whole, for the sake of most people's health or survival and to addressing the socially extremely uneven repercussions that result from multiple forms of oppression that forces many people to live in harm's way. What such changes need to be or will be depends on the outcomes of multiple forms of social struggles.

Since the 1800s, a main way in which alternatives to capitalism were envisaged and fought for go under the banner of socialism (herein encompassing communist movements for the sake of simplicity). There have been and continue to be diverse forms of socialism, some of which succeeded in toppling capitalist regimes, starting with the Russian Revolution of 1917. Succeeding decades saw the repression of socialist movements in capitalist countries and socialist countries under siege or invaded by imperialist liberal democracies until many of the socialist revolutions were defeated by the late 1980s. Meanwhile, the socio-ecological crisis has continued intensifying while conventional environmentalism has repeatedly shown its capitalism-friendly bankruptcy. Socialism suffered major setbacks globally, but it persists, arguably even at the helm of certain states. One socialist scion emerging in liberal democracies during the 1980s looked for ways to combine what they have deemed a necessary renewal of socialism with radicalised forms of environmentalism. This is, in broad strokes, what has come to be called ecosocialism.

Ecosocialism is a way to gather socialist and environmentalist principles and objectives together. As Jin Xue states, following Löwy (2005):

> Ecosocialism is an ideological and theoretical field associated with Marxism and thus fundamentally challenges capitalist productivism. … ecosocialism also criticizes productivist socialism that ignores the biosphere's limits. (Xue 2016b, 200)

The debatable thesis of "productivist socialism", considering the fact that Cuba, for instance, hardly fits such a description (Moran et al. 2008), is taken up subsequently (mainly in Chapter 8). Ecosocialism is, nevertheless, socialist in the sense of identifying capitalist relations as the ultimate and systemic cause of structural inequalities and environmental destruction. Politically, this means struggling for social equality by establishing the social control of the means to life. This includes decolonisation and cross-generational justice as well as overcoming patriarchal relations and developing respect for differing knowledge systems, striving to combine them to the benefit of all. Ecosocialism is environmentalist in calling attention to the biophysically destructive character of currently conventional ways of living and in premising the understanding of biophysical processes on diverse forms of systematic knowledge and inquiry, institutional and otherwise. Ecosocialism, in other words, stands for the development of biophysically sustainable egalitarian (classless) communities worldwide (Kovel 2014; Löwy 2015; Brownhill and Turner 2020).

A Summary of the Book

A brief review of socialist histories comprises the following chapter to provide some background useful towards understanding the substance of ecosocialism and its internal diversity. Chapter 3 traces the history of ecosocialism proper, including a description of existing tendencies. In Chapter 4, the geography and political forms of ecosocialist currents are described and analysed. Chapters 5 and 6 are dedicated to the two primary theoretical frameworks guiding ecosocialism, Ecological

Marxism and Ecofeminism, along with the schools of thought and debates therewith associated. In Chapter 7, since ecosocialists frequently claim affinity with First Peoples, the relationship of ecosocialism with the self-determination struggles and political philosophies of First Peoples is put under critical scrutiny. Chapter 8 is another critical assessment of ecosocialist thought, but this time in relation to the evolving Ecological Civilisation project in the People's Republic of China. The penultimate chapter involves a description of ecosocialist political strategy and accomplishments, with particular attention to Bolivarian Venezuela. Finally, in the concluding chapter, a discussion of prospects for ecosocialism is offered.

References

Angus, Ian, and Simon Butler. 2011. *Too Many People? Population, Immigration, and the Environmental Crisis*. Chicago: Haymarket.

Arora, Naveen K. 2019. "Impact of Climate Change on Agriculture Production and Its Sustainable Solutions." *Environmental Sustainability* 2: 95–96

Barry, Andrew, and Mark Maslin. 2016. "The Politics of the Anthropocene: A Dialogue." *Geography and Environment* 3 (2): e00022.

Bereiter, Bernhard, Sarah Eggleston, Jochen Schmitt, Christoph Nehrbass-Ahles, Thomas F. Stocker, Hubertus Fischer, Sepp Kipfstuhl and Jerome Chappellaz. 2015. "Revision of the EPICA Dome C CO_2 Record from 800 to 600 kyr Before Present." *Geophysical Research Letters*. https://doi.org/10.1002/2014GL061957.

Brownhill, Leigh, and Terisa E. Turner. 2020. "Ecofeminist Ways, Ecosocialist Means: Life in the Post-capitalist Future." *Capitalism Nature Socialism* 31 (1): 1–14.

Ceballos, Gerardo, Paul R. Ehrlich, and Rodolfo Dirzo. 2017. "Population Losses and the Sixth Mass Extinction." *Proceedings of the National Academy of Sciences* 114 (30): E6089–E6096.

Ceddia, M. Graziano. 2020. "The Super-Rich and Cropland Expansion Via Direct Investments in Agriculture." *Nature Sustainainability* 3: 312–318.

Ciais, P., C. Sabine, G. Bala, L. Bopp, V. Brovkin, J. Canadell, A. Chhabra, R. DeFries, J. Galloway, M. Heimann, C. Jones, C. Le Quéré, R.B. Myneni, S. Piao, and P. Thornton. 2013. "Carbon and Other Biogeochemical Cycles." In *Climate Change 2013: The Physical Science Basis. Contribution of Working Group I to the Fifth Assessment Report of the Intergovernmental Panel on Climate Change*, edited by T.F. Stocker, D. Qin, G.-K. Plattner, M. Tignor, S.K. Allen, J. Boschung, A. Nauels, Y. Xia, V. Bex and P.M. Midgley.

Cambridge, United Kingdom and New York, NY, USA: Cambridge University Press.

Crutzen, Paul J., and Eugene F. Stoermer. 2013 [2000]. "The "Anthropocene". How Can We Live in a World Where There Is No Nature Without People?" In *The Future of Nature: Documents of Global Change*, edited by Libby Robin, Sverker Sörlin, and Paul Warde, 479–490. New Haven: Yale University Press.

Di Virgilio, Giovanni, Jason P. Evans, Stephanie A. Blake, Matthew Armstrong, Andrew J. Dowdy, Jason Sharples, and Rick McRae. 2019. "Climate Change Increases the Potential for Extreme Wildfires." *Geophysical Research Letters* 46 (14), 8517–8526.

Engels, Friedrich. 1972 [1875–1882]. *Dialectics of Nature*. Translated by Clemens Dutt. Moscow: Progress Publishers.

Engels, Friedrich. 1876 [2010]. "The Part Played by Labour in the Transition from Ape to Man." In *Marx and Engels Collected Works, Volume 25*, 452–464. London: Lawrence & Wishart.

Finney, Stanley C., and Lucy E. Edwards. 2016. "The 'Anthropocene' Epoch: Scientific Decision or Political Statement?" *GSA Today* 26 (3–4): 4–10.

Foster, John Bellamy, and Brett Clark. 2021. "The Capitalinian: The First Geological Age of the Anthropocene." *Monthly Review* 73 (4): 1–16.

Franta, Benjamin. 2018. "Early Oil Industry Knowledge of CO_2 and Global Warming." *Nature Climate Change* 8: 1024–1025.

Gare, Arran. 1995. *Postmodernism and the Environmental Crisis*. London: Routledge.

Gibbons, Ann. 2017. "Oldest Members of Our Species Discovered in Morocco." *Science* 356 (6342): 993–994.

Global Witness. 2020. "Defending Tomorrow. The Climate Crisis and Threats against Land and Environmental Defenders." https://www.globalwitness.org/documents/19938/Defending_Tomorrow_EN_high_res_-_July_2020.pdf. Accessed 26 October 2020.

Gore, Tim. 2020. "Confronting Carbon Inequality. Putting Climate Justice at the Heart of the COVID-19 Recovery." https://oxfamilibrary.openrepository.com/bitstream/handle/10546/621052/mb-confronting-carbon-inequality-210920-en.pdf. Accessed 9 October 2020.

Haines, Andy J., and Jonathan A. Patz. 2004. "Health Effects of Climate Change." *Journal of the American Medical Association* 291 (1): 99–103.

Herring, Stephanie C., Nikolaos Christidis, Andrew Hoell, Martin P. Hoerling, and Peter A. Stott. 2020. "Explaining Extreme Events of 2018 from a Climate Perspective." *Bulletin of the American Meteorological Society* 101: S1–S140.

IEA. 2022. *CO_2 Emissions from Energy Combustion and Industrial Processes, 1900–2021*. Paris: IEA. https://www.iea.org/data-and-statistics/charts/co2-emissions-from-energy-combustion-and-industrial-processes-1900-2021. Accessed 11 January 2023.

Jacob, Daniel J., and Darrell A. Winner. 2009. "Effect of Climate Change on Air Quality." *Atmospheric Environment* 43 (1): 51–63.
Isla, Ana. 2018. "Climate Chaos Mother Earth Under Threat." In *Climate Chaos. Ecofeminism and the Land Question*, edited by Ana Isla, 1–12. Toronto: INANNA Publications and Education.
Kaplan, Jed O., Kristen M. Krumhardt, Niklaus Zimmermann. 2009. "The Prehistoric and Preindustrial Deforestation of Europe." *Quaternary Science Reviews* 28 (27–28): 3016–3034.
Kovel, Joel. 2007. *The Enemy of Nature. The End of Capitalism or the End of the World?* London: Zed Press.
Kovel, Joel. 2014. "Ecosocialism as a Human Phenomenon." *Capitalism Nature Socialism* 25 (1): 10–23.
Lewis, Simon L., and Mark A. Maslin. 2015. "Defining the Anthropocene." *Nature* 519: 171–180.
Lewontin, Richard, and Richard Levins. 2007. *Biology Under the Influence: Dialectical Essays on the Coevolution of Nature and Society.* New York: Monthly Review Press.
Li, Yi, Jiahui Shang, Chi Zhang, Wenlong Zhang, Lihua Niu, Longfei Wang, Huanjun Zhang. 2021. "The Role of Freshwater Eutrophication in Greenhouse Gas Emissions: A Review." *Science of The Total Environment* 768: 144582.
Löwy, Michael. 2005. "What Is Ecosocialism?" *Capitalism Nature Socialism* 2: 15–24.
Löwy, Michael. 2015. *Ecosocialism. A Radical Alternative to Capitalist Catastrophe*. Chicago: Haymarket Books.
Luciano, Eugenio. 2022. "Is 'Anthropocene' a Suitable Chronostratigraphic Term." *Anthropocene Science* 1: 29–41.
Malhi, Gurdeep Singh, Manpreet Kaur, and Prashant Kaushik. 2021. "Impact of Climate Change on Agriculture and Its Mitigation Strategies: A Review." *Sustainability* 13 (3): 1318.
Malm, Andreas. 2016. *Fossil Capital: The Rise of Steam Power and the Roots of Global Warming*. Verso Books.
Malm, Andeas and Alf Hornborg. 2014. "The Geology of Mankind? A Critique of the Anthropocene Narrative." *The Anthropocene Review* 1 (1): 62–69.
Marks, Robert B. 2017. *China. An Environmental History*. Lanham: Rowman & Littlefield.
Marx, Karl. 1867 [1992]. *Capital. A Critique of Political Economy. Volume 1.* Translated by S. Moore and E. Aveling. New York: International Publishers.
McConnell, Joseph R., Wilson, Andrew I., Stohl, Andreas, Arienzo, Monica M., Chellman, Nathan J., Eckhardt, Sabine, Thompson, Elisabeth M., Pollard, A. Mark, Steffensen, Jørgen Peder. 2018. "Lead Pollution Recorded in Greenland Ice Indicates European Emissions Tracked Plagues, Wars, and Imperial

Expansion During Antiquity. *Proceedings of the National Academy of Sciences* 115 (22): 5726–5731.

McNeil, Cameron L, David A. Burney, Lida P. Burney. 2010. "Evidence Disputing Deforestation as the Cause for the Collapse of the Ancient Maya Polity of Copan, Honduras." *Proceedings of the National Academies of Science* 107 (3): 1017–1022.

Meena, Ram Swaroop, Sandeep Kumar, Rahul Datta, Rattan Lal, Vinod Vijayakumar, Martin Brtnicky, Mahaveer Prasad Sharma, Gulab Singh Yadav, Manoj Kumar Jhariya, Chetan Kumar Jangir, Shamina Imran Pathan, Tereza Dokulilova, Vaclav Pecina, and Theodore Danso Marfo. 2020. "Impact of Agrochemicals on Soil Microbiota and Management: A Review." *Land* 9 (2): 34.

Moore, Jason W. 2017. "The Capitalocene, Part I: On the Nature and Origins of Our Ecological Crisis." *The Journal of Peasant Studies* 44 (3): 594–630.

Moran, Daniel D., Mathis Wackernagel, Justin A. Kitzes, Steven H. Goldfinger, Aurélien Boutaud. 2008. "Measuring Sustainable Development—Nation by Nation." *Ecological Economics* 64 (3): 470–474.

NOAA. 2022. "Trends in Atmospheric Carbon Dioxide." https://gml.noaa.gov/ccgg/trends/. Accessed 18 January 2023.

O'Connor, James R. 1998. *Natural Causes: Essays in Ecological Marxism*. New York: Guilford Press.

Parsons, Howard. 1977. *Marx and Engels on Ecology*. Westport: Greenwood Press.

Preunkert, Susanne, Joseph R. McConnell, Helene Hoffmann, Michel Legrand, Andrew I. Wilson, Sabine Eckhardt, Andreas Stohl, Nathan J. Chellman, Monica M. Arienzo, and Ronny Friedrich. 2019. "Lead and Antimony in Basal Ice from Col du Dome (French Alps) Dated with Radiocarbon: A Record of Pollution During Antiquity." *Geophysical Research Letters* 46: 4953–4961.

Ripple, William J., Christopher Wolf, Thomas M. Newsome, Phoebe Barnard, William R. Moomaw. 2020. "World Scientists' Warning of a Climate Emergency." *BioScience* 70 (1): 8–12.

Ritchie, Hannah. 2002. "The World Has Lost One-Third of Its Forest, But an End of Deforestation Is Possible." https://ourworldindata.org/world-lost-one-third-forests. Accessed 9 January 2023.

Ritchie, Hannah, Max Roser and Pablo Rosado. 2022. "Energy." https://ourworldindata.org/energy. Accessed 17 January 2023.

Ritchie, Hannah, Max Roser and Pablo Rosado. 2024. "CO_2 and Greenhouse Gas Emissions." https://ourworldindata.org/co2-and-other-greenhouse-gas-emissions. Accessed 15 November 2024.

Sánchez-Bayo, Francisco, and Kris A.G. Wyckhuys. 2019. "Worldwide Decline of the Entomofauna: A Review of Its Drivers." *Biological Conservation* 232: 8–27.

Schwartzman, Peter, and David Schwartzman. 2019. *The Earth Is Not for Sale: A Path Out of Fossil Capitalism to the Other World That Is Still Possible*. Singapore: World Scientific Publishing.

Şekercioğlu, Çağan H., Richard B. Primack, Janice Wormworth. 2012. "The Effects of Climate Change on Tropical Birds." *Biological Conservation* 148 (1): 1–18.

Solomon, Susan, Kane Stone, Pengfei Yu, D.M. Murphy, Doug Kinnison, A.R. Ravishankara, and Peidong Wang. 2023. "Chlorine Activation and Enhanced Ozone Depletion Induced by Wildfire Aerosol." *Nature* 615 (7951): 259–264.

Stolarski, Richard S., Anne R. Douglass, Luke D. Oman, and Darryn W. Waugh. 2015. "Impact of Future Nitrous Oxide and Carbon Dioxide Emissions on the Stratospheric Ozone Layer." *Environmental Research Letters* 10 (3): 034011.

Tian, Hanqin, Rongting Xu, Josep G. Canadell, Rona L. Thompson, Wilfried Winiwarter, Parvadha Suntharalingam, Eric A. Davidson, Philippe Ciais, Robert B. Jackson, Greet Janssens-Maenhout, Michael J. Prather, Pierre Regnier, Naiqing Pan, Shufen Pan, Glen P. Peters, Hao Shi, Francesco N. Tubiello, Sönke Zaehle, Feng Zhou, Almut Arneth, Gianna Battaglia, Sarah Berthet, Laurent Bopp, Alexander F. Bouwman, Erik T. Buitenhuis, Jinfeng Chang, Martyn P. Chipperfield, Shree R. S. Dangal, Edward Dlugokencky, James W. Elkins, Bradley D. Eyre, Bojie Fu, Bradley Hall, Akihiko Ito, Fortunat Joos, Paul B. Krummel, Angela Landolfi, Goulven G. Laruelle, Ronny Lauerwald, Wei Li, Sebastian Lienert, Taylor Maavara, Michael MacLeod, Dylan B. Millet, Stefan Olin, Prabir K. Patra, Ronald G. Prinn, Peter A. Raymond, Daniel J. Ruiz, Guido R. van der Werf, Nicolas Vuichard, Junjie Wang, Ray F. Weiss, Kelley C. Wells, Chris Wilson, Jia Yang, and Yuanzhi Yao. 2020. "A Comprehensive Quantification of Global Nitrous Oxide Sources and Sinks." *Nature* 586 (7828): 248–256.

Tilly, Charles. 1990. *Coercion, Capital, and European States, AD 990–1990*. Cambridge, USA: Basil Blackwell.

Timperley, Jocelyn. 2020. "Who Is Really to Blame for Climate Change?" BBC, June 18. https://www.bbc.com/future/article/20200618-climate-change-who-is-to-blame-and-why-does-it-matter. Accessed 22 June 2020.

Tucker, Michael. 1995. "Carbon Dioxide Emissions and Global GDP." *Ecological Economics* 15: 215–223.

Wallace, Rodrick, Luis Fernando Chaves, Luke R. Bergmann, Constância Ayres, Lenny Hogerwerf, Richard Kock, and Robert G. Wallace. 2018. *Clear-Cutting Disease Control Capital-Led Deforestation, Public Health Austerity, and Vector-Borne Infection*. Cham: Springer.

Wallis, Victor. 2018. *Red-Green Revolution: The Politics and Technology of Ecosocialism*. Toronto: Political Animal Press.

Weart, Spencer R. 2008. *The Discovery of Global Warming*. Cambridge, USA: Harvard University Press.

Wiedmann, Thomas, Manfred Lenzen, Lorenz T. Keyßer, and Julia K. Steinberger. 2020. "Scientists' Warning on Affluence." *Nature Communications* 11: 3107.

Williams, Michael. 2003. *Deforesting the Earth: From Prehistory to Global Crisis*. Chicago: University of Chicago Press.

WWF. 2020. *Living Planet Report 2020—Bending the Curve of Biodiversity Loss*. Edited by R.E.A. Almond, M. Grooten and T. Petersen. Gland, Switzerland: WWF.

Xue, Jin. 2016a. "Arguments for and Against Economic Growth." In *Crisis System. A Critical Realist and Environmental Critique of Economics and the Economy*, edited by Petter Næss and Leigh Price, 69–88. London and New York: Routledge.

Xue, Jin. 2016b. "China at the Crossroad. Ecological Modernization or Ecosocialism?" In *Crisis System. A Critical Realist and Environmental Critique of Economics and the Economy*, edited by Petter Næss and Leigh Price, 192–208. London and New York: Routledge.

Zillman, John W. 2009. "A History of Climate Activities." *World Meteorological Organisation Bulletin* 58 (3): 141–150.

CHAPTER 2

A Historical Sketch of Socialism

Socialism and Communism

So as not to presume too much prior knowledge, a sketch of what socialism and capitalism are about could be helpful, so a bit of room is dedicated to explaining these concepts. For socialism, there is no single definition, yet, at a minimum, it refers to a type of society whose purpose is meeting everyone's needs and maximising everyone's personal potentials, in contrast to capitalism. At bottom, socialism is a response to the many social injustices and horrors inhering capitalism. It is an alternative to capitalism that is often described as society in general controlling the means of production, means that are to be developed and put to use for the benefit of everyone. As such, socialism is a different kind of society (not just a form of government) to replace capitalism systemically, in its entirety.

Socialism is definable in relation to that to which it is historically and specifically opposed, which is capitalism. The principal characteristics of capitalism include a systematic division of people according to classes and an economy based on the endless accumulation of capital (gaining wealth to get even more wealth and so on to infinity). Historically and crucially, capitalism emerged through the violent separation of people according to racial categories, including by means of mass abductions for slavery-based economic gain and multiple genocides to secure resources (including the means of production) necessary for capital accumulation

S. Engel-Di Mauro, *Ecosocialism*, Marx, Engels, and Marxisms,
https://doi.org/10.1007/978-3-031-74288-0_2

even to be possible. By such sordid deeds and by reinforcing intergenerationally their socially horrific and divisive outcomes, the means of production are owned or controlled by capitalist and landowning classes (the ruling classes) and, to survive, workers must sell their ability to work (labour-power).[1]

The surplus produced by workers is systematically appropriated by ruling classes in diverse ways (the wage system, rent, interest on loans, international trade, etc.). That is, the workers altogether, whether paid or unpaid, get less than what they produce or contribute to producing, as otherwise there would be no profits for the ruling class few. This is what is technically called exploitation in Marxist terms and the state plays a central role in establishing and enforcing the mechanism of exploitation (e.g., through legal frameworks privileging private property, taxation systems, police and militaries to apply capitalist prerogatives within and between countries, etc.). Getting control over the surplus is key if any capital is to be accumulated, so there is a lot at stake. It should not surprise anyone that expansionism and warfare are endemic to capitalism. Ruling classes, as they compete ruthlessly over control of the surplus, are bound to clash and by doing so they bring ruin mainly to working classes and colonised peoples generally, as well as environmental devastation.

Notably, businesses that succeed in capitalist competition (also known as the free market) will be able to concentrate increasing amounts of the total surplus (the wealth created by society) into their hands in one or several economic sectors and centralise their control over parts of an economy. They can in this way exercise effective control over the most productive or most key labour processes at national or even at international levels (e.g., agro-chemically intensive farming, computer chips manufacture, high-tech military equipment, etc.). This is how some businesses exert decisive influence over governments and state apparatuses and most such businesses (also called monopolies) are concentrated in a handful of typically liberal democratic countries (or parliamentary democracies), those that inflicted or inflict colonial dictatorships elsewhere and/or benefited and continue to benefit from past or present colonial expansion. These are also known as imperialist countries. Imperialism is thus also a tendency intrinsic to capitalism (Lenin 1917; see also King 2021).

[1] To clarify, workers may own some means of production, like home computers, or possess monetary savings or even invest in stock exchanges, but all such economic resources are insufficient to live on.

There is a simultaneous acculturation component to capitalism into a specific kind of belief system that hampers the development of socialist consciousness and ecosocially sustainable practices. Whatever does not fulfil the objective of capital accumulation (profitability, the bottom line, marketability, etc.) is considered or treated as relatively unimportant or worthless, like most activities necessary for the biological reproduction of our species, our relationship with other organisms and to ecosystems, and anyone who is not contributing to making profits for capitalists and landlords. As these exclusions intimate, capitalism is deeply anti-social and anti-ecological, based on sponging off the labour of those who enable society to exist but are not directly involved in the capital accumulation process (e.g., unwaged workers, most women careworkers, etc.).

Structurally, working classes have material interests at odds with ruling classes. A more obvious example of how this is manifested is when trade unions strike for more wages and better working conditions or renters organise for rent strikes, which means getting more of the share of the surplus produced and reducing profit margins or rent revenue for businesses. However, just as in any kind of class-differentiated society, class polarisation is a tendency in capitalism, not any static condition. There are intergrade classes, like small business owners, for example. The form that capitalism takes can also differ substantively according to social context and histories. There are often other—transitional or vestigial—forms of social stratification from preceding non-capitalist kinds of society, like clerics, lords, and monarchies. Especially in settler colonial countries and countries under (neo)colonial yokes (i.e., most countries), there can often co-exist more egalitarian non-capitalist structures that may include such structures as leadership posts in confederacies and clan chiefs. All such non-capitalist social relations are in tension with and/or waning under capitalist pressures and affect how capitalism develops in different places. What complicates matters further is that all classes have internal divisions and conflicts. In their incessant race for capital accumulation (e.g., gaining wealth to get more wealth), ruling classes vie for global or national market share or the means to outcompete each other. Within capitalist classes there are widely differing levels of influence or power such that there can develop political affinities across class positions on the basis of socialised (cultural, identitarian) differences, including those preceding capitalism (e.g., some forms of patriarchy, ethnicity, religion, etc.) or resulting from or associated with capitalism (e.g., nationhood, racialisation, settler colonialism, and the like).

Socialists struggle for a society that is free of capitalist exploitation, capital accumulation processes, and all the trappings of capitalism and oppressive relations thereto related. For communists, the sort of society sought is one that is free of classes altogether, as well as states, and all forms of oppression, with all means of production held communally. How socialism is envisioned and is to be achieved is often contentious, at times to deadly degrees. Socialist achievements historically have frequently been incomplete or severely curtailed due to permanent attacks by capitalist powers and generally inimical reactionary forces within countries where socialism has taken root. Perhaps more generally, in terms of more immediate political aims, socialism could be defined as a struggle for and ideally a politics of substantive instead of formal ("free market") democracy.[2] It is not about state control or greater government spending, although in part this can be as part of strategies or policies to realise socialist objectives over different time horizons. Socialism has been thought about by its proponents in terms of addressing, resolving, preventing the social disasters endemic to capitalism, like workplace deaths, mass poverty, hunger, structural unemployment, systematically unpaid work in the home, houselessness, and nearly constant warfare. These and many other ills are what have persuaded masses of workers, peasants, and allied intellectuals over the past couple of centuries to organise themselves to fight for and build socialism and, for many, a communist future (Newman 2015; Van Der Linden 2023a).

[2] Formal democracy, for example, includes rights for abstract individuals (rather than communities or nationalities, or individuals in their specific situations), but without any economic security (substantive rights), those quickly degenerate into rhetorical devices justifying freedom for the wealthy few and social inequality. Similarly, "free and fair elections" in such countries is a self-serving slogan for capitalist ruling classes to conceal their control over the economic resources needed to get elected or to hold on to office and their overwhelming influence over electoral processes and governing institutions to ensure anti-capitalist political parties have at most marginal power (otherwise, coups become the favoured option). This is the problem with liberal or parliamentary democracies, which are founded on capitalist relations and histories of genocide, slavery, and conquest. Aside from omitting such inconvenient facts, champions of "free market" democracy will also deny the substantively capitalist foundations of countries like Nazi Germany, Fascist Italy, Saudi Arabia, the Gulf Monarchies, Brunei, etc.

Origins of Socialism

The eventual development of socialism was in part nourished by the enduring influences and memories of peasant revolts, millenarian movements, philosophies, and beliefs within Europe as far back as the 1300s. The sheer existence of classless societies, the constant anti-colonial revolts and liberation struggles, the relative successes of marronage (the formation of independent polities of self-freed slaves, often with First Peoples), and the 1791–1804 Revolution in Ayiti (Haiti) may also have shaped the political imaginary leading to the formation of socialist movements. The extent of this remains nebulous, but Marx and Engels are among those who were clearly influenced not only by past and contemporary popular struggles within Europe, but also by reports of non-capitalist societies outside the Europe-centred capitalist world. Notions of social equality in some currents of the 1789 Bourgeois Revolution in France, especially the 1796 Conjuration des Égaux (Conspiracy of the Equals), were also of major influence (Birchall 1997; Krader 1974, 5–6; Meisenhelder 1995; Merchant 1980; Van Der Linden 2023a).

Still, as a politics and ideal, socialism emerged specifically as a response to "the social question" wrought through industrial capitalism mainly in Northwestern and Central Europe. It was a way to envision alternatives to what multitudes were suffering through, the massive and widening inequalities, with many peasants and artisans displaced and living in squalor. The increasingly disastrous material conditions of the many were also undermining community ties, cooperative and associative structures, and related cultural values. The name to describe such alternatives, socialism, only came into use during the first half of the 1800s. By the 1860s, socialism denoted associationism, mutualism, cooperativism, or collectivism, and had started to spread to other places, especially Japan.

Arguably, notions of socialism arrived several centuries late if one understands the basis of capital accumulation. Near-incessant social turmoil and warfare, mass displacements and exterminations, slavery systems, and continent-sized episodes of grand larceny, known also as settler colonialism, had already been in full force and further intensifying in much of the rest of the world by the time the word socialism was first uttered. The lasting scourge of European colonialism, slavery systems, and its mystifying liberal ideology (Losurdo 2011) had created the very conditions for the development of socialism. Socialist concern over the historical and recurring basis of capital accumulation took time to develop. At first, there was widespread insensitivity. For instance, those who elected to form intentional communities in the Americas, such as the Icarian and Owenite socialists did not demonstrate any compunction regarding settler colonialism, nor any support for any contemporary First People's resistance to settler colonialism. Apparently, Saint Simonians and Fourierists even thought of establishing "socialist" colonies in Algeria by taking over land stolen from local communities (Losurdo 2011, 180). Similar forms of settler colonial mindsets or supremacism pervaded the 1870 Commune of Algiers and tinged exiled communards' encounters with Kanak communities (Plaetzer 2021).[3] The articulation of a systematic rejection of and socialist alternative to capitalism was a very gradual process.

It came about from within multifaceted modes of oppression preceding and negatively revolutionised through the dictatorship of capital, including racialism and patriarchy. Because of the diversity of forms of oppression, degradation, and havoc generated through capitalism, socialism has always been polyvocal and variable in content and emphasis. Its multifariousness is grounded in the diverse and historically shifting material conditions, whether as proletarians, peasants, artisans, gatherer-hunters, herders, or other kinds of social strata and modes of being. What today is called socialism has changed substantively in content and scope, yet the egalitarianism-seeking core and classes-demolishing aims persist, even if expressed in diverse ways. While the historical and current horrors caused by capitalist societies are not lost to the majority socialist world, especially now, there remains an oblivious, sometimes even justifying (Eurocentric) socialist minority. Ecosocialism is not immune to this enduring minoritarian trend, one that is usually associated with material beneficiaries of (neo-)colonialism and imperialism. It is a self-contradictory offshoot of what Losurdo (2011, 107) refers to as the "master race" democracy, an ideology that subtends liberal thought.

This is one aspect of what divided early socialists. Some currents, very few at first, were inclusive and sensitive to multiple forms of oppression, and some were masculinist, chauvinistic, and even justified colonialism and racism. There were those, such as Cabet, Fourier, and Owen, who

[3] That this supremacist attitude was part of a settler colonial system, not confined to French colonists or exiled communards, is evident in the exiled Kabyle anti-colonial fighters' readiness to join French colonial forces in the massacres to squash the 1878 Kanak uprising against French colonialism in New Caledonia.

imagined and at times set up egalitarian communities or workers' cooperatives that, more often than not, fell apart within a few years. As mentioned above, when setting up such communities especially in North America, they apparently paid no mind to the settler colonial basis of the land they purchased and made no attempts to connect to First Peoples nearby or support their struggles. Some of these communities were religious in character, basing their socialist quest on dissenting interpretations of holy scriptures. More recent movements like Liberation Theology and the Catholic Worker have these as historical precedents. Other contemporary socialists in the 1800s advocated for technocratic centralist rule by producers, who included industrialists. Saint Simon was the foremost exponent of this view. His followers pressed the state for the establishment of a technocratic order based on the large-scale scientific management of the economy.

Syndicalists instead focused on worker power as the way to reach socialism and formed unions to that effect. The visionary Inka-French feminist Flora Tristán introduced the concept and practices of working-class unity in the struggle to improve workers' conditions at both work and home (Beik and Beik 1993). Most syndicalists, however, largely accepted the legitimacy of the existing social order, at least in economically prosperous years. Finally, Jacobins, a prominent grouping during the French revolution, strove to turn state institutions into organs for the gradual extension of political rights to all. Shared by all the major socialist movements was the belief that everyone involved in production could eventually cooperate to improve society for mutual benefit, which assumed a harmonious relationship between, for instance, peasants, industrialists, and workers (Cole 1953).

The delineation of the now familiar labels of socialist, communist, and anarchist took roughly a century to emerge. Anarchism and socialism would start assuming clearly separable politics in the 1860s, largely splitting not too long thereafter. A demarcation between socialism and communism would not appear until the 1920s splits within socialist and social democratic parties, although self-described communist organisations already existed by the 1840s. Karl Marx and Friedrich Engels so identified their politics at that time (Braunthal 1967a; Graham 2005; Guérin 1970).

In the 1830s, with the spread of workers' mutual-assistance associations, Pierre Joseph Proudhon developed the concept of anarchism as an alternative kind of social order based on mutualism and federated communities. These ideas would be further elaborated on, modified, and diffused by Mikhail Bakunin, Élisée Reclus, Pëtr Kropotkin, and the Owenite William Benbouw, among others. They stressed the inseparability of state and capitalism and therefore the necessity to overcome both simultaneously. Imbued with these ideas, revolutionary syndicalist and anarcho-communist movements organised revolts in places like Ukraine, México, and Spain in the 1900s (Eckhardt 2016; Graham 2005; Guérin 1970). At the same time, Kropotkin and especially the communard Reclus, whose work includes an early form of bioregionalism, started to call attention to the importance of the environment in a non-deterministic fashion (Reclus 2013).

Marx's and Engels' analyses of capitalism as a system of dispossession and exploitation and their development of dialectical historical materialism started maturing in the 1840s. The *Communist Manifesto* (1948), written for the Communist League (1847–1852) and revived in the 1870s, was an early distillation of this framework. It was part of a new approach eventually called Scientific Socialism, grounded on studying social relations, both past and present, to arrive at political strategies attuned to existing possibilities. Such an overarching worldview offered an alternative to a tendency in prior and contemporary perspectives, deemed Utopian, to formulate pre-packaged blueprints devoid of analyses of social context and hence unable to find effective applicability. This was attested multiple times by the negligible political effects of that kind of socialism. Noteworthy are Engels' early studies involving attention to environmental degradation and Marx's scattered, laconic, but clear analyses of the environmentally destructive propensities of capitalism (Dickens 1992; Foster 2023). This aspect is discussed in depth in the next chapter.

Neither Scientific Socialism nor historical and dialectical materialism were ever unified perspectives under any one theoretical or party line. Much effort would be exerted from the very beginning, especially by Engels, to counter various undialectical, economically deterministic (vulgar materialist), and idealist perspectives claiming scientific standing. The struggle for scientific rigour and dialectical materialist open-endedness was often unsuccessful, as rigid, closed, and sometimes doctrinaire worldviews have often prevailed within Marxism, institutional and otherwise.

Revolutions and State Power

In much of Europe and, partly, in European settler colonies, the ideas of Marx and Engels came to prominence especially through the International Workingmen's Association (the First International, 1864–1876), where the intrepid duo played key roles, alongside Proudhonists, nationalists, syndicalists, and collectivist anarchists like Bakunin. The First International was, among other things, an early instance of working-class organising and coordination within and across countries, but also a focal point of decisive political bifurcation among socialist forces. The First International fell apart under the strains and disarray created by relentless police infiltration and repression.

A final blow was the heightened persecution of Internationalists, falsely presumed to have organised the agitation and uprisings that flowered into the formation of communes in many French cities in 1871, as a response to French government capitulation to invading Prussian armies and in the shadow of the murderous squashing of worker revolts in the 1840s. The communes, of which the Paris commune is the most famous, involved takeovers of entire cities and their political transformation into centres for the development of direct democracy. All communes were promptly crushed within a period of several months by colluding French and Prussian armies, who murdered thousands of communards (Gaillard 1971; Lejeune 2016).

To recover organisational capacity after the exile, imprisonment, and dispersal of many members of the First International, Engels, Marx, and their allies (including some communards) tried to centralise decision-making processes under their direction. The aim was the formation of parties to be coordinated by means of a renewed International. Anarchists and some syndicalists responded by forming an alternative International or federation, while Bakuninists continued to form shadow and conspiratorial groups. The schism that resulted from the increasing divide, exacerbated by state repression, fractured the first International irreparably, but its demise was overshadowed by the rise of social democratic parties (Braunthal 1967a, 175; Cole 1961; Eckhardt 2016).

Recurring state violence targeting socialists, preceding and during the First International's existence, led some socialist formations to establish secret societies to achieve their aims, using force if deemed necessary. This route became particularly easier to justify after the 1871 massacres and persecution of communards. Other tendencies opted for gradualist approaches, working within existing institutions. Some socialist formations developed both kinds of strategies, or shifted from one to the other. This diversity of strategies has since characterised all socialist movements in one way or another.

Parliamentarism, that is, seeking the achievement of socialism by means of government institutions within capitalism, became progressively the more prominent political strategy from the late 1800s onwards. The founding of the Second International in 1889 was dominated by political parties, especially the German Social Democratic Party, the largest of all. In 1893, anarchist formations and trade unions were essentially barred, leading to a permanent severance of anarchist and some syndicalist organisations from what became mainstream socialist institutions. Out of these changes also came greater affinity between socialist and labour parties and parliamentarism, and then reformism (understood as prioritising the modification over the overcoming of capitalism). Anarchists and syndicalists would thenceforth form their own independent international organisations and rarely interact with other socialist groups, often in conflictual ways (Braunthal 1967b; van der Linden 2023b).

Communism was integral to Marxism early on but also to some forms of anarchism, where the term was inflected as anarchist communism. The differences from Marxism lay mainly in seeing the state as a fundamental obstacle to be overcome immediately and in refusing any intermediate stage to achieving communism. Marx and Engels, in contrast, upheld strategies of movement centralisation, predicated on striving for working-class gains, and the conquest and use of the state to dissolve it. They defined the process as proceeding in two steps: first establishing worker majority rule ("the dictatorship of the proletariat"), and then, through that transitional stage, the eventual achievement of communism. State socialism, arguably, conforms to this notion of a transitory phase.

Marx and Engels would later see a potential for skipping directly to communism in non-capitalist societies, but this would be contingent on successful revolutions in the most powerful capitalist countries (Rodney 2011). Those writings would not be widely known until the 1960s, but communist movements would arrive at similar conclusions independently out of practical considerations, as in revolutionary Russia and Mongolia. The means to achieve communism—a state-free, classless society—would nevertheless be highly contested within Marxism, ranging from conforming to and/or using liberal democracy to develop socialism to organising underground to foment the violent overthrow of any form of capitalist regime and its replacement with a socialist system (Cole 1960; Graham 2005).

For both Marx and Engels, and not a few other socialist thinkers (including anarchist-communist Reclus), understanding what would now be called ecological principles (a "mastery of nature") was crucial to developing an eventually communist society. They warned against what would be much more recently called environmentally unsustainable practices. These proclivities persisted within Marxist thought (e.g., Rosa Luxemburg, Vladimir Ilyich Ulyanov "Lenin"). They furnished precedents and continuities that would form socialist states' environmentalist currents, nourished by histories of struggles against colonial depredations, the material effects of industrialisation and recurring and, mainly by the 1980s, rediscoveries of insights from Marx and Engels outside of socialist countries (Dickens 1992; Gare 1993). This still under-appreciated aspect of socialist thought is explored further in the next chapter.

In contrast, with respect to politics, some socialist, social democratic, and labour parties adopted narrow and gradualist objectives and, in some respects, even liberal principles, confining political work mainly to parliamentarism as an end in itself. In the early 1900s, anti-Marxist and Marxist revisionist currents arose that converged over a disavowal of class struggle. Communism remained the province of still-influential segments of social democratic parties. In the case of the Russian Social Democratic and Labour Party (founded in 1898), such a segment was embodied by what came to be called the Bolshevik wing, who definitively split from the Mensheviks by 1912 and founded their own party.

The Bolsheviks were ahead of their time. The contradictions within social democratic parties became a centrifugal force that, in the wake of World War I, tore them apart on questions of nationalism and support for war. Most social democratic parties (including labour and socialist parties) opted on the side of their respective national governments in a crude display of chauvinism, flatly contradicting their 1907 anti-militarism resolution. A minority within the Second International, convened at Bolsheviks' insistence, met in 1915 to organise opposition to the war, but it was ultimately in vain. Nationalist binges in mass slaughter prevailed upon the Second International, which was officially defunct by 1916.

Excepting Russia and Mongolia, revolutionary attempts like the 1919 German Spartacist revolt and Hungarian Soviet Republic were crushed militarily within months. Newly formed or reconstituted capitalist regimes subsequently incarcerated, executed, or exiled thousands of socialists of all stripes. Similar convulsions in other parts of the world preceded or roughly coincided with the 1917 Russian Revolution, even if not as influential at the time. In México, the Mexican Revolution of 1910–1920 included anarchism-leaning Magonismo and the pro-peasant agrarian socialist Zapatismo. The institutions that came about would later enable Cuban revolutionaries to set up their main base in México in the 1950s.

In Russia, a militarily weakened Czarist dictatorship gave way, in October 1917, to the rising tides of workers' councils (soviets) occupying factories and mines, peasant communes and land appropriations, national liberation movements (especially in Siberia and Central Asia) and a variety of socialist formations, mainly the Bolsheviks, Mensheviks, Social Revolutionaries and, to some degree, anarcho-communists (e.g., Makhno-led revolutionaries in Ukraine). The Bolsheviks succeeded not only in gaining popularity in the industrialised, urban, and especially western parts of Russia, but also in centralising decision-making processes and consolidating their own political power. This they achieved by providing workable alternatives enabling many to survive the multiple wars and famines, repressing antagonistic socialist formations, and crushing revolts (e.g., at Kronstadt and Tambov), in the process reducing soviet decision-making authority.

Developing a new army (the Red Army), the Bolsheviks, jointly with Left Socialist Revolutionaries and with indirect help from other formations (e.g., the Makhnovists), were able to defeat, by 1922, multiple armies (the White Guard) allied to the Czarists. Notably, the feat included the repulsion of invading foreign troops from 14 countries supporting the White Guard, including thousands sent by the US Wilson government (the American Expeditionary Force, Siberia). These liberal democratic and subsequent Nazi-fascist invasions in World War II would permanently scar and shape the politics of the new socialist state, the USSR. A major outcome was greater political centralisation, the privileging of military investment to ensure self-defence, the rapid industrialisation that would provide such military capacity, and the contradictory environmental practices and ramifications that would follow.

In China under the Qing Dynasty, which at the time included present-day Mongolia, there were more promising developments for socialist ends. The 1911 overthrow of the monarchy (the Xinhai Revolution) facilitated the flourishing and spread of socialist ideas, including Marxism, in the newly founded Republic of China and the Bogd Khanate of Mongolia. It would first be in herder-majority Mongolia that a communist party would be founded—the Mongolian People's Party (MPP)—in 1920. Out of the chaos ensuing from first Republican Chinese military and then Russian White Guard invasions, the MPP led the June 1921 Mongolian People's Revolution, aided by a fledgling USSR decimated by civil wars and multiple foreign military interventions. The Mongolian People's Republic, founded on 7 June 1924, became the second successful socialist revolution in world history. Like the USSR and all other successful socialist revolutions, the Mongolian People's Republic would be besieged militarily and economically, already by the 1930s having to repel Imperial Japanese armies with the help of the USSR.

It would take more than a couple of decades for the next successful socialist revolutions to take place. Marxist and assorted socialist (even anarchist) ideas had also infused the 1919 May Fourth Movement in the Republic of China, leading to the founding of the Communist Party of China (CPC) in 1921. This included such subsequently famous co-founders and thinkers as Máo *Zédōng*, eventual CPC leader by 1943. But it would not be until the People's Liberation Army (founded in 1927) defeated the Guomindang (the nationalists and their warlord allies), the Imperial Japanese armies, and US and satellite states' incursions that the Chinese Revolution would succeed in establishing the People's Republic of China in 1949.

Elsewhere, out of the mayhem of World War I and schisms within most social democratic and socialist parties, leftist currents often broke away to form communist parties, either as the war was ending or shortly thereafter. Most of those communist parties, but also socialist and syndicalist formations, faced state repression and were often forced underground soon after forming. In Italy this occurred by 1926 under the fascist regime and was soon followed elsewhere in Europe, reaching particularly horrific intensity in Nazi Germany with death camps (the first, Dachau, was built in 1933 for political opponents). In countries like the US the violent repression of any socialist organisation and deportation of socialists of any stripe, especially communists, was already in full swing by 1919.

The Russian and Mongolian Revolutions

Communist fortunes changed radically with the Russian and Mongolian Revolutions, for the better for some and for the worst for others. One-party Bolshevik rule was resisted or denounced by dissidents on the left from the very beginning. There were nevertheless major social gains and a flowering of egalitarian culture. It was during the 1920s that ecologists, naturalists, and conservationists exerted increasing influence and major strides were made in environmental protection (Gare 1993; Weiner 1999). These developments were curtailed in some respects but expanded in other ways by the late 1930s, under Stalin's leadership, in the aftermath of internecine struggle among Bolshevik factions over economic policies, political strategies, and the level and kind of centralisation. Political purges occurred multiple times, thousands of people were disappeared or interned, and many went through labour camps. Several million perished or suffered greatly, sometimes the same people who had inflicted the suffering. The horrific developments were not solely due to government repression, but to diffuse levels of violence and regional power dynamics that at times flatly contradicted government policies. In the end, it must also be appreciated, the majority had their lives improved despite a World War, a civil and revolutionary war involving foreign military invasions, and a war within the Bolshevik party. The carnage and trauma from the largely externally inflicted horrors would reach its apogee with yet another capitalist conflagration, World War II. Recovering from the terrifying and incalculable long-term repercussions was stymied by permanent military encirclement and economic sabotage by the most powerful states on the planet (D'Mello 2009; Getty and Naumov 1999; Kokosalakis 2023).

The repressive apparatuses evolved from the wars and famines of the 1920s enabled the country's further existence but also met with criticism from within as well as without. Prominent among those levelling the sharpest criticisms was Trotsky, erstwhile Menshevik turned Bolshevik in 1917 and founder of the Red Army. As an organiser of actions outside the party to counter the rising influence of Stalin's faction, he was expelled from the party in 1927 and assassinated in México in 1940. He was hardly the only one to suffer such a fate. He would be in the company of many prominent Bolsheviks, such as Bukharin, Zinoviev, and Galiev, subjected to trials of dubious legality, imprisoned, and executed. While living in exile, Trotsky founded the Fourth International in 1938 as response to the Third or Communist International (the Comintern, 1919–1943).

Critiques of the USSR under Stalin would vary with time and according to political persuasion, but largely centred on the lack of substantive worker control over the economy and the suppression of legitimate dissent. It is from these perspectives that the first critical studies of socialist states emerged. Developing and applying Marxist theories, Trotsky and his allies concluded that the USSR and similar societies had become degenerated workers' states, with later Trotskyists going further in regarding such social systems as bureaucratically deformed, ruled by a new and increasingly entrenched bureaucratic class, or, in other renditions, as state capitalist (Chase-Dunn 1982). Criticism eventually included environmental issues from within the USSR. It was often confined to scientific communities and naturalist organisations, but such organisations numbered millions of USSR citizens. Unlike other forms of political dissent, environmentalism was generally not met with repression. It was often expressed openly, including within the highest decision-making organs (Brain 2011; Weiner 1999).

The social convulsions and ecological destruction of multiple wars combined with revolutionary changes were intensified by the striving to reach industrial production levels. Within two decades, the USSR was transformed forcefully, feverishly, and violently into an industrialised society, organised into federated socialist republics under the rule of what came to be called the Communist Party of the Soviet Union (CPSU). Decision-making was centralised, often within the CPSU's Politburo, sweeping aside, more by force of circumstances than intention, the early Bolshevik internal policy of freedom of discussion, unity of action (democratic centralism). Periodic policies of forced requisitions and farm subordination to cooperative or state management ended private control over most farmland. All other land, surface waters, and offshore areas were for the most turned into state property under different ministerial jurisdictions, including mines, fishing areas, and the rapidly expanded ecological preserves (zapovedniki). Mixtures of economic coercion and enticement induced ever-greater employment in factories and cooperative and state farms. While it became increasingly repressive for many Bolsheviks as well, the USSR quickly began to rival economically and militarily the most advanced industrialised capitalist countries. This proved crucial to gaining popular consent, to curry leftists' support abroad, to rout the Nazi and imperial Japanese invaders from the west and east during World War II, and to help liberate many countries from German and Japanese colonial expansion.

Many communists and socialists drew great inspiration from the Russian Revolution and the Bolsheviks in particular. They began forming communist parties and/or aligning themselves with the Bolsheviks through the Moscow-centred Comintern. There, communist and allied organisations would meet periodically and coordinate activities, at least in the first years, to promote global proletarian revolution. Decisions made through the Comintern came to have far-reaching consequences, some of them unintended. For example, strategies for the colonies and relative to oppressed peoples were to involve the cultivation of solidarity among workers across different communities and alliances with liberal elements in national liberation movements. And the various affected parties dutifully aligned their activities accordingly in their respective countries. In China, this is what led to the early collaboration between the CPC and the Guómíndǎng until, after the death of the party founder, Sun Yat-sen (Sun Zhongshan), the Guómíndǎng purged and slaughtered communists in 1927. The reverberations were, among other places, felt by the communist anti-colonial Thanh Nien movement, which folded as a result of losing their primary base of operations in China and their main leader Nguyen Ai Quoc (Ho Chí Minh), who found safety in the USSR.

From 1928 to 1935 the Comintern called on communist parties to undermine moderate leftist parties. Thereafter, with the rise of fascism and Nazism, the policy shifted to a popular front strategy, characterised by forming broad alliances to fight fascism. Such abrupt shifts created confusion, demoralised or marginalised activists, and created conditions conducive to conformism and personal vendettas. Worker-led revolutions, as in Spain (1936), were sacrificed in the name of protecting the USSR, in a sense making a universalising socialist virtue out of the particularity of Bolshevik self-preservation necessity. Perhaps this, along with the trauma of the 1927 massacre, eventually contributed to the schism with the People's Republic of China years later.

Those allied to or taking inspiration from Bolshevik strategy turned ever more towards a framework oriented towards vanguard party structure and the takeover of the state to promote working-class interests. The objectives and techniques, to simplify, amounted to founding and expanding a rigidly disciplined party structure, often in clandestine form (mimicking experiences in Czarist Russia, where such strategies were essential for sheer survival). The next steps entailed building and severing alliances as necessary to appropriate state institutions, involving infiltration of the military or at a minimum the formation of a well-trained military wing. The capture of state apparatuses would be followed by the consolidation of power and elimination of bourgeois opposition and antagonistic leftist groups. The ensuing governing socialist institutions would then embark upon wealth redistribution (such as land reform and full employment), establishing social provisions (such as free education, healthcare, and public libraries), and industrialisation. The objective was to create the conditions for socialism, understood as a preliminary phase towards communism.

Capitalist Encirclement, Evolving Socialist States, and the Revolution in China

As socialist states developed, forced industrialisation became an attractive way in many countries, especially former colonies, to break from economic dependence on former colonising powers, while consolidating power and raising the material well-being of most people. Industrialisation also proved pivotal for self-defence against the constant threats of invasion or coups by liberal democracies or other capitalist states. The matter of self-defence may have been useful and still can be for propaganda purposes, especially in marshalling patriotic sentiment, but it was and is also a legitimate preoccupation. It must be recalled, for example, that the US invaded Russia in 1918–1920 and China in 1900–1901 to try and suffocate revolutionary movements and that such policies as military build-up and alliances like the Warsaw Pact, founded in 1955, were responses to global US military expansionism and to the 1948 formation and subsequent expansion of NATO (Berend 1996).

The very existence of the USSR (Mongolia remains perplexingly ignored by most) had also greatly modified the world economy and the political contexts of countries where leftist forces were emergent. At the same time, new socialist states established through Bolshevik-modelled and supported revolutions were expected to be coordinated with the CPSU Politburo or related institutions. To some extent, such an arrangement was established with communist parties in Western Europe, but in looser form. The linkage became more a formality by the early 1970s. Actual subordination was temporary and confined to a few central and eastern European countries, as important buffers against encroachment by what would become NATO powers. This arrangement unfolded through the USSR Red Army's liberation from Nazi and fascist regimes and subsequent military presence, which pre-empted political reversals in favour of the enemy camp but also stifled independent socialist state action. Salient examples of the latter are the military suppression of the revolts in Hungary in 1956 and in Czechoslovakia in 1968 (Berend 1996). The longer-term consequence felt to this day has been, arguably, the exacerbation of a pre-existing and diffuse anti-communism.

The degree of the USSR's grip over such states was always tenuous. The Red Army withdrew in 1958 from Romania, where the government pursued an independent path, essentially disengaging from the Warsaw Pact by 1963 and becoming among the US government's "Most Favoured Nation" trading partners in 1975. In 1973, after the capitalism-leaning reforms of 1968, the Hungarian government acceded to the General Agreement on Tariffs and Trade (the precursor of the World Trade Organisation), and then the IMF in 1983. These cases hardly countenance the notion of the USSR as an empire with satellite states, an ideological construct that pervades part of the left. Instead, those are instances illustrating the progressive subordination of such socialist countries to the core capitalist bloc during the USSR's very existence (Berend 1996). Nor was there the sort of neocolonial dependence rife among formerly colonised societies. By the late 1970s the USSR was basically subsidising its supposed "satellites" to the tune of $20 and $80 billion through an unequal exchange of raw materials and energy for manufactures (Turnock 2006, 284).

There were also socialist states forged out of autochthonous forces expelling the fascist invaders, like Albania and Yugoslavia. Those countries were always independent of and repeatedly clashed with the USSR. Overall, in environmental matters, as well as domestic social policy and even internal government politics (which could even involve multiple parties, as in the German Democratic Republic), socialist states were independent of the USSR (Berend 1996). Eventually, the USSR would succumb to the same core capitalist bloc under the combined weight of the liberal democracies' military-economic siege and the centrifugal effects of internal strife and anti-socialist forces.

The 1949 founding of the PRC played a major role in shifting influence away from Europe, as it eventuated into rivalry with the USSR over political leadership among communist formations worldwide. The schism happened by the late 1950s, leading to an undeclared border war in 1969 and rapprochement with the US government, made obvious with US President Nixon's 1972 visit. The Mongolian government, though allied with the USSR, played an important role as mediator between the USSR and PRC to quell tensions. There was also much political contentiousness within the PRC, for reasons not too dissimilar from those in the USSR during the 1930s. A major manifestation of the conflicts was the Chinese Cultural Revolution (1966–1976). The outcome of that conflict, the ousting of the Máo faction, eventually yielded a reformed CPC under

Deng Xiaoping's leadership. The sea change would pave the way for the expansion of capitalist relations under state control of key economic sectors and the political command of the CPC (Li 2009, 2016; Xu 2018).

The intensifying economic dependence of socialist states on capital from liberal democracies translated politically into a reform phase by the 1960s and 1970s in nearly all socialist countries (Frank 1989). During this period, party affiliation, especially in central and east European state-socialist countries, started to become less economically important than technical ability. A dual system came into temporary existence characterised by military and political dependence on the USSR by most socialist states, especially excepting the PRC, and economic dependence on the most powerful capitalist states (Böröcz 1992). By the 1980s, other, more recently established socialist states in Southeast Asia that endure today would follow a path more similar to that of the PRC (e.g., Vietnam's 1986 *Đổi Mới reforms*).

The Spread and Contraction of Socialist Successes

Socialist revolutionary movements were successful in many other countries as well, overwhelmingly in Africa and other parts of Asia. Just to illustrate the enormous diversity of political-economic contexts, cultures, and histories involved, self-defined socialist governments or states emerged, often briefly, in, for example, Afghanistan (1978–1992), Benin (1975–1990), Burkina Faso (1984–1987), Ethiopia (1974–1991), Grenada (1979–1983), Iraq (1958–2003), Libya (1969–2011), Nicaragua (1979–1990), Sudan (1969–1985), and Yemen (1969–1990) (see also Takriti and Ieddine 2023). Almost all such successes, whether taking the form of socialist states or one-party socialist governments, were consolidated between the 1960s and 1970s, during or after the USSR–PRC schism and the independent lines taken by Yugoslavia and Albania. These include the so far most durable socialist countries of Cuba, Laos, and Vietnam, founded between the 1950s and 1970s. The exception of the Democratic People's Republic of Korea (DPRK) is discussed briefly below. In some countries, like Bolivia and Venezuela, socialist governments have come to power in the aftermath of the dissolution of most socialist states, and therefore with little possibility of any support in the face of capitalist reaction internally and imperialist pressures from NATO countries.

Most of the successful socialist revolutionary formations were grounded in anti-colonial national liberationist movements and ideologies and deeply etched by long colonial histories of racialised segregation. Revolutionary struggles were deeply intertwined with decolonisation and involved alliances with nationalist and liberal democratic elements. Furthermore, revolutionary movements depended heavily on established socialist countries for military and economic assistance. That there would be conflicts among ruling socialist formations within and between newer socialist countries should therefore not be surprising. For example, in 1979 Vietnam and the PRC were at war in relation to the Vietnamese government, with USSR support, invading Cambodia to oust the Khmer Rouge, which had historically developed out of the Vietnamese-dominated Communist Party of Indochina, founded in 1930 by the above-mentioned Hồ Chí Minh (also known as Nguyễn Sinh Cung).[4]

The situation and trajectories in Africa, Asia, and to minor degree in the Americas could not be more divergent from those in Europe, and the USSR had even less influence in such places. Harsh colonial repression induced secretive organising, making Bolshevik strategies more attractive and effective. In colonies and semi-colonies in North Africa and much of Asia, Marxist social democratic parties had already formed since the days of the Second International and then Bolshevik parties by the 1920s. Most of them were repressed or essentially destroyed by governments aligned with core capitalist states. There are also cases of sheer desperation leading to draconian political measures. In the wake of genocidal, city-annihilating, and infrastructure-flattening US aggression, enormous reconstruction efforts, and following prolonged power

[4] The Khmer Rouge removed the US-backed Lon Nol dictatorship after the US carpet bombed the country (1965–1973) and killed hundreds of thousands of civilians (with the excuse of destroying Việt Nam Cộng-sản military corridors). In power, the Khmer Rouge, composed mainly of non-communists, engaged in mass atrocities as they robbed an already starved people of even the basic means of survival, tortured and murdered thousands, and imposed forced labour to accumulate capital for eventual industrialisation (Tyner 2017; Vltchek 2015, 610–611). Following liberation with the crucial assistance of socialist Vietnam, dissident Khmer Rouge took over the government and, after more than a decade of one-party rule, eventually came to be characterised by the currently dominant Cambodia People's Party, in the framework of a representative democracy and constitutional monarchy. The ousted Khmer Rouge regrouped, took refuge in Thailand, and had the institutional support of both the US and PRC governments until the 1990s. The Vietnamese army withdrew in 1989.

struggles within the Workers' Party of Korea, DPRK featured the intensification of decision-making centralisation as the Kim Il Sung Kapsan faction seized power. By 1972, they would achieve improvements in material living conditions to rival or supersede most countries at similar economic levels and steer the country towards a greater emphasis on Juche or "self-reliance". The latter policy has proven important in developing the wherewithal to withstanding the combined impacts of economically crippling US sanctions and successive bouts of prolonged extreme weather (droughts, floods) that led to untold numbers of deaths, including by starvation (Baer 2018; Cummings 2004; David-West 2011; Gills 1992; Robinson 2007).

Socialist Struggles in Africa

In much of Africa, socialist independence movements successfully overthrowing colonial rule were, on the whole, negatively impacted by the USSR-PRC split, as already stated. The newly independent states and socialist/communist formations had to rely on fleeting linkages and support. Successive and changing USSR and PRC governments lent a hand to varying factions or states, even anti-socialist ones like the Mobutu regime in the case of the PRC, according to what was deemed advantageous relative to their world position and perceived self-preservation needs. This was not a one-way process. Excepting the marginalised Pan-Africanist and Afrocentric Communist movements, who promoted African forms of collectivism, self-described socialist or communist parties were rather fluid politically, aligning themselves with liberal democracies, the USSR, or the PRC according to changing circumstances and power-retention prerogatives. It should be appreciated more that the possibility of losing state power usually meant the annihilation of socialist forces and the establishment or restoration of neocolonial control by Euro-American imperialist powers.

To some extent, what transpired internationally largely reflected the societal diversity negated by boundaries and states imposed by colonial dictatorships and the widely different histories of African communities. As in the Americas, many African peoples had no backgrounds of highly centralised states or sometimes any state formation at all. Most communities were, and in some cases continue to be, largely agrarian, with peasant or herder majorities. In the Americas, however, genocidal settler colonialism had prevailed along with anti-communism and early industrialisation developed on the backs of millions of enslaved Africans and on stolen lands. In contrast, African societies had undergone little to no industrialisation and often peasant communities had enough land access for self-reproduction, so that Bolshevism had arguably greater difficulty spreading compared to places in Africa affected by settler colonialism and industrialisation such as South Africa (Rodney 1972; see also Guevara 2022).

Pre-existing communalistic institutions in many Asian and African societies could be used towards developing and diffusing socialist ideas or practices, but the capacity for self-subsistence in most peasant, pastoralist and gatherer-hunter communities has also often made socialist notions or policies redundant. Hence, a diffuse radicalised social base, supporting otherwise well-organised political groups, was often absent and this also impeded any firm and lasting political footing in most countries. A consequence was at times a mix of socialist and liberal developmentalist policies characterising most socialist governments and states, as in Tanzania under Nyerere's administration (Campbell 1975). The Somali Barré administration is a more extreme paragon, switching sides and ideological discourse according to which government offered a better military and economic deal.

Furthermore, socialist states or one-party socialist governments in African countries, but also in many Asian countries, could not emulate Bolshevik strategies because they lacked the economic and military means to do so and, as a consequence of the USSR's diminishing capacity for international assistance and policies of détente, their economies remained under the orbit of colonial powers (the PRC, still very far from an industrialised economy, were hardly in a position to provide substantive aid). When gaining state power, communist and socialist parties, much like their counterparts in contemporary and, by the 1970s, increasingly indebted socialist states (Berend 1996), were easily undercut through raw material pricing (uneven terms of trade), direct financial pressures (e.g., extortionate interests on loans) and other such tactics from core capitalist countries or former colonial powers, which were often the same.

Otherwise, as in the cases of Angola and Mozambique, the new socialist states were under constant military attack and invasions by the Portuguese colonial militaries and the South African racist state, supported by the US and allies, as well as by local proxies. Had it not been for Cuban government help (not only in terms of military intervention), the fate of the independence movements in Angola and Mozambique would have been even worse and the prospects for the defeat of Apartheid in South Africa much diminished (Brittain 1988; Castro 1996; Saney 2023; Villafaña 2017). Other countries suffered a fate akin to that of the above-listed, short-lived Burkinabé socialist state (1983–1987), which was struck down through a murderous pro-French coup before any lasting positive social and environmental impact could even start to materialise. Such horrific post-liberation conditions made it virtually impossible to carry out any sensible environmental policy or to start reversing the long-term biophysical damage of colonial dictatorship. Excluding the relatively ephemeral Burkinabé case, the average lifespan of socialist states in Africa was 17.3 years, barely enough time to leave much of a mark (Engel-Di Mauro 2021). Rather, the environmentally ruinous mass extraction of raw materials to benefit core and semi-peripheral capitalist economies continued virtually unabated through neocolonial relations, facilitated through supranational institutions like the IMF and World Bank.

Burkina Faso

The Burkinabé Revolution merits further discussion, as it can be arguably understood as a historical example from which ecosocialists could draw inspiration, but also ideas for self-education regarding what can be achievable where and when. In fact, some even go as far as claiming that the Burkinabé Revolution is an example of ecosocialism.[5] The name Burkina Faso was introduced under the Sankara government in 1983. Until then, it was known as Upper Volta. Formal independence was gained from the French dictatorship in 1960. Until the Burkinabé Revolution, the country was mostly ruled by French government puppets. The revolutionary leader, Thomas Sankara (1949–1987), was a Marxist Pan-Africanist and military officer. He became prime minister in 1983 (after a coup that had countered a prior coup), but he was arrested during yet another a French-supported coup because he had been pushing for egalitarian policies. With popular support, he was freed the same year and led the Burkinabé Revolution.

[5] See, for instance: https://www.thomassankara.net/burkina-faso-de-sankara-experience-ecosocialiste-africaine/.

The Sankara government that emerged from the process was soon brutally brought down by means of assassinating most of the political leadership, including Sankara, and the violent repression of the Revolution's supporters. The Compaoré dictatorship, allied to France, lasted until 2014, when it was was ousted through a popular revolt. The convulsions of civil war and invasion enveloping the Sahel region since 2011 have deeply affected Burkina Faso, leading to sudden and frequent changes in government. Eventually, a military coup in 2022 rid the country of French troops, but Burkina Faso remains a beleaguered country where the current government has little to no control over much of the country.

The aims of the Burkinabé Revolutionaries were to achieve self-reliance on the basis of utilising local resources (notice the affinity with the DPRK's policies), the democratisation of society, establishing gender equality, and breaking with existing models of development. These objectives had internationalist and national dimensions simultaneously in that they entailed the end to "aid" dependency, reducing the political influence of landlords and petit bourgeoisie (bureaucrats, merchants, etc.), and, most ambitiously, restructuring and democratising the United Nations. The main policies that followed the Revolution fused social and ecological questions by combining the development of the conditions to raise living standards with the restoration of community and ecosystem health, ravaged as they had been by French colonial extractivism and undermined by persisting feudal relations. Land nationalisation (including mines), the promotion of gender equality at all levels, farmland redistribution to peasants, and the suppression of peasants' tribute and forced labour for landlords were policies that went hand in hand with reforestation programmes, mass literacy, and vaccination campaigns (assisted by socialist Cuba), the halting of unplanned wood cutting and livestock movement, and a reforestation programme to fight desertification (Biney 2018).

It may be tempting to see ecosocialism in such commitment to both social equality (land redistribution, support for women's empowerment, etc.) and ecological sustainability (reforestation and self-sufficiency), but it must be recalled that the Burknabé revolutionaries, especially their leader, Sankara, were of a Marxist-Leninist persuasion that was at odds with the kinds of anti-state-socialist "New Left" and other political seeds out of which ecosocialism sprouted. Such contemporary movements would have frowned upon military coups, instances of violent repression of upper-class forces and even trade unions, and ties to the socialist state of Cuba, among the crucial elements that made the Burkinabé Revolution possible, even if struck down within a few years by French and imperialist intervention with local reactionary support (Jaffré 1989; Sankara and Jaffré 2017).

American and Asia-Pacific Predicaments and Revolutionary Cuba

While in much of Africa socialist revolutionary movements faced formidable impediments especially on account of historical conjuncture, elsewhere prospects were and are even grimmer. The Americas and Asia-Pacific, dominated by settler colonists through racist systems, have been rough terrain for working-class unity of action through national liberation and decolonisation. Aside from draconian policies to separate people according to racial categories, the annihilation of many First Peoples and the appropriation of their lands necessitated an entirely different approach to socialism, one that never really came about as a result of racism or Eurocentrism among at least segments of settler colonial socialists (Bedford and Irving-Stephens 2000).

As discussed above, the various early Utopian socialist communities formed in the US by the likes of Cabet and Owen in the 1800s attest to a widespread indifference among socialists of the day to the racism, genocide, and settler colonial land theft that enabled the establishment of such Utopian socialist communities. The revolution in Haiti (1791–1804)—a yoke-shattering, world-transforming freedom struggle—was and still is little appreciated in socialist movements (Blackburn 2006). Environmental impacts in the Americas are thus overwhelmingly tied to shifts in various types of capitalist regimes. This regrettable situation has been abetted by the chronic failures of most socialist movements to develop a truly anti-colonial politics and prioritise the fight against the settler colonial regimes they inhabit (see Chapter 6). Most, but certainly not all socialist formations have been this defective politically. Hundreds of years of Indigenous and Afro-descendent peoples' liberation struggles have also nurtured movements that have led to some socialist successes in taking over the reins of government. The pluri-national state of Bolivia and the Bolivarian state of Venezuela (Bansart 2012) are the main illustrations of socialist rule over a still capitalist economy, but much improved policies with respect to dismantling settler colonialism.

Mixed-heritage peasant movements have also played a major role, as in the establishment of the Sandinista government in Nicaragua. But these are short-lived, limited, or very recent turns of events and have not had the chance to overturn centuries of settler colonialism. A veritable contrast is Cuba, where socialist revolution and sustained state policies have been eroding the foundations of US neo-colonialism and a heritage of genocide and slavery. Since sloughing off US colonial dominance in 1959, Cuba exemplifies a continuity of socialist politics despite huge odds. There have been instances of repression, not too unlike in other state-socialist contexts and in some respects not too different from situations like Burkina Faso, and in part due to military invasion and permanent military-economic siege (importantly, the US state continue to impose their military presence in Cuba by means of a naval base in Guantánamo). State socialism was a decisive pathway out of neocolonial stagnation and led to massive social gains. The environmental repercussions of socialism have eventually been especially fruitful, making Cuba one of the most environmentally sustainable countries on the planet, as will be further discussed in Chapter 8.

Concluding Considerations

Stemming from widespread discontent primarily spread across parts of industrialising capitalist Europe, movements that eventually came to be described as socialist brought together a burgeoning proletarian class with displaced artisans and peasants, among others radicalised by the increasingly horrific conditions of life forced upon masses of people by capitalist ruling classes. Out of persistent efforts at local to international levels of organising, socialists overcame the murderous and repressive apparatuses of capitalism and its disaggregating effects to mount a political force that to this day preoccupies ruling classes everywhere. Though fragmentary from the beginning and often beset by violent internecine strife, socialist movements continue to emerge and consolidate out of the fundamental contradictions created by a tendency for the division of society into a minority of owners of the means of production and capital and the masses of people forced to sell their labour-power, colonisers and colonised, slavers and enslaved, paid and unpaid, and by the continuous and violent private appropriation of socially produced wealth and of the labour of social reproducers.

The multiple forms and often simultaneously disconnected struggles ensuing from such contradictions have been and continue to be fierce and deadly, especially when they took on international proportions, starting with the Russian and then Mongolian Revolutions. Entire wars, including World War II, have been planned and waged by capitalist governments, including to try to annihilate socialism. What has resulted from such permanent upheaval and the many convulsive social developments prior to, during, and following socialist revolutions is a record that ranges from enormous and lasting advances in the material conditions of and life chances for the many to cases of terrible hardship, repression, and mass violence. Ecologically, despite fast-paced industrialisation and catastrophic instances, the overall record of socialism has been and continues to be positive compared to any kind of capitalist regime. In this evaluation, it is important to keep in mind the permanently altering effects and planetary-level environmental destruction wrought by capitalist forces during the centuries prior to the formation of the first socialist movements in the 1800s. Overall, the histories and development of socialist states and their ecosocial impacts are incomprehensible without factoring in historical and changing ecosocial conditions, internal struggles, capitalist legacies, and constant pressures from belligerent capitalist forces. Ecosocialists, whether they are willing to acknowledge it or not, inherit therewith the sum total of socialist movements' gigantic strides and heart-wrenching setbacks and the institutions thereby built, relative to which ecosocialists will have to measure their own achievements and prospects.

References

Baer, Hans A. 2018. *Democratic Eco-Socialism as a Real Utopia: Transitioning to an Alternative World System*. New York: Berghahn Books.

Bansart, Andrés. 2012. *Hacia el Ecosocialismo*. Caracas: Gobierno Bolivariano.

Bedford, David, and Danielle Irving-Stephens. 2000. The Tragedy of Progress: Marxism, Modernity, and the Aboriginal Question. Halifax, NS: Fernwood.

Berend, Iván T. 1996. *Central and Eastern Europe, 1944–1993: Detour from the Periphery to the Periphery*. Cambridge: Cambridge University Press.

Beik, Doris, and Paul Beik. 1993. *Flora Tristan, Utopian Feminist. Her Travel Diaries and Personal Crusade*. Bloomington and Indianapolis, Indiana University Press.

Biney, Ama. 2018. "Madmen, Thomas Sankara and Decoloniality in Africa." In *A Certain Amount of Madness. The Life, Politics and Legacies of Thomas Sankara*, edited by Amber Murrey, 127–146. London: Pluto Press.

Birchall, Ian. 1997. *Spectre of Babeuf*. Houndmills: Palgrave Macmillan.

Blackburn, Robin. 2006. "Haiti, Slavery, and the Age of the Democratic Revolution." *The William and Mary Quarterly*. 63 (4): 643–674.

Böröcz, József. 1992. "Dual Dependency and Property Vacuum: Social Change on the State Socialist Semiperiphery." *Theory and Society*. 21: 77–104.
Brain, Stephen. 2011. *Song of the Forest: Russian Forestry and Stalinist Environmentalism, 1905–1953*. Pittsburgh: University of Pittsburgh Press.
Braunthal, Julius. 1967a. *History of the International—Volume 1: 1864–1914*. New York: Praeger.
Braunthal, Julius. 1967b. *History of the International—Volume 2: 1914–1943*. New York: Praeger.
Brittain, Victoria. 1988. "Cuba and Southern Africa." *New Left Review* 172: 117–124.
Campbell, Horace. 1975. "Socialism in Tanzania: A Case Study." *The Black Scholar* 6 (8): 41–51.
Castro, Fidel. 1996. "Cuba and the End of Apartheid." *Socialism and Democracy* 10 (1): 1–5.
Chase-Dunn, Christopher. 1982. "Socialist States in the Capitalist World Economy." In *Socialist States in the World-System*, edited by Christopher Chase-Dunn, 21–55. Beverly Hills: SAGE.
Cole, George D.H. 1953. *A History of Socialist Thought: Volume I—The Forerunners 1789–1850*. London: Macmillan.
Cole, George D.H.. 1960. *The Second International 1889–1914, Part II*. London: Macmillan.
Cole, George D.H.. 1961. *Marxism and Anarchism 1850–1890*. London: Macmillan.
Cummings, Bruce. 2004. *North Korea: Another Country*. New York: The New Press.
David-West, Alzo. 2011. "Between Confucianism and Marxism-Leninism: Juche and the Case of Chŏng Tasan." *Korean Studies* 35: 93–121.
D'Mello, Bernard. 2009. "What Is Maoism?" *Economic and Political Weekly* 44 (47): 37–48.
Dickens, Peter. 1992. *Society and Nature. Towards a Green Social Theory*. Philadelphia: Temple University Press.
Eckhardt, Wolfgang. 2016. *The First Socialist Schism: Bakunin vs. Marx in the International Working Men's Association*. Translated by Robert M. Homsi, Jesse Cohn, Cian Lawless, Nestor McNab and B. Moreel. Oakland: P.M. Press.
Engel-Di Mauro, Salvatore. 2021. *Socialist States and the Environment. Lessons for Ecosocialist Futures*. London: Pluto: Press.
Foster, John B. 2023. "Engels and the Second Foundation of Marxism." *Monthly Review* 72 (2): 2–12.
Frank, Andre Gunder. 1989. "The Development of Underdevelopment." *Monthly Review* 41 (2): 37–51.

Gaillard, Jeanne. 1971. *Communes de Province, Commune de Paris (1870–1871)*. Paris: Flammarion.

Gare, Arran. 1993. "Soviet Environmentalism: The Path Not Taken." *Capitalism Nature Socialism* 4 (4): 69–88

Getty, J. Arch., and Oleg Naumov. 1999. *The Road to Terror: Stalin and the Self-Destruction of the Bolsheviks, 1932–1939*. New Haven: Yale University Press.

Gills, Barry. 1992. "North Korea and the Crisis of Socialism: The Historical Ironies of National Division." *Third World Quarterly* 13 (1): 107–130.

Graham, Robert, ed. 2005. *Anarchism: A Documentary History of Libertarian Ideas. Volume 1: From Anarchy to Anarchism (300 CE to 1939)*. Montréal: Black Rose Books.

Guérin, Daniel. 1970. *Anarchism: From Theory to Practice*. New York: Monthly Review.

Guevara, Ernesto Che. 2022. *Congo Diary: Episodes of the Revolutionary War in the Congo*. San Francisco: Seven Stories Press.

Jaffré, Bruno. 1989. *Burkina Faso. Les années Sankara, de la révolution à la rectification*. Paris: L'Harmattan.

King, Sam. 2021. *Imperialism and the Development Myth*. Manchester: Manchester University Press.

Kokosalakis, Yiannis. 2023. *Building Socialism: The Communist Party and the Making of the Soviet System, 1921–1941*. Cambridge: Cambridge University Press.

Krader, Lawrence. 1974. *The Ethnological Notebooks of Karl Marx*. Assen: Van Gorcum.

Lejeune, Dominique. 2016. *La France des Débuts de la IIIe République: 1870–1896*. Paris: Armand Colin.

Lenin, Vladimir Ilich. 1917 [1963]. "Imperialism, the Highest Stage of Capitalism." In *Selected Works*, vol. 1, pp. 667–766. Moscow: Progress Publishers. https://www.marxists.org/archive/lenin/works/1916/imp-hsc/. Accessed 6 March 2024.

Li, Minqi. 2009. *The Rise of China and the Demise of the Capitalist World Economy*. New York: Monthly Review Press.

Li, Minqi. 2016. *China and the 21st Century Crisis*. London: Pluto Press.

Losurdo, Domenico. 2011. *Liberalism. A Counter-History*. Translated by Gregory Elliott. New York: Verso.

Meisenhelder, Thomas. 1995. "Marx, Engels, and Africa." *Science & Society* 59: 197–205.

Merchant, Carolyn. 1980. *The Death of Nature: Women, Ecology, and the Scientific Revolution*. San Francisco: Harper.

Murrey, Amber. 2018. "Introduction." In *A Certain Amount of Madness. The Life, Politics and Legacies of Thomas Sankara*, edited by Amber Murrey, 1–18. London: Pluto Press.

Newman, Michael. 2015. *Socialism: A Very Short Introduction.* Oxford: Oxford University Press.
O'Connor, James R., ed. 1998. *Natural Causes: Essays in Ecological Marxism.* New York: Guilford Press.
Plaetzer, Niklas. 2021. "Decolonizing the 'Universal Republic': The Paris Commune and French Empire. *Nineteenth-Century French Studies* 49 (3): 585–603.
Reclus, Elisée. 2013. *Anarchy, Geography, Modernity: Selected Writings of Elisée Reclus.* Introduced and edited by Camille Martin and John Clark. Oakland: PM Press.
Robinson, Michael E. 2007. *Korea's Twentieth-Century Odyssey: A Short History.* Honolulu: University of Hawai'i Press.
Rodney, Walter. 1972. "Tanzanian Ujamaa and Scientific Socialism." *African Review* 1 (4): 61–76.
Rodney, Walter. 2011 [1972]. *How Europe Underdeveloped Africa.* Baltimore: Black Classic Press.
Saney, Isaac. 2023. *Cuba, Africa, and Apartheid's End: Africa's Children Return!* New York: Rowman & Littlefield.
Sankara, Thomas, and Bruno Jaffré. 2017. *La liberté contre le destin.* Paris: Syllepse.
Takriti , Abdel Razzaq, and Hicham Saf Ieddine. 2023. "Arab Socialism." In *The Cambridge History of Socialism, Volume II*, edited by Marcel van der Linden, 474–516. Cambridge: Cambridge University Press.
Turnock, David. 2006. *The Economy of East Central Europe, 1815–1989: Stages of Transformation in a Peripheral Region.* London: Routledge.
Tyner, James A. 2017. *From Rice Fields to Killing Fields: Nature, Life, and Labor under the Khmer Rouge.* Syracuse: Syracuse University Press.
van der Linden, Marcel. 2023a. "Introduction to Volume I." In *The Cambridge History of Socialism, Volume I*, edited by Marcel van der Linden, 1–25. Cambridge: Cambridge University Press.
van der Linden, Marcel. 2023b. "Introduction to Volume II." In *The Cambridge History of Socialism, Volume II*, edited by Marcel van der Linden, 1–44. Cambridge: Cambridge University Press.
Villafaña, Frank. 2017. *Cold War in the Congo: The Confrontation of Cuban Military Forces, 1960–1967.* London: Routledge.
Vltchek, Andre. 2015. *Exposing Lies of the Empire.* Jakarta: Badak Merah Semesta.
Weiner, Douglas R. 1999. *A Little Corner of Freedom: Russian Nature Protection from Stalin to Gorbachëv.* Berkeley: University of California Press.
Xu, Zhun. 2018. *From Commune to Capitalism: How China's Peasants Lost Collective Farming and Gained Urban Poverty.* New York: Monthly Review Press.

CHAPTER 3

A History of Ecosocialism

"Ecosocialism is socialism made ecologically rational." (Kovel 2005, 2)

Rooting Ecosocialism

Though ecosocialism is principally a 1980s phenomenon, some contend that the roots go back much further. According to such a view, what would now be regarded as ecological concerns from a socialist perspective were already expressed in the writings of Marx, Engels, Reclus, Kropotkin, Morris, Luxemburg, and Lenin, among others (Benton 2023; Foster 2000; Gare 1993; Pepper 1993; Pepper 2010b; Wall 2010; Williams 1982). Roelofs (1993) deems Fourier a precursor to the red-green movement, thereby arguing for parallels with Utopian socialism (see also Pepper 2010b; Vettese and Pendergrass 2022).[1] Buhle (2000) notes a couple of examples of socialists involved in forestry and parks management in the US drawing together environmental conservation and

[1] Suing (2018), among others, would be among those who might agree with such research explorations into historical antecedents, but for different reasons. He sees in much of ecosocialism the same sort of limitations Engels and Marx had identified in their socialist predecessors and many contemporaries, in contrast to the actual and practical progress made on ecological matters in socialist countries, without recourse to developing ecosocialist perspectives.

S. Engel-Di Mauro, *Ecosocialism*, Marx, Engels, and Marxisms,
https://doi.org/10.1007/978-3-031-74288-0_3

socialist values during the early twentieth century. These are among the examples of seeing a foreshadowing of ecosocialism in the ideas of past influential or lesser-appreciated socialist figures. A presumed suppression of environmentalism within socialist movements or formations, so the narrative goes, led to failures in developing ecologically sustainable alternatives. This, along with the anti-ecological and ultimately destructive nature of capitalism, is a main justification for ecosocialism.

On the surface, this recounting appears sensible, but it misses the social context out of which ecosocialism has emerged. The scouring for and exhumation of ecological thought in socialist writings is a historiographical (perhaps at times hagiographical) practice of relatively recent vintage. It is largely traceable to the rethinking among leftists that was stimulated by social movements, including environmentalism, in 1960s liberal democracies. The rethinking often culminated in rejecting most prior forms of socialism and sometimes even embracing liberalism (prominent examples being Daniel Cohn-Bendit and Christopher Hitchens).

Rather than an inheritance of some socialist undercurrents of yore, ecosocialism's historical antecedents are to be found in the leftist movements and neo-Marxist intellectual circles arising out of two general and intertwined political processes. One is the spread of environmentalism and so-called new social movements, including civil rights, anti-racist, feminist, and what eventually would be identified as LGBTQ formations. All were and mostly remain for the most part eviscerated of structural analysis or devoid of awareness regarding capitalism as internally heterogeneous totality (Wallis 2020, 71–72). Another process is rooted in the much longer-standing repercussions of splits within socialism that occurred, as illustrated in the previous chapter, multiple times and for differing reasons. The most recent iteration, the one out of which ecosocialism emerged, is represented in the dissatisfaction with and often rejection of post-revolutionary, that is, socialist countries, and all those movements and parties therewith associated. Much of ecosocialist thought thus bears the imprint of aspects and combinations of broadly Marxist "New Left" perspectives (e.g., Baer 2018; Benton 2023; Kovel 2007; O'Connor 1998; Saito 2023; Wallis 2020; Wolf and Löwy 2007), peppered with Socialist Feminist and Trotskyist tendencies (e.g., Brownhill and Turner 2020; Löwy and Tanuro 2022) and various neo-Marxist schools of thought, like the Frankfurt ("Critical Theory") School (Gare 1993; Vettese and Pendergrass 2022) and the Baran-Sweezy Monopoly Capital framework (e.g., Foster 2023b).

The Influential Background of the 1960s Movements in Liberal Democracies

The global 1968 revolts were a particularly important juncture in this recovery and elaboration process. Anarchist, feminist, neo-Maoist, and Trotskyist organisations were influential in those revolts, an outpouring of liberation movements long excluded or marginalised. Multiple kinds of social upheavals and national liberation struggles provided the foundations to the formation and development of ecosocialism, as part of other anti-systemic non-capitalist perspectives. These include the movements that form and rapidly dissolve following episodic eruptions of popular demands for social justice and the kinds of social provisions that have been the hallmark of socialist states, like guaranteed employment, workplace rights, unemployment benefits, national healthcare and much else that is denied or provisional in capitalist countries.

In areas of the world where struggles for self-determination or sheer survival involve the protection of ecosystems, ecosocialism is rarely found. Where it is found, ecosocialism is rooted in more advanced ecological understandings associated with socialist or communist movements. A foremost current example is the convergence with ecosocialist ideas of the alliances with Lakota communities fighting against impositions of pipelines. More closely associated with ecosocialism is the Seringueiros movement led by Chico Mendes (assassinated in 1988). He exerted much effort in joining trade unionism with Indigenous communities and the liberation theology movement to create the Forest Alliance and mount thereby more effective struggles against firms involved in deforestation. This is deemed to be a first in combining ecological and socialist approaches in an explicit manner outside of a core capitalist country (Löwy 2015).[2] Cooperative Jackson in Mississippi (US) is another, though rare, example in core capitalist countries of forging locally based alliances to lay the groundwork for livelihood improvements and political empowerment for the majority Black residents, on an ecosocialist basis and drawing from the historical experiences of the Civil Rights and Black Power movements (Akuno and Nangwaya 2017). Further details about Cooperation Jackson are offered below.

[2] The relationship between ecosocialism and First Peoples' highly varied politics is explored in Chapter 5.

By and large, social movements and alternative socialist formations rising primarily out of the social protest movements of the 1960s constituted the seeds of ecosocialist thought and politics. With exceptions (e.g., Labour Parties in Australia, New Zealand, and the UK), those alternative socialist currents tended to situate themselves outside parliamentary representation processes or to refuse such institutional politics altogether. This includes leftist elements that exist within contemporary environmentalism. A defining characteristic, carried over into much of ecosocialism now, is such socialist groups' self-distancing or severance from communist parties aligned in any way with socialist states. Drawing principally from those kinds of socialist frameworks, closer to or in line with democratic socialism, ecosocialists, to reiterate, have been motivated not only by the rejection of capitalist relations, but also, especially in the case of ecosocialists in liberal democracies, by the critique of, if not repudiation of existing socialism. In some instances, such as Ryle (1988, 25), liberal democracy is openly preferred to "actually existing socialism."

In sum, all ecosocialists deem capitalism as the ultimate cause of the socio-ecological crisis and as inherently inadequate to the task of addressing the crisis. However, many ecosocialists extend the latter to "first epoch" socialism as well, or they deny the socialist character of socialist countries. Perhaps unsurprisingly, ecosocialism is a phenomenon delimited to capitalist countries, especially those ruled through the globally most powerful, imperialist states, which are liberal democracies.

Developing Red-Green Perspectives

Initially, ecosocialism was a way to designate the project of drawing labour and environmental movements together. This has also been referred to as Red-Green politics, a different label for what essentially points to Ecosocialism. The bicoloured designation describes the attempt to blend socialist (Red) with environmentalist (Green) perspectives. Presumably, Ecosocialism would signify the amalgam being consummated.[3] The term Ecosocialism also appears in both hyphenated (Eco-Socialism) and unified (Ecosocialism) form, but there seems to be no denotational difference

[3] Out of such colour mixture one could flippantly make a Brown, a symbol that could evoke a pledge to prioritise the struggles of racially minoritised peoples. As it turns out, this direction has not been particularly prominent in attempts to combine Red with Green movements, which is unfortunate.

intended (e.g., Hayward 1990; Mulberg 1992; Pepper 1993; Ryle 1988; Weston 1986).[4]

This striving for combined labour-environmentalist politics appears formally as "Eco-Socialism" in a 1980 comic-strip pamphlet of the Socialist Environment and Resources Association (SERA), an organisation affiliated with and campaigning for the UK's Labour Party (SERA 1980[5]; see also Albritton 2019, 6). Founded in 1973, SERA remain in existence, continuing in their attempts to influence what since the 1990s have become essentially and anti-socialist political party. From its inception, Ecosocialism has emerged from and is grounded primarily in the politics of liberal democratic Anglophone European and European-descendent settler colonial capitalist countries.

The Environmentalism that could cohere with socialism is, for obvious reasons, egalitarianism leaning. Examples are Ecofeminist groups and organisations like Green Peace, Friends of the Earth, and possibly Earth First! and leftist currents within some Green parties. As Pepper has explained, a major challenge in uniting Red and Green is that the former is mainly Marxist in orientation, while the latter is underlain by a predominant anarchist influence and often by claims of transcending both capitalism and socialism (Pepper 1993, 3). Hence, an Ecosocialist project often becomes an attempt to overcome Green anti-socialism and to bring together socialist movements (if one includes anarchism in socialist history) that have been divergent and mutually hostile. At least, for the most part, all socialist currents typically shun Malthusian, technocratic, and other environmentalist currents close to or supportive of capitalism. The example of SERA tends to be exceptional, but unfortunately even unconventional Greens may harbour Malthusian or like sentiments. More recently, Green parties have amply demonstrated instead how well they can conform to or represent capitalist interests, including by colluding with right-wing forces and imperialist institutions.

[4] Perhaps the hyphen is to signal a fusion of labour and environmental movements (as for SERA) or a rupture or difference with other and usually past forms of socialism (e.g., Löwy 2005), but the rationale has not been made explicit. I prefer the unified, unhyphenated version because I see largely a continuity with forms of socialism and radical environmentalism prevalent in capitalist countries, especially of the liberal democratic variety. This is likely a minoritarian view within Ecosocialism.

[5] Thanks to Ian Angus, the main force behind the website "Climate and Capitalism" (https://climateandcapitalism.com/), the pamphlet has been scanned and made widely available.

The Red component is relatively wide-ranging and mainly undergirded by one or another sort of socialism, often Marxism in particular (Pepper 1984; Wallis 2018). However, in part as a consequence of its origins in socialist movements within liberal democracies, Red politics developed out of, if not repudiation of existing and past state-socialist systems, regarded as authoritarian or dictatorial. Most leftists in liberal democracies have progressed little from the kind of peremptory judgement featured in the above-cited SERA comic-strip, where British male industrial workers lecture caricatured communist officials on developing "a planned economy with a human face." The accompanying caption reads:

> The Soviet experiment went off the rails long ago... Today it's more or less a parody of socialism...a centralised, authoritarian, bureaucratic and energy-hungry society riddled with nukes. (SERA 1980, 17)

A similar argument is repeated in Ryle (1988, 39). One could exchange the compliment to all such socialists dismissive of the "Soviet experiment," given their spectacular failure to establish any "planned economy with a human face" in their own countries. The record becomes even more embarrassing for the likes of SERA, who persist in aligning themselves with a warmongering political party whose policies are "riddled with nukes" and high energy consumption policies.

The sensitivity to authoritarianism is anyway not always consistent among such Reds. In much of the ecosocialist imagination, Red may include unions, vaguely leftist political parties (such as labour parties), and movements that are not necessarily anti-capitalist or democratic (in the sense usually understood by leftists in liberal democracies). Concepts like authoritarianism are anyhow relatively flexible or contingent on the specific leftist point of view considered. Authoritarianism may be a smear about the state generally, democratic centralism, or even centralism in any form. It depends on which kind of socialist or communist one asks. Assessments of political systems in Cuba or the People's Republic of China (PRC) can be one way of discerning what sort of socialist or communist current is represented. Noteworthy is the rarity of ecosocialists defining themselves as communist, as Schwartzman has noted (2022, 286). This is because most ecosocialists do not regard ecosocialism as a transitional phase towards communism, abandoning thereby a core understanding of long-term political strategy developed by Marx and Engels.

In contrast, there is more room for religious currents that align or intersect with leftist anti-capitalist causes, like the Liberation Theology movement and the Catholic Worker. To Anarchists, of course, much if not all kinds of socialism are hopelessly authoritarian, so they tend to reject political formations denoted by the colour Red. Antagonistic sentiments between anarchists and socialists are, often, reciprocated still today, but there are nevertheless points of convergence with respect to struggling for a stateless, classless society (Hayward 1990; Pepper 1993). Akuno and Nangwaya (2023, 64–65) see worker-owned cooperatives as especially suitable training grounds for the construction of an egalitarian world, and as a way to integrate the ideas of Marx (communism) and Bakunin (collectivist anarchism). This is yet another crucial departure from Marx and Engels that gets too little attention but should merit scrutiny and debate among ecosocialists.

Be that as it may, at first, use of the term ecosocialist was not necessarily complimentary. Some perspectives from liberal democratic and socialist countries, though for differing reasons, converged in describing ecosocialism as ingenuous politics exemplified by the leftist environmentalism of the likes of Barry Commoner and André Gorz (Emel and Peet 1989, 60–61; Ostrovityanov 1987; Pepper 1984). Others, such as anarchist academics, would predictably regard it as too constraining in focusing only on capitalism (Atkinson 1991). Anarchist environmentalists, often closer to what are sometimes called ecocentric perspectives, still generally find little resonance in ecosocialism, in continuity with the above-described historical splits established by the 1870s (Pepper 1993).

Dialogue and debate over socialism's relationship to environmentalism spread and became more sustained by the early 1990s, mainly with respect to Marxism. This is covered in detail in subsequent chapters. Suffice it here to state that the discussions occurred both within leftist parties or movement coalitions, where ecosocialism first emerged, and subsequently scholarly journals and the academic left. Eventually, especially since the early 2000s, ecosocialism reached wider political and academic traction, though globally of marginal consequence. Its formal articulation as a platform was reflected in the 2001 *Ecosocialist Manifesto*, which was revised and expanded in 2007 for its presentation at the 2009 World Social Forum (Löwy 2015). Combined with an increasing number of currents within leftist political parties and other organisations adopting some form of ecosocialism, there have been several institutions where ecosocialism is being incorporated as part of policy or advocacy. Examples range from the Bolivian and Venezuelan governments to parties or coalitions like the Left-Green Movement (Iceland), the Nordic Green Left Alliance, and the Partido Socialismo e Libertade (Brazil). More about the geography of ecosocialist politics will be discussed in Chapter 8. What is emphasised, aside from care for working class and ecological issues, is attentiveness to and promotion of gender-egalitarian, anti-racist, and decolonial or post-colonial perspectives (Baer 2018; Bansart 2012, 23–24; Barca 2019).

Main Ecosocialist Currents

At the time of writing, ecosocialist formations are largely confined to currents within leftist political parties and loose academic networks. Several political parties in different parts of the world have adopted ecosocialist platforms over the past two decades. As the ecosocialist component is only one among various kinds of advocacy frameworks or internal party currents, it would be a misrepresentation to define such parties as ecosocialist. The only substantial institutional presence is in Venezuela, where there is a Ministry dedicated to environmental protection on the basis of ecosocialist principles. This is the Ministerio del Poder Popular para el Ecosocialismo (MINEC, http://www.minec.gob.ve/). More will be discussed in Chapters 4 and 8 about this and the degree of ecosocialism's political reach in different parts of the world.

Like socialism has been historically, ecosocialism is a heterogeneous politics. It draws also from different, sometimes contrasting perspectives beyond Marxism, including Ecofeminism, certain strands of Utopian socialism, and even anarchism, the sort often represented implicitly in radical environmentalism (Hayward 1990). Generally, the main theoretical manifestations of ecosocialism are Eco-Marxism and socialist or ecosocialist Ecofeminism (to some, interchangeable with Ecofeminist ecosocialism). They provide the philosophical planks for much of ecosocialist ideology and politics. Subsequent chapters are dedicated to each of them because of this.

There are several kinds of ecosocialist politics that draw directly or indirectly from those frameworks, aside from ecosocialist Ecofeminism, which stands as its own. Proponents of socialist or ecosocialist Ecofeminism are usually involved in or supportive of grassroots activism or social movements and suspicious of centralism. Political parties and typical forms of socialist organising through such institutions as trade unions tend to be shunned or regarded as inadequate for the task of bringing about social change. This may at first seem to accord more with anarchist frameworks and politics, though it is not coextensive with them on account of capitalist patriarchy and social reproduction being the focus of critical scrutiny more than the state itself (Bansart 2012, 18–20; Barca 2023, 8–9; Brownhill and Turner 2020; Salleh 1997; Shiva and Mies 1993).

Other ecosocialists who may or may not be associated with any definite current share this view, even as they may advocate for political party formation (e.g., Kovel 2007; Wall 2010). Eco-Marxists, on the other hand, are primarily academics who may or may not be associated with any ecosocialist formation or collective but tend to advocate for some sort of state involvement or takeover, and/or some sort of wide social movement that is not necessarily averse to state intervention (e.g., Akuno and Nangwaya 2023, 75–76; Magdoff and Williams 2017, 326; O'Connor 1998, 310). A prominent expression of this is to be found in the ideas proposed by self-described democratic ecosocialists (Baer 2018, 136–153).

A predominant influence on ecosocialism, especially in political parties espousing ecosocialism, is to be found in what is often called the "New Left" of the 1960s, as adumbrated above. It is associated with the development and promotion of socialist democracy, or what is also frequently named democratic socialism, as an alternative to state socialism. "New Leftists" have argued that "first epoch" socialism failed at achieving socialist aims. Some reject aspects of Marxism. This can be in the form of a perceived omission of social reproduction and inadequate treatment of social relations that are understood as being beyond class. Some point to what they deem original flaws in Marxism that should be corrected, such as a dearth of ecological thought and a tendency to exalt technological progress.

Often, socialist states are the main target of critique and sometimes they are used as illustrations of the problems with Marxism. Some democratic socialists recognise the importance of the many historical challenges faced by socialist states, but they remain critical of and politically aloof from socialist countries. Reasons usually given for this are a lack of workers' control over the economy and what democratic socialists deem to be an authoritarian system. How socialist revolutions degenerated to such undemocratic conditions is variably explained. Generally, in "New Left" readings, underdevelopment and inimical external and internal pressures are indicated as factors that prevented the development of socialist democracy after successful revolutions. There have been different interpretations given for the lack of development in socialist countries of what "New Leftists" view as socialist democracy. A main critique was and continues to be about the absence of substantive worker control of the economy and the existence of the so-called one-party state, which socialist democrats see as intrinsically undemocratic (Ali 2009; Baer 2016; Foster 2000; Magdoff and Magdoff 2005; Miliband 1994; Wallis 2020).

For the most part, ecosocialism's basis lies in the combination of socialist politics with pacifist and parliamentary strategies within a liberal democratic order (Baer 2016; Löwy 2015; Wolf and Löwy 2007). According to this vision, ecosocialism must be underpinned by a set of democratic institutions for genuine socialism to emerge and presumably ecologically sustainable economies. By democratic is meant political pluralism (including multi-party elections), freedom of speech and assembly, access to the media, the right to form trade unions, and cultural liberty. Such political institutions and guarantees are to be underlain or supported by a dominant public sector characterised by partial economic decentralism with only small-medium firms and mainly worker cooperatives. In other words, socialist democracy is about building on liberal democratic institutions to attain workplace democracy and popular participation in decision-making over matters of direct concern (Akuno 2023; Baer 2018, 132–136; Magdoff and Williams 2017; Wall 2010; Wallis 2020).

Democratic eco-socialist narratives illustrate, in particular, the ways in which ecosocialism is internally differentiated as well as contrasted to other forms of socialism. Already the term itself, democratic eco-socialism denotes the existence of and distancing from forms of ecosocialism considered undemocratic. The identity of such currents of undemocratic ecosocialism is curiously not divulged. Some clues can be gleaned from democratic eco-socialist recounting of ecosocialism's history.

According to democratic eco-ecosocialist Satgar (2018, 11), there have been five historical phases since the 1970s that have led to the emergence of specifically democratic eco-socialism. They involve both critique and reinterpretation of Marxism, the development of neo-Marxism, as well as the rise of anti-systemic movements. The first three phases are respectively reflected in the Marxist ecology critique of "first epoch" ("productivist") state socialism. Elements of the first are to be found in the works of the critical Utopian Marxist turned conservative spiritualist Rudolf Bahro (Amberger 2023) and the fascist turned communist Manuel Sacristán (Tello 2016). The second phase was modifying Marxism by including ecology concerns, as done by the likes of James O'Connor, Ted Benton, Joel Kovel, and Michael Löwy. The third phase involved bringing out the supposedly neglected ecological principles in the works of Marx and Engels, such as in the work of John B. Foster. The fourth and fifth historical phases have involved using a form of historical materialist "ecology" and capitalist crisis theory to address current environmental problems (e.g., Nimmo Bassey, Naomi Klein, David Layfield) and the rise of ecosocialist or "green left" forces championing systemic alternatives (e.g., La Via Campesina, the Forest Alliance, the World Social Forum), as also identified by Derek Wall (2010). What is noticeable in such self-defining genealogy is the adherence to eco-Marxism and implicit rejection of classical Utopian socialism (cf. Pepper 2010b) and downplaying or omission of Ecofeminism or Ecofeminist concerns. The latter two are replaced by combinations of social movements and political party organisation to take over the reins of the state by means of electoral avenues or parliamentary roads.

Evidently, ecosocialists part ways with each other over a number of principles and strategies, something adduced by what is emphasised and what is omitted. One major issue is the degree of political and/or economic decentralisation or centralisation, which implies ideas and policies about the mode of decision-making. Put differently, a salient difference is among those who see state power or centralisation as essential, those who find such strategy anathema, and those who strive to find complementarity between grassroots and state institutions. The issue of the state impinges on how ecosocialists envision strategies or what are regarded as feasible objectives and/or effective strategies in efforts to promote and bring about ecosocialism. Among the strategies there are often implicit differences over the role of parties, trade unions, movements, and other ways of organising.

The question of the state and the main political vehicles to develop for effecting social change is at times introduced according to whether strategies are reformist or revolutionary, but these terms are unhelpful in representing actual differences. Just as in the case of socialism generally, what is revolutionary from one's vantage point is reformist from another's. Circumstances also change in the same place and the political situation also differs from one location to another. It is revolutionary to call for universal suffrage when a handful of propertied white men can vote, but it is reformist to campaign for legal means to ensure everyone can vote when only anti-communist parties are effectively to present candidates. In either case, both forms of political action can be useful, though in different ways, towards building power to overcome capitalist dictatorship.

Sometimes, there are also points of contention among ecosocialists regarding the relative importance of forms of oppression aside from class relations, including patriarchy, racism, homophobia, transphobia, and speciesism. The relative emphasis over certain modes of oppression affects which subject positions are interpreted as possessing revolutionary potential. For instance, as will be discussed in subsequent chapters, Ecofeminist Ecosocialists see patriarchy as ultimate target, with capitalist relations as its latest incarnation, so that the social position of womanhood is a main, though not sole focus of attention. People involved in ecologically regenerative, non-capitalist relations may also be included. Eco-Marxists see capitalism as a totality, including patriarchy, rather than a derivation of patriarchy. They will tend to privilege proletarians as the revolutionary subject, although the definition of proletarian is not exclusive of unwaged workers and has become very elastic in some theoretical frameworks.

Nevertheless, there are commonalities to all ecosocialist perspectives that should always be borne in mind. At a minimum, like all socialists, ecosocialists share a commitment to social change away from capitalism and towards a world based on social equality. A commitment to internationalism seems to be shared as well among ecosocialists. Solidarity and cooperation feature among basic principles, although what they mean in practice is not always made clear. Notably, most ecosocialist currents converge in their negative appraisal of socialist countries or what they often deem as "first epoch" socialism. Usually, "first epoch" socialism is defined as insufficiently attentive to environmental issues or to be associated with destructive environmental practices. This is understood to be intimately linked to authoritarianism and, according to many ecosocialists, an over-emphasis on economic growth or the development of the forces of production to the detriment of ecosystem and human health. Such forms of socialism are to be superseded by an ecologically based kind of socialism, according to these viewpoints. The quest for socialist economies that are ecologically sustainable usually unites ecosocialists.

Returning to the splintered nature of ecosocialism, notwithstanding substantive grounds for convergence and possibly concerted action, ecosocialists are divided on several major concerns and this may impede the creation of a unified and coordinated ecosocialist politics. Anitra Nelson identifies two major tensions within ecosocialism that roughly correspond to political tendencies or philosophies (Nelson 2022, 32–33). According to Nelson,

> ecosocialism has been ridden with frictions between democratic, even parliamentary, ecosocialist advances and post-development ecosocialist grassroots activism at a determinedly local level of governance and market-free production and exchange. (Nelson 2022, 35)

This can be one way of characterising the differences within ecosocialism that accords with what is illustrated above. Another perspective on the range of ecosocialist politics is to define the tendencies according to relative affinity with differing degrees and forms of political-economic centralisation and coordination, as discussed in the case of socialist histories, in the previous chapter.[6]

The issue can be analysed politically and/or economically, but the matter can also be framed in terms of what sort of and how to build social relations that can overcome capitalism. Politically, ecosocialist currents oscillate between calling and organising for political parties, attaining state or government power, and the like, compared to those privileging or identifying with grassroots movements and decentralised decision-making processes. A related tension is over the economic means to achieve ecosocialism, whether by modifying existing structures (setting up local currencies, cooperatives, state firms, etc.) or by constituting entirely new or prefigurative ones (communal housing, collective land ownership, barter systems, etc.).

This perspective on ecosocialism reproduces the broad historical lineaments of ideological fractures among socialists, which may or may not

[6] This is usually described in the dichotomous terms of "top-down" and "bottom-up" frameworks. I resist using such categories because they foreclose the possibility of both forms co-existing or making ecosocialism viable by means of combining them, as Wallis (2020) and Akuno and Nangwaya (2023) have pointed out. This is a major feature of the struggles ongoing within Venezuela, for example, in terms of making the formation and coordination of worker-managed communes cohere with state policies. Furthermore, perhaps more prosaically, it should be recognised that there also exists a "top-down" organising with "bottom-up" acquiescence or relative agreement, for otherwise "top-down" approaches are impossible (one could also argue this in Gramsci's terms of coercion and consent processes, but the claim herein made is rather banal in comparison). The discussion, in other words, should be centred about political strategy effectiveness in relation to context or the current balance of forces and power dynamics, not about abstract principles.

cohere with actual socialist practices or wider political processes. Principles must inform strategy but must also have their day of reckoning with social context and material conditions. This may be what, among other arguments, animates Michael Albert's concern over ecosocialist thought tending towards visions lacking in strategy ("abstract utopianism") and forethought (Albert 2023, 12). Otherwise stated, what are often missing are discussions about how to get to ecosocialism (Albert's "transition pathway"), what is likely to happen once ecosocialism is reached, and what to do to ensure the survival of an ecosocialist society.

Once more, virtually the same questions arise that have been historically confronted by socialists generally, especially by communist formations who have led successful revolutions. It is in some ways a continuation of the differentiation and relationship between utopian and scientific socialism as identified long ago by Engels and Marx as well as between socialism, anarchism, and communism and their internally diverse, intertwining, and evolving threads. It will remain a challenge to reconcile them to each other, especially under global social conditions mainly under capitalist ruling classes' sway.

Principles and Aims

Though substantial issues divide them, ecosocialists converge on the main guiding principles of meeting basic social needs, attaining social equality, and preserving ecological sustainability (Löwy and Tanuro 2021, 28–29; Somerville 2022). The divergence is over how these principles are to be put into practice and how the ultimate objective of reaching an ecosocialist society is to be achieved. The issue of strategy is discussed further in Chapter 8. One way of viewing these differences as already stated is in terms of the degree and role of centralism and coordination, whether in the shorter-term activities to build movements and institutions or in the ways that ecosocialist communities are to be organised or structured. According to some, the role of market exchange is not necessarily denied as part of ecosocialism (Mulberg 1992, 341), but this seems to be a minoritarian view that depends on understanding "market" as exchange relations that may or may not be capitalist.

This general assessment about what unites ecosocialists rests on the assumption that the final objective is shared among all ecosocialists. This is not necessarily clear. Certainly, for Ecofeminist Ecosocialists, the endpoint is not ecosocialism, but an ecofeminist society or whatever one would call a society freed of patriarchy once and for all (see Chapter 6). For others, it is democratic eco-socialism. For most, it is ecosocialism (or eco-socialism, the reason for a hyphen never explained) without accompanying qualifiers. It can be difficult to tell whether these terms are mere differences of inflection or substantive departures about what is envisioned as a future society. Those who prefer to call themselves democratic eco-socialists do not differ substantively from most other ecosocialists, especially those, like Wallis, who are trying to find a way to mix centralism and decentralism by means of "common ownership," by which is meant mainly state (public) ownership, at least for what pertains to proximate objectives (Baer and Singer 2022, 86). In contrast, Schwartzman regards ecosocialism as a phase towards solar communism (Schwartzman 1996; 2009, 10–12; 2022), while others see the horizon as an unqualified communism (Pepper 2010a, 234). These latter views regarding relative centralism and phases of ecosocialist construction are arguably more consistent with the ideas of Marx and Engels and most Marxist understandings. Presumably, all these labels for some future, ideal kinds of society point to the same sort of characteristics, but to date there have been no systematic, direct discussions, or debates about this.

For self-described democratic eco-socialists, "The construction of eco-socialism needs to be based upon a commitment to long-term sustainable balance between sociocultural systems and the natural environment" (Baer and Singer 2022, 86). Basic principles and political strategies are to be deployed to arrive at a steady-state economy or an economy characterised by net zero growth. In the latter, democratic eco-socialism bears similarities with degrowth advocates who are not averse to some forms of Marxism, insofar as they conform to their worldview (e.g., De Maria et al. 2013, 206). The idea of a steady-state economy has existed for decades, especially as part of discussions within what came to be known as ecological economics. It rests on making a distinction between growth and development. These notions, along with the formation of institutions based on democratically self-governing associated producers, seem agreeable to all ecosocialists (Wallis 2020, 34).

The former growth describes the endless use of resources in the current economic system. It is often argued by democratic eco-socialists that this focus on growth was rife in state-socialist countries as well as among most socialist movements and is among the reasons that "first epoch" socialism is alleged to have failed with respect to ecological issues. The latter, development, is redefined in terms of meeting people's needs (cf.

Nelson 2022). The argument regarding growth and development is that there must be a worldwide redistribution of resources so that everyone's needs can be fulfilled worldwide. That implies reducing resource consumption where there currently is overconsumption and the converse is to take place where people are deprived of healthcare, nutritious foods, housing, education, and other such aspects of well-being. In democratic eco-socialist Hans Baer's words:

> A serious redistribution of the world's resources would ensure an adequate living standard for everyone on the planet, but this shift would require serious discussion of how much is enough and, with the elimination of poverty, the recognition that the global population would begin to dwindle, easing strain on the ecosystem. (Baer 2021, 165)

That is, what constitutes the fulfilment of needs compared to unnecessary consumption must be specified and sensitised to people's individual conditions. The issue being enormously complex, democratic eco-socialists, as other ecosocialists, defer to future discussions and debates that would be presumably enabled by a global coordination of democratic eco-socialist parties in power. This would be important in setting up institutions to devise and refine criteria and bring about effective redistribution mechanisms.

To return to environmental policies, democratic eco-socialists propose several major shifts in society to arrive at a steady-state economy. One is raising energy efficiency during a transition to renewable sources of energy and reducing per capita resource consumption levels "among the world's affluent while allowing the poor to draw on these new energy sources to access basic resources" (Baer 2021, 165). Another is the development of public transportation to replace as much private vehicular travel as feasible, relative to people's circumstances. This entails redesigning and redeveloping urban and regional areas to enable such arrangements and reduce overall distances for everyday activity. A third shift is relative to the food and forestry systems. Low to no-meat diets and agroecological and organic farming techniques would be promoted, including in urban areas, along with the integration of food production and forestry (agroforestry) where applicable. Such production techniques would eventually supplant existing conventional methods, which are capital intensive and highly energy inefficient while emitting large amounts of greenhouse gases and relying on fossil fuels and pollution-inducing agrochemicals. Food and, arguably, fibre and wood production systems would also be reorganised into collectives, based on "decentralized planning and respect for regional variation within a country" (Baer 2021, 168). Related changes would be imposed on world trade so that consumption of faraway resources would be minimised and product conveyance occurs through means of transport and packaging that are the least environmentally impacting. Importantly, all these shifts are to be facilitated by a "planned centralized economy" that includes transferring workers to forms of employment that are ecologically sustainable (ibid., 166). These proposals are in tune with many other ecosocialists (e.g., Devine 2022; Wallis 2020).

The centralisation aspect would be contested by some ecosocialists as ultimately deleterious to the construction of ecosocialism. The Cooperation Jackson project, founded on 1st of May 2014 and located in Jackson (Mississippi), exemplifies this alternative viewpoint. Theirs are efforts at building ecosocialism by means of cooperativism in a Black-majority city. It is in some ways a throwback to the Owenite and Fourerist currents of Utopian socialism. Not coincidentally, the Mondragón Cooperative,[7] where tens of thousands of workers have relative job security and can vote on matters of management, production, and leadership, is among the main inspirations (Akuno and Nangwaya 2023). Whether the Mondragón Cooperative constitutes a paragon of worker self-management is a contentious matter, especially over whether it can even be shown to be an alternative to capitalism. This is openly acknowledged by Cooperation Jackson leadership (Akuno and Nangwaya 2023, 74). Some view it instead as a place-specific capitalist strategy to neutralise worker militancy (e.g., trade unionism is banned), aided by a radical nationalist ideology to assuage workers (Amado-Borthayre 2009; Kasmir 1999).

Cooperation Jackson, in contrast, is an attempt to achieve local Black worker ownership of the means of production, foster ecologically sound production, and transform the region's politics and economy by democratic means. These are understood as avenues to bring to fruition the

[7] Started initially as a small factory in 1941 in Arrasate (Mondragón, in Spanish), Euskal Herria (part of Basque Country under Spanish rule), the cooperative was set up as such in 1956 according to a mixture of Catholic, Marxist, and Owenite principles, combining factory production with schooling and a credit union, all locally owned and, to a large extent, a worker-managed process. Thus, the cooperative is typically considered an example of associated producers (whether freely so, in a capitalist context, is another question).

Jackson–Kush Plan, which involves Black self-determination in the US southeast as part of decolonising the entire US (Akuno 2023, 11–12). These aims overlap with democratic eco-socialism with respect to the insistence on democratic means, although without the party form as a means for organising and struggling for social change. In fact, there is an internal tension to the Cooperation Jackson project stemming from a mixture of Bakuninist anarchist and Utopian socialist principles with a vague form of Marxism that invokes the likes of Cabral but implicitly rejects his democratic centralism (exemplified in the African Party for the Independence of Guinea and Cape Verde). There is therefore more substantive convergence with the electoral politics of democratic eco-socialism, even if, to reiterate, without a party structure. This is evident in the history of the project. Initially, Cooperation Jackson was envisioned as part of the late Jackson mayor Chokwe Lumumba's policies aiming to facilitate the formation and development of cooperatives (Akuno and Nangwaya 2023, 59–60). Chokwe Lumumba's sudden departure to join the ancestors ended such a prospect of complementing municipal and party politics with decentralised cooperativism. This is just an example of how Cooperation Jackson are not averse to electoral and other institutional avenues to achieve the wider Jackson–Kush Plan.

Degrowth-Ecosocialism

Among a faction of ecosocialists there has developed recently a keen interest in the degrowth framework (e.g., de Kadt 2023; Foster 2023b; Löwy et al. 2022; Saito 2022). Differing and possibly mutually antagonistic perspectives are ensconced within the degrowth paradigm, but essentially it is an academic activist current that deems infinite economic growth as inherently devastating and ultimately self-destructive, given that constant economic expansion entails raising resource consumption and ever-increasing throughputs contradict the reality of finite resources on this planet. To start on the road to a worldwide contraction of resource use, without relying on socially devastating downturns, requires substantive or deep social change. What sort of social change and how it is to be achieved politically are among the more contended aspects of degrowth. Degrowth proponents offer little clarity on this beyond abstract conceptualisation, much like ecosocialists. The tendency is to prefer government intervention and non-violent consent-based avenues, provided planetary catastrophe is avoidable (see also Hickel et al. 2022).

Be that as it may, the degrowth current that overlaps with ecosocialism is the one that insists on a differential pathway and adopts a green left or ecosocialist platform that includes sensitivity to First Peoples' worldviews (Andreucci and Engel-Di Mauro 2019). Global South countries, the world's majority, are to continue economic growth in the most ecologically sustainable manner possible, while the minority of humanity in the Global North must rapidly reduce consumption levels to enable the Global South to reach liveable standards mainly denied at present. Implied is a massive transfer of technologies and wealth redistribution within and between countries and regions. Paradoxically, such proponents of degrowth with a sense of social justice find that they must promote growth in most of the world in order to achieve their ultimate aims of global degrowth. As the focus here is on ecosocialism, the diversity of degrowth perspectives is not discussed here, only versions of degrowth finding similarities with ecosocialist positions (see Hickel 2020; Kallis et al. 2020).

This emerging ecosocialist affinity with degrowth is, however, disconnected from the outcomes of dialogues on the topic that occurred half a century ago. Hence, a brief detour could be useful on the historical relationship between Marxism and the concept of physical limits to economic growth. At least by the early 1970s, within the USSR, preoccupations arose on the question of economic growth in relation to resource limits. The discussions that took place were politically influential and evinced a general split. Some advocated for major changes to reduce consumption and pollution levels, while others insisted that the matter is resolvable with the worldwide spread of socialism. Another viewpoint held that the USSR was exceptionally capable of addressing the contradiction through planning (Gare 2021; Kelley 1976; Ostrovityanov 1987; Voprosy Filosofii Editors 1974).

In socialist countries more broadly, environmental concerns informed government policies to differing degrees and were or have been expressed in diverse ways, depending on conjuncture, including levels of belligerence from liberal democracies and the prevailing balance of forces within socialist countries. This would be a general background to current environmental policies in the remaining socialist countries. The concept and mandate of building Ecological Civilisation in the People's Republic of China is a salient example of the evolution of concerns over ecological sustainability within a ruling communist party framework (see Chapter 8).

Within liberal democracies in the Global (figurative) North, where much of ecosocialism emerged, a few Marxists drew the conclusion from the dour "Limits to Growth" forecast (Meadows et al. 1972) that both abundance and industrial production must be excised from the socialist repertoire. Accordingly, "palaeo-Marxists will have to learn to love the peasants after all" (Foster-Carter 1974, 93). A renewed appreciation for peasant-based socialist revolutions was among the implications of this call for a progressive ruralisation and de-industrialisation of society. This Marxist undercurrent gained more traction in liberal democracies with the first attempts to join environmentalist movements (see Chapter 5), though shorn of enthusiasm for peasant lifestyles (see, e.g., Amberger 2023; cf. Sarkar 1999).

In some ways, the undercurrent became a precursor to what is now called Ecological Marxism, one of the main schools of thought on which ecosocialism is founded. Within socialist feminist circles the understanding of growth limits was critically incorporated into a form called Ecofeminism, another major pillar of ecosocialism. The issue has been addressed by way of, for instance, the subsistence perspective and more recently the concept of eco-sufficiency (see Chapter 6). There have thus been several streams of thought that have made ecosocialism sensitive to the question of economic growth and ecologically unsustainable resource use.

However, it has only been over the past couple of decades that ecosocialists have engaged with the largely intellectual and academic movement called degrowth, a movement at first focused exclusively with ways to reduce resource and energy consumption. Like Marxist forms of environmentalism in core capitalist countries, this movement is also traceable to the 1970s, developing mainly in France and Spain (Kallis 2023). Writings on and advocating for degrowth have grown tremendously. It has evolved in a sort of parallel fashion with the relatively more recent ecosocialist movement. It is beyond the scope of the present work to reflect on degrowth's history, evolving ideas and findings, and political diversity, but the reader is encouraged to explore such matters further and in relation to ecosocialism by means of a modest overview offered here, pointing to key publications that have led to an ecosocialism-degrowth convergence.

Ecosocialists' dialogues with degrowth proponents were initially rare but picked up by the 2010s. Since then, they have yielded mixed outcomes ranging from mutual critique to calls for close collaboration (Andreucci and Engel-Di Mauro 2019; Correia 2012; de Kadt 2023; Pineault 2019; Schwartzman 2012). The anti-Marxist trappings of early degrowth writings were an initial impediment, such as those expressed by Serge Latouche (2012). These have been partially surmounted as a degrowth current has emerged that understands class differentiation (though not necessarily surplus value extraction; see de Kadt 2023) as intrinsic to economic growth in capitalism and that also takes multiple modes of oppression into account, as well as embracing aspects of Marxism (Kallis 2019) and, to an inadequate extent still, Ecofeminism (Barca 2023; Brownhill et al. 2012). On the other hand, Xue (2016, 201) sees overlaps between certain forms of ecosocialism and degrowth that embrace "a multi-scalar transformation" and state intervention strategy.

The development of a basis for convergence has been accomplished in part through debates and symposia published in the ecosocialist journal *Capitalism Nature Socialism* over the past two decades (saed 2012; Andreucci and Engel-Di Mauro 2019). The Eco-Marxist metabolic rift school (see Chapter 6) has also become more receptive to degrowth, after an initial dismissiveness (De Maria et al. 2013; Foster 2011), provided it sticks to certain (socialist) central planning prescriptions (Foster 2023b) or to a specific interpretation of Marxian text (Foster 2023a; cf. Saito 2022). Among the planks drawing ecosocialism and degrowth together is a shared aversion for "Productivist 'socialism'" (Kallis 2019; Löwy et al. 2022). In other words, the burgeoning ecosocialist degrowth current blends particular strands of the two frameworks (state-socialist inimical ecosocialism and anti-capitalist degrowth) to the inevitable exclusion of other currents within both perspectives.

Several areas could be further tread in this new coupled degrowth-ecosocialism adventure. A systematic treatment of imperialism and colonialism, aside from inklings of applications of world-systems perspective (such as unequal ecological exchange theory), seems yet to occur in relation to the politics of and strategies to implement degrowth ecosocialism, but this is a problem in ecosocialist discourse more widely as well. Another challenge of possibly greater weight is to develop greater precision and accuracy with respect to physics concepts, biophysical processes, and ecosystem dynamics, the latter of which tend to be treated as unchanging entities. Imprecision and inaccuracy in the use of biophysical science concepts and towards nonhuman processes lead to misinterpreting the ecological impacts of diverse forms of economic growth by underestimating (e.g., lack of accounting for tree species diversity and biodiversity potentials in forest biomes) or exaggerating problems (e.g., counting soil erosion by means of sediment yield values without examining the landscape-wide sediment distribution or the mitigating effects of enhanced soil formation in downslope areas due to enhanced sediment influx) (Schwartzman and Engel-Di Mauro 2019).

The Continuing Quest for Ecologically Sustainable Socialism

Ecosocialism came out of the historical experiences of leftist formations within a limited number of capitalist countries, especially liberal democracies (or, in a Leninist sense, the imperialist countries[8]). Historical and geographical contexts can be useful to keep in mind when reading ecosocialist texts or considering ecosocialist politics. Just as socialism in general, ecosocialism is also a product of specific social conditions. This ecosocialists understand with respect to socialism broadly, but without so far much critical self-reflection regarding ecosocialism's own roots (e.g., Wallis 2020, 37–38).

Influenced by an expanding environmental movement and self-described anti-authoritarian versions of socialism, many Marxists predominantly in the imperialist countries felt compelled, and some still do, to turn to the writings within the socialisms of the 1800s and early 1900s to justify a defence of Marxism and a call for a fundamental change in what they see as a degraded form of socialism or "first epoch socialism." Such scholarship sprouted especially in the 1990s as a response to the combination of an intellectually stultifying capitalist triumphalism and a long-standing anti-socialism within much of the environmentalist movement.

The search for an alternative kind of socialism that overcomes capitalism and that is ecologically sustainable underlies the plethora and diversity of ideas propounded by ecosocialists. They vary considerably

[8] By this is meant countries where there are capitalist institutions with a global monopoly over the most technologically advanced and productive labour processes (this implies the existence of non-monopoly economic sectors within imperialist countries and in the rest of the world, with varying degrees of global political influence, but much less than monopoly capital). This is also known as financial monopoly capital, since such monopolies include finance. Monopolies result from the tendency in capitalism for the concentration and centralisation of capital, a process identified by Marx (Lenin 1917; see also King 2021).

between calls for the building of parallel institutions from below to envisioning the eventual capture of states to usher an ecosocialist world system. Ecosocialists have also laid out a large array of policy ideas, including technically feasible ones that would, for example, account for the challenges of ending the energy poverty of the world's majority while phasing out fossil fuels (Schwartzman and Schwartzman 2019). For the most part, this wealth of proposals, similar to the degrowth framework to which some ecosocialists have been gravitating, remains part of a wish list of what could be done if ecosocialists gained power, whether by means of grassroots organising or by capturing institutions or both. It is ecosocialists' collective actions, organising effectiveness, and consequent political reach that will convince more people of the desirability of ecosocialism. After more than forty years, however, it seems that instead much more ink continues being spilled compared to efforts aimed at organising to gain the wide social mobilisation ecosocialists themselves recognise as necessary to confront the socio-ecological crisis. For the moment, with few important national-level influential exceptions, much of ecosocialism inhabits the realm of theory. The geographical and political reach of ecosocialism is explored in the next chapter. The theoretical frameworks underlying and diversifying ecosocialism are confronted in Chapters 5 and 6.

References

Akuno, Kali. 2023. "Build and Fight: The Program and Strategy of Cooperation Jackson." In *Jackson Rising Redux. Lessons on Building the Future in the Present*, edited by Kali Akuno and Matt Meyer, 12–55. Oakland: PM Press.

Akuno, Kali, and Ajamu Nangwaya. 2023. "Toward Economic Democracy, Labor Self-Management, and Self-Determination. In *Jackson Rising Redux. Lessons on Building the Future in the Present*, edited by Kali Akuno and Matt Meyer, 56–90. Oakland: PM Press.

Albert, Michael J. 2023. "Ecosocialism for Realists: Transitions, Trade-Offs, and Authoritarian Dangers," *Capitalism Nature Socialism*, 34 (1): 11–30.

Albritton, Robert. 2019. *Eco-Socialism for Now and the Future*. Geneva: Springer International Publishing.

Ali, Tariq. 2009. *The Idea of Communism*. London: Seagull Books.

Amado-Borthayre, Lontzi. 2009. "Enjeux de gouvernance dans une coopérative multinationale: Fagor Electrodomésticos face à la globalisation et à la critique syndicale." *Pôle Sud* 2 (2009): 87–102.

Amberger, Alexander. 2023. *Dissident Marxism and Utopian Eco-Socialism in the German Democratic Republic*. Leiden: Brill.

Andreucci, Diego, and Salvatore Engel-Di Mauro. 2019. "Capitalism, Socialism and the Challenge of Degrowth: Introduction to the Symposium." *Capitalism Nature Socialism* 30 (2): 176–188.

Atkinson, Adrian. 1991. *Principles of Political Ecology*. London: Belhaven.

Baer, Hans A. 2016. "Toward Democratic Eco-Socialism as the Next World System." https://seed.uw.edu/wp-content/uploads/baer-eco-socialism-2016.pdf. Last accessed 4 September 2022.

Baer, Hans A. 2018. *Democratic Eco-Socialism as a Real Utopia: Transitioning to an Alternative World System*. New York: Berghahn Books.

Baer, Hans A. 2021. "Toward Democratic Eco-socialism as the Next World System: A Vision for the Future." In *The New Systems Reader: Alternatives to a Failed Economy*, edited by James Gustav Speth, and Kathleen Courrier, 127–144. New York and London: Routledge.

Baer, H.A., and M. Singer. 2022. Can Ecological Modernisation Contain Climate Change: An Eco-Socialist Perspective. *Journal of Australian Political Economy* 90: 75–91.

Bansart, Andrés. 2012. *Hacia el Ecosocialismo*. Caracas: Gobierno Bolivariano.

Barca, Stefania. 2019. "Labour and the Ecological Crisis: The Eco-Modernist Dilemma in Western Marxism(s) (1970s–2000s)." Geoforum 98: 226–335.

Barca, Stefania. 2023. "Forces of Reproduction." In *A Map to Ecosocialism, Emancipations: A Journal of Critical Social Analysis* 2 (1, Article 8): 7–9.

Benton, Ted. 2023. "Socialism and Ecology." In *The Cambridge History of Socialism, Volume II*, edited by Marcel van der Linden, 664–686. Cambridge: Cambridge University Press.

Brownhill, Leigh, Terisa E. Turner, and Wahu Kaara. 2012. "Degrowth? How About Some 'De-Alienation'?" *Capitalism Nature Socialism* 23 (1): 93–104.

Brownhill, Leigh, and Terisa E. Turner. 2020. "Ecofeminist Ways, Ecosocialist Means: Life in the Post-capitalist Future." *Capitalism Nature Socialism* 31 (1): 1–14.

Buhle, Paul. 2000. "First Wave Eco-Socialism." *Capitalism Nature Socialism* 11 (3): 38–40.

Correia, David. 2012. "Degrowth, American Style: *No Impact Man* and Bourgeois Primitivism." *Capitalism Nature Socialism* 23 (1): 105–118.

De Kadt, Maarten. 2023. "Ecosocialists and Degrowth Advocates Should Work Together." *Capitalism Nature Socialism* 34 (2): 1–7.

De Maria, Federico, François Schneider, Filka Sekulova, and Joan Martinez-Alier. 2013. "What Is Degrowth? From an Activist Slogan to a Social Movement." *Environmental Values* 22 (2): 191–215.

Devine, Pat. 2022. "Ecosocialist Economics," In *The Routledge Handbook on Ecosocialism*, edited by Leigh Brownhill, Salvatore Engel-Di Mauro, Terran Giacomini, Ana Isla, Michal Löwy, and Terissa Turner, 271–279. New York: Routledge.

Emel, Jacque, and Richard Peet. 1989. "Resource Management and Natural Hazards." In *New Models in Geography, Volume 1*, edited by Richard Peet and Nigel Thrift, 51–81. London: Routledge.

Foster, John B. 2000. *Marx's Ecology*. New York: Monthly Review Press.

Foster, John B. 2011. "Capitalism and Degrowth—An Impossibility Theorem." *Monthly Review* 62 (8): 26.

Foster, John B. 2023a. "Engels and the Second Foundation of Marxism." *Monthly Review* 72 (2): 2–12.

Foster, John B. 2023b. "Planned Degrowth: Ecosocialism and Sustainable Human Development." *Monthly Review* 75 (3): 1–29.

Foster-Carter, Aidan. 1974. "Neo-Marxist Approaches to Development and Underdevelopment. In *Sociology and Development*, edited by Emanuel De Kadt and Gavin Williams, 67–105. Cambridge: Tavistock.

Gare, Arran. 1993. "Soviet Environmentalism: The Path Not Taken." *Capitalism Nature Socialism* 4 (4): 69–88.

Gare, Arran. 2021. "The Eco-socialist Roots of Ecological Civilisation." *Capitalism Nature Socialism* 32 (1): 37–55.

Hayward, Tim. 1990. "Ecosocialism—Utopian and Scientific." *Radical Philosophy* 56: 2–14.

Hickel, Jason. 2020. *Less Is More: How Degrowth Will Save the World*. New York: Random House.

Hickel, Jason, Giorgos Kallis, Tim Jackson, Daniel W. O'Neill, Juliet B. Schor, Julia K. Steinberger, Peter A. Victor and Diana Ürge-Vorsatz. 2022. "Degrowth Can Work—Here's How Science Can Help." *Nature* 612: 400–403.

Julia K Steinberger Peter A Victor Diana Ürge-Vorsatz 2022 Degrowth Can Work—Here's How Science Can Help Nature 612 400 403

Kallis, Giorgos. 2019. "Capitalism, Socialism, Degrowth: A Rejoinder." *Capitalism Nature Socialism* 30 (2): 267–273

Kallis, Giorgos. 2023. "Degrowth and the Barcelona School." In *The Barcelona School of Ecological Economics and Political Ecology: A Companion in Honour of Joan Martinez-Alier*, edited by Villamayor-Tomas, Sergio, and Roldan Muradian, 83–90. Cham: Springer International Publishing.

Kallis, Giorgos, Susan Paulson, Giacomo D'Alisa, and Federico Demaria. 2020. *The Case for Degrowth*. New York: Wiley.

Kasmir, Sharryn. 1999. "The Mondragon Model as Post-Fordist Discourse: Considerations on the Production of Post-Fordism." *Critique of Anthropology* 19 (4): 379–400.

Kelley, Donald R. 1976. "Economic Growth and Environmental Quality in the USSR: Soviet Reaction to 'The Limits to Growth.'" *Canadian Slavonic Papers* 18 (3): 266–283.

King, Sam. 2021. *Imperialism and the Development Myth*. Manchester: Manchester University Press.

Kovel, Joel. 2005. "The Ecofeminist Ground of Ecosocialism." *Capitalism Nature Socialism* 16 (2): 1–8.

Kovel, Joel. 2007. *The Enemy of Nature. The End of Capitalism or the End of the World*. London: Zed Books.

Latouche, Serge. 2012. "Can the Left Escape Economism?" *Capitalism Nature Socialism* 23 (1): 74–78.

Lenin, Vladimir Ilich. 1917 [1963]. "Imperialism, the Highest Stage of Capitalism. In *Selected Works*, vol. 1, 667–766. Moscow: Progress Publishers. https://www.marxists.org/archive/lenin/works/1916/imp-hsc/. Accessed 6 March 2024.

Löwy, Michael. 2005. "What Is Ecosocialism?" *Capitalism Nature Socialism* 2: 15–24.

Löwy, Michael. 2015. *Ecosocialism. A Radical Alternative to Capitalist Catastrophe*. Chicago: Haymarket Books.

Löwy, Michael, and Daniel Tanuro. 2021. "Introduction. Le XXIe Siècle Sera le Siècle de la Question Écosociale." In *Luttes Écologiques et Sociales dans le Monde. Allier le Vert et le Rouge*, edited by Michael Löwy and Daniel Tanuro, 7–35. Paris: Les Éditions Textuel.

Löwy, Michael, and Daniel Tanuro. 2022. "The Fourth International's Contribution to Ecosocialism." In *The Routledge Handbook on Ecosocialism*, edited by Leigh Brownhill, Salvatore Engel-Di Mauro, Terran Giacomini, Ana Isla, Michal Löwy, and Terissa Turner, 346–356. New York: Routledge.

Löwy, Michael, Bengi Akbulut, Sabrina Fernandes, and Giorgos Kallis. 2022. "For an Ecosocialist Degrowth." *Monthly Review* 73 (11): 56–58.

Magdoff, Harry, and Fred Magdoff. 2005. "Approaching Socialism." *Monthly Review* 57 (3): 19–61.

Magdoff, Fred, and Chris Williams. 2017. *Creating an Ecological Society: Toward a Revolutionary Transformation*. New York: Monthly Review Press.

Meadows, Donella H., Dennis L. Meadows, Jørgen Randers, and William W. Behrens III. 1972. *The Limits to Growth; A Report for the Club of Rome's Project on the Predicament of Mankind*. New York: Universe Books.

Miliband, Ralph. 1994. *Socialism for a Sceptical Age*. Cambridge: Polity Press.

Mulberg, Jon. 1992. "Who Rules the Market? Green Versus Ecosocialist Economic Programmes: A Response to Eckersley." *Political Studies* 40 (2): 334–341.

Nelson, Anitra. 2022. "Ecosocialism from a Post-Development Perspective." In *Post-Capitalist Futures: Paradigms, Politics, and Prospects*, 31–40. Singapore: Springer Singapore.

O'Connor, James R., ed. 1998. *Natural Causes: Essays in Ecological Marxism*. New York: Guilford Press.

Ostrovityanov, Yuri. 1987. *Where Do We Come from? What Are We? Where Are We Going? Marxism versus Sociological Pessimism and Its Neo-Marxist Version*. Moscow: Novosti Press Agency Publishing House.

Pepper, David. 1984. *The Roots of Modern Environmentalism*. London: Croom Helm.

Pepper, David. 1993. *Eco-Socialism. From Deep Ecology to Social Justice*. London: Routledge.

Pepper, David. 2010a *Eco-Socialism. From Deep Ecology to Social Justice*, 2nd ed. London: Routledge.

Pepper, David. 2010b. "On contemporary Eco-Socialism." In *Eco-Socialism as Politics: Rebuilding the Basis of Our Modern Civilisation*, edited by Qingzhi Huan, 33–44. Dordrecht: Springer.

Pineault, Eric. 2019. "From Provocation to Challenge: Degrowth, Capitalism and the Prospect of 'Socialism Without Growth': A Commentary on Giorgios Kallis." *Capitalism Nature Socialism* 30 (2): 251–266.

Roelofs Joan. 1993. "Charles Fourier: Proto-Red Green." *Capitalism Nature Socialism* 4 (3): 69–88.

Ryle, Martin. 1988. *Ecology and Socialism*. London: Radius.

saed. 2012. "Introduction to the Degrowth Symposium." *Capitalism Nature Socialism* 23 (1): 26–29.

Saito, Kohei. 2023. *Marx in the Anthropocene: Towards the Idea of Degrowth Communism*. Cambridge: Cambridge University Press.

Salleh, Ariel. 1997. *Ecofeminism as Politics: Nature, Marx and the Postmodern*. London: Zed Books.

Sarkar, Saral K. 1999. *Eco-Socialism or Eco-Capitalism? A Critical Analysis of Humanity's Fundamental Choices*. London: Zed Books.

Satgar, Vishwas. 2018. "The Climate Crisis and Systemic Alternatives.' In *The Climate Crisis: South African and Global Democratic Eco-Socialist Alternatives*, edited by Vishwas Satgar, 10–16. Johannesburg: Wits University Press.

Schwartzman, David. 1996. "Solar Communism." *Science & Society* 60 (3, Fall): 307–331.

Schwartzman David. 2009. "Ecosocialism or Ecocatastrophe?" *Capitalism Nature Socialism* 20 (1): 6–33.

Schwartzman, David. 2012. "A Critique of Degrowth and Its Politics." *Capitalism Nature Socialism* 23 (1): 119–125.

Schwartzman, David. 2022. "Solar Communism." In *The Routledge Handbook on Ecosocialism*, edited by Leigh Brownhill, Salvatore Engel-Di Mauro, and Terran Giacomini, Ana Isla, Michael Löwy, and Turner Terisa, 280–289. Oxon and New York: Routledge.

Schwartzman, David, and Salvatore Engel-Di Mauro. 2019. “A Response to Giorgios Kallis’ Notions of Socialism and Growth.” *Capitalism Nature Socialism* 30 (3): 40–51.

Schwartzman, Peter, and David Schwartzman. 2019. *The Earth Is Not for Sale: A Path Out of Fossil Capitalism to the Other World That Is Still Possible*. New York: World Scientific Publishers.

SERA [The Socialist Environment and Resources Association]. 1980. *Eco-Socialism in a Nutshell*. London: Writers and Readers Publishing Cooperative, also available at https://climateandcapitalism.com/wp-content/uploads/2013/11/Ecosocialism-in-a-Nutshell.pdf. Last accessed 9 February 2024.

Shiva, Vandana, and Maria Mies. 1993. *Ecofeminism*. London: Zed Books.

Somerville, Peter. 2022. “Problems with Ecosocialism.” *The Ecologist*, December 5. https://theecologist.org/2022/dec/05/problems-ecosocialism. Accessed 7 January 2022.

Suing, Guillaume. 2018. *L’Écologie Réelle. Une Histoire Soviétique et Cubaine*. Paris: Éditions Delga.

Tello, Enric. 2016. “Manuel Sacristán at the Onset of Ecological Marxism After Stalinism.” *Capitalism Nature Socialism* 27 (2): 32–50.

Vettese, Troy, and Drew Pendergrass. 2022. *Half-Earth Socialism: A Plan to Save the Future from Extinction, Climate Change and Pandemics*. New York: Verso Books.

Voprosy Filosofii Editors. 1974. “Foreword: ‘Man and His Environment’.” *Soviet Studies in Philosophy* 13 (2–3): 3–6.

Wall, Derek. 2010. *The Rise of the Green Left: Inside the Worldwide Ecosocialist Movement*. London: Pluto.

Wallis, Victor. 2018. *Red-Green Revolution: The Politics and Technology of Ecosocialism*. Toronto: Political Animal Press.

Wallis, Victor. 2020. *Socialist Practice: Histories and Theories*. London: Palgrave Macmillan.

Weston, Joe, ed. 1986. *Red and Green: A New Politics of the Environment*. London: Pluto Press.

Williams, Raymond. 1982. *Socialism and Ecology*. London: Socialist Environment and Resources Association.

Wolf, Frieder O., and Michael Löwy. 2007. “Party-Building for Eco-Socialists: Lessons from the Failed Project of the German Greens.” In *Coming to Terms with Nature: Socialist Register*, edited by Leo Panitch and Colin Leys, 310–336. London: The Merlin Press.

Xue, Jin. 2016. “China at the Crossroad. Ecological Modernization or Ecosocialism?” In *Crisis System. A Critical Realist and Environmental Critique of Economics and the Economy*, edited by Petter Næss and Leigh Price, 192–208. London and New York: Routledge.

CHAPTER 4

A Geography of Ecosocialism

Locating Ecosocialist Hearths

Ecosocialist organisations and/or intellectual currents exist primarily in parts of Europe and the Americas, along with a presence in South Africa, Australia, and the Philippines. Notable is the ecosocialist content of the government and state institutions in Bolivarian Venezuela. There is otherwise a virtual absence in the most populous regions of the world, like South and East Asia, and in socialist countries. Some ecosocialist currents or activists are on the fringes of Green parties and others are at the margins of Leftist anti-capitalist or social democratic parties (see also Wall 2010, 81–104). For the most part, ecosocialist currents or organisations have had very limited public reach or have been short-lived, whether centralist or decentralist (Tanuro 2020, 250–251). Centralist ecosocialists have yet to demonstrate the ability to form parties or institutions that would build any political power. They are, at most, members of a pre-existing party, which may have adopted an ecosocialist platform. Otherwise, they are ecosocialists at large, involved in loose networks.

Centralist forms of ecosocialism tend to enjoy more political visibility as a result of being associated with often better-resourced and internally coordinated parties and/or internationalist organisations. Though there exists no ecosocialist party (except Zabalaza for Socialism, a small, recently established one in South Africa) or something akin to such a political party, ecosocialists have become involved in a variety of different

S. Engel-Di Mauro, *Ecosocialism*, Marx, Engels, and Marxisms,
https://doi.org/10.1007/978-3-031-74288-0_4

parties, ranging from environmentalist parties like the US Green Party to self-described revolutionary Trotskyist formations and even within the pro-capitalist UK Labour Party. Greater details about the political reach and accomplishments are discussed further below. First, the degree of influence attained through parties in different parts of the world is discussed, interspersed by observations about more decentralist forms of ecosocialism, followed by greater details on cases of international and primarily decentralist organising.

Ecosocialism in Africa and Potentials for a Pan-Africanist Orientation

The geographical distribution of ecosocialism tends to be concentrated in a relatively few countries, mainly in the Americas and Europe. Nevertheless, in South Africa, a small ecosocialist movement has developed out of the historical environmental degradation by the apartheid regime affecting mainly Black African communities. As leftist parties have tended to ignore environmental problems, ecosocialists have concentrated on linking existing activists who draw together anti-racist, socialist, and environmentalist struggles. But it was not until 2008 that the linkages were formalised in the Green Socialist Coalition to address social inequality and environmental issues simultaneously. It was a merger of the anti-privatisation Operation Khanyisa Movement and Ecopeace (founded in 1995), as well as Socialist Group, Keep Left, Socialism from Below, the Democratic Socialist Movement, and the Workers Party. This list already demonstrates the fractious character of South African leftist politics, so the building of such a coalition is no mean achievement in itself. The Coalition participate in elections as an alternative to the neoliberal African National Congress and South African Green Party (see also Satgar and Cock 2022).

Elsewhere in Africa there does not seem to be any ecosocialist movement emerging. Even in the Burkinabé case, about which some claim as a form of ecosocialism, there has been no ecosocialist movement following the 2014 ousting of the France-aligned Compaoré dictatorship. More about Burkina Faso is related below, but the recent resurgence of Pan-Africanism could be a motivational force for ecosocialists to focus their attention on becoming relevant to and supportive of pre-existing and developing political forces that have combined socialism with continent-wide liberation struggle.

Connecting ecosocialism to Pan-Africanism may be a recent turn confined to few ecosocialists' imagination so far, but there is a much longer history and much greater social reach to the intertwining of Pan-Africanist and socialist perspectives and struggles. As in the case of First Peoples' self-defence from settler colonialism and then self-determination struggles (see Chapter 6), there is a rough relationship between Pan-Africanism and socialism and a multitude of ways of linking the two. Some forms of Pan-Africanism find much affinity or complementarity with socialism. Other Pan-Africanist and socialist formations have instead arrived at mutual aversion, even hostility. Broadly speaking, the two movements, also because they are internally quite diverse, have at times converged and at other times diverged over fundamentals in political philosophy, strategy, and even ultimate objectives. Overlaps alternate with incompatibilities. The same can be said more broadly about socialist and African or Black national liberation movements.

Though views on what constitutes Pan-Africanism vary, it is relatively uncontroversial to claim that perspectives associable with a Pan-Africanist outlook arose from within the African Diaspora, possibly by the late 1700s (Adi 2018, 7–8; James 1969; Malisa and Missedja 2020, 36). Clarke (1988, 31) sees its earliest verifiable emergence thousands of years ago with "the unification of the Upper and Lower civilizations" and the more recent form since 1400s, originating with the resistance to the cross-oceanic trafficking of captives and to colonialism, including within Africa. Regardless of controversies over chronology, most would agree that Pan-Africanism stands for

> the unity, common history and common purpose of the peoples of Africa and the African diaspora and the notion that their destinies are interconnected. In addition, many would highlight the importance of the liberation and advancement of the African continent itself, not just for its inhabitants, but also as the homeland of the entire African diaspora. (Adi 2018, 2)

Pan-Africanism, in a sense, is the expression of transcontinental Africanness, with the unification and liberation of all people of African descent as watchwords. This inflection mainly emerged by the late 1800s and crystallised in 1900 with the first Pan-African Conference in London. Pan-Africanist perspectives evolved and diversified also in struggle with internalised forms of racism and colonialism in some currents, which included claims of modernising or bettering Africa through Diasporic

African interventions or resettlement and that eventually yielded to notions of self-determination and national independence (Malisa and Missedja 2020).

The latter objectives found special affinity with the communist movement, formed by the 1920s. Pan-Africanists had uneasy relations with socialist or social democratic organisations or became disillusioned with socialist forces that turned out to be unconcerned with, if not even supportive of colonialism. There was also growing dissatisfaction with approaches advocating for resettlement or repatriation (e.g., the Universal Negro Improvement Association) or focused on petitioning colonial and international institutions for more self-governance in Africa and Caribbean, an end to racism, and other pressing issues. In many respects, the outrage of the 1935 Italian Fascists' invasion of Ethiopia, following the two decades of mass-murderous US military occupation of Haiti (1915–1934), radicalised or further radicalised Diasporic and Continental Africans, whether or not they were active in the struggle against colonialism and for self-determination. Pan-Africanists were among the many people of African descent who were drawn to and contributed much to the development and spread of communist parties, as leaders or active members. These included major influential intellectual figures like Claudia Jones, a member of Communist Party USA (CPUSA) and Communist Party of Great Britain, and W. E. B. Du Bois, who joined the CPUSA later in life (1961), and visited, and expressed much praise for the USSR (Adi 2018).

Intertwining Pan-Africanism with ecosocialism builds on much precedent, drawing from and in continuity with the efforts of Diasporic and Continental African socialists and communists. Yet so far, in practice, there is no Pan-African ecosocialism in the making in terms of a systematised and an implementable worldwide or continental strategy. There are, however, a few ideas on the topic.

Christelle Terreblanche identifies in particular Ubuntu, a caring and sharing ethics paradigm (applied to both social and ecological relations) widespread in Africa, as complementary to Sumak Kawsay (see Chapter 6) and a basis to "inspire endogenous [African] eco-socialist imaginaries" (Terreblanche 2018, 170). Ubuntu, among other cultural and ethical frameworks, could be a conduit for a unified, Pan-African struggle against colonial powers, as well as the means whereby to build social equality within and beyond the continent. A difficulty in the specific case of

Ubuntu is that its meaning and practices are highly variegated, as Terreblanche points out (ibid., 169). The paradigm has been deployed for many kinds of political ends that can be and have been mutually antagonistic. The effort to recuperate a form of Ubuntu that can be put into dialogue with ecosocialism seems impractical without first establishing organisations capable of mass mobilisation across the continent. If the matter is confined to a dialogue in the abstract, as in a piece of academic writing, not much can be expected. It is not a promising start to define socialist struggles in Africa as "mostly disastrous" (ibid., 170), while omitting the Burkinabé Revolution as well as the recurring colonial violence of liberal democratic Euro-American powers (multiple military invasions, economic sabotage, coups, among other means of subjugation) that represses any form of socialism.

It is rather more the concrete situation and history of Burkina Faso than an attempt to appropriate a paradigm that could function as an example for a Pan-African ecosocialism. As described in Chapter 2, the coup that enabled revolutionaries to take over the state was essential in the effort to break away from neocolonial oppression and to usher in major changes towards gender equality and ecologically sustainable practices. Though short-lived, the Burkinabé revolution lives on as inspiration to many Africans beyond Burkina Faso, and it serves as symbol for the current attempts to kick out Euro-American forces out of the Sahel countries once and for all. Regrettably, ecosocialist formations are virtually absent in this process. There have appeared explicit discussions on Sankara's "ecosocialism" in some leftist Francophone circles, such as the French Parti de Gauche (https://www.lepartidegauche.fr/?p=5063), aside from Wall (2010). But these are contrived readings superimposing ecosocialism on what was essentially an anti- and decolonial Marxist-Leninist framework.

Ecosocialist Organising in Southeast Asia and Australia

In Southeast Asia, the only noteworthy appearance of ecosocialist ideas is in Lanao del Norte, Mindanao, the Philippines. There, the Revolutionary Workers Party-Mindanao (RPM-P in Mindanao)[1] left the Revolutionary Workers Party in 2001 eventually to join the Fourth International (about which more below) in 2003 and follow a sort of ecosocialist orientation.[2] Among their policies are the provision of contraceptives, promotion of peasant women into administrative posts, and the fight against gender-based discrimination. While in the RPM-P, they have been trying to establish an autonomous zone since 1994 over an area of 1200 hectares in ways similar to what the EZLN have been establishing in Chiapas (México) since roughly the same time. In the areas under their control, the RPM-P Mindanao have been transforming local farming practices into ecologically sustainable ones. Agroecologically-based farms, often under women peasant management, provide for local needs and serve as demonstration centres for educational and training purposes. To accomplish this, they have received support from the National Rice Institute of the University of the Philippines and various environmental NGOs, but they face pressures from mining and logging firms as well as the Moro Islamic Liberation Front. To fend off military attacks, the RPM-P Mindanao have an armed self-defence force, the Revolutionary Proletarian Army or RPA (Colle 2022). It is unclear the extent to which the wider population embraces ecosocialism and the extent of popular support in general for the party.

In the case of the settler colonial regime of Australia, there is a precedent in 1970s trade unionism being combined with environmentalism. Trotskyists Alan Roberts and Nick Origlass were main organisers and influenced what later became a still small, but very active ecosocialist movement. Democratic Socialist Perspective was one ecosocialist current in the Australian Socialist Alliance. In 2010, they merged into the Socialist Alliance (founded 2001).[3] As far as known, they are a Trotskyist party with a national membership of less than 2000, they have been elected to

[1] The Rebolusyonaryong Partido ng Manggagawà ng Pilipinas (RPM-P) themselves had split from the Communist Party of the Philippines in 1995.

[2] https://internationalviewpoint.org/spip.php?article6574, accessed 10 April 2024.

[3] https://socialist-alliance.org/, accessed 10 April 2024.

and still hold a few municipal and regional offices. They exert some influence at local levels, but much of their presence is in grassroots and trade union activism, outdoing the Greens especially in climate justice activism. Elsewhere in the Asia-Pacific region there appears to be no ecosocialist presence of political significance.

European Ecosocialism

In Europe, home of NATO and major colonial powers, ecosocialist activism exists mainly as networks and currents within parties. The extent of political underdevelopment is interesting, given the early introduction of ecosocialism within the UK's Labour Party, led by Raymond Williams, as discussed in Chapter 3. Such precocious start has not materialised in the establishment of any ecosocialist party. Instead, the likes of SERA remained rigorously within Labour Party bounds, regardless of the Blairite neoliberal capitalist turn and later suppression of the Corbynite current.

For the most part, ecosocialists have not been able of building much beyond affinity groups or networks and political currents within parties, such as the Alliance for Green Socialism (UK; https://green-socialist.com/). At times, such groups are seemingly disconnected from each other even within the same country and it is unclear how many people are involved. An example known to the author is the Movimento EcoSocialista (https://movimentoecosocialista.org), revolving about a new-ager and self-published individual who lives in London, and founded apparently in 2020. They have no relation to the more recently formed Rete Ecosocialista Nazionale (Italy; https://www.retecosocialista.it/), who in turn have entirely ignored antecedents like the defunct Ecologia Politica (established in 1991), an organisation that was founded and headed by the late Giovanna Ricoveri and was associated with the journal *Capitalism Nature Socialism*. Without direct contact with those involved, it is difficult to assess the significance of networks or groups whose presence is little more than online.

An example of a political party current is the Green Left within the Green Party of England and Wales. Several political parties with ecosocialist platforms also exist with parliamentary representation, including at the level of the European Union (EU). The politics they represent are quite varied and not necessarily anti-capitalist. GroenLinks (Green Left, Holland, founded 1991) is a pro-EU social democratic formation of the sort prevalent since the First World War, that is, with a vision of a reformable, humane capitalism, or market democracy.

If it is understood that the EU is an imperialist organisation based on capitalist principles, one could be forgiven for regarding any pro-EU party as, at bottom, anti-ecosocialist, but this is where the term socialism itself is what is implicitly in dispute. Since especially the 1980s, political parties that call themselves socialist (e.g., the Socialist Parties of France, Italy, and Spain) have shed themselves of any semblance of socialism or express no attempt to create a socialist system, so it can follow that ecosocialism would have a similar fate under such influence.

This is how there are contrasting forms of socialism in Europe that diverge over the most basic principles about even the meaning of socialism. In contrast to Groenlinks, for example, Bloco de Esquerda (Left Bloc, Portugal, founded 1999) is a feminist democratic socialist and decidedly not pro-EU formation. Vinstri Græn (Green Left, Iceland, founded 1999) is another feminist democratic socialist, anti-EU, and anti-NATO party. It was part of a ruling coalition in 2009 and 2017. Some of these viewpoints are represented as minoritarian currents at the EU level. The Left in the EU Parliament is a major illustration. It is a coalition formed in 2021 of communist and red-green parties (including the Nordic Green Left, founded in 1996). They have little representation and no influence at the highest decision-making levels, like the EU Commission (see also Huan 2014).

Ecosocialism in the Americas

It is rather in South America that ecosocialism has had much greater fortunes, with a veritable flowering in Venezuela. For this reason, a special section about Venezuela is provided in Chapter 9, focusing on the achievements of Bolivarian ecosocialism (see also Wall 2010, 95–98). This version of ecosocialism is based on coordinating the decentralist approach of communal councils with the more centralist framework of government rule. It is the only country in the world where ecosocialism has become an essential part of state policies and has attained the status of a government ministry.

A different ecosocialist current is traceable to early efforts at joining various movements and organisations outside of state institutions. As already noted, Michael Löwy identifies the movement headed by Chico

Mendes in the Brazilian Amazon as a first manifestation of ecosocialism, the Alliance of Forest Peoples. It combined syndicalism and First Peoples' struggles with leftist religious formations like Liberation Theology. Though a dubious characterisation to make of the Alliance an ecosocialist organisation, it certainly is a precursor to Brazilian ecosocialism and an inspiration for ecosocialists more broadly, especially in exemplifying what ecosocialist strategy could be.

Ecosocialism in South America is thereby much inspired by First Peoples' long-standing struggles for survival and for environmental protection. This is critically discussed at length in Chapter 7, where it is made evident that ecosocialism and First Peoples may intersect, but only contingently, and the heterogeneity between and within First People communities does not make for any easy formation of alliances or convergence of political interests or principles. In Bolivia and Ecuador, nevertheless, aspects of First Peoples' struggles have led to constitutional laws to defend Nature (e.g., the "Pachamama" Law in Bolivia; the attempt to prevent oil extraction at Yasuní-ITT in Ecuador). These are the sort of developments that are consonant with ecosocialist aims. However, US government belligerence as well as internal and external pressures for resource extraction make it very difficult to put such policies or ideas into practice.

Still, outside of Venezuela, the ecosocialist presence is slim across the region. In Brazil, for instance, there was an ecosocialist network with some presence in the Workers Party (PT), but with very little political influence (erstwhile ecosocialist Marina Silva eventually shifted to mainstream positions). The status of such a network is presently unknown to the author. In Central America, there seems not to have been a notable ecosocialist presence until the 2010s, mainly in México. This is composed of intellectual currents within universities, and a few, small political formations identifying with ecosocialism, such as the Coordinadora Socialista Revolucionaria. It is mainly by way of the Fourth International that ecosocialism has made inroads in South America.

Notwithstanding a relatively more successful diffusion in South America, ecosocialism has not been taken up by mass organisations, excepting, again, Venezuela. This may be due to political redundance (or maybe objectionability) with a movement's existing politics or to earlier historical existence, as in the case of the Landless Workers Movement (MST) in Brazil, founded in 1984 and with about a million and a half members. The MST, who have adopted agroecological principles, are structured in coordinating units that range from households, within settlements, to national bodies representing the entire organisation, including by way of political coalitions and electoral activities. At the time of writing, the MST, unlike, for example, the PSOL, have ministries in the Lula government (see Engel-Di Mauro et al. 2023).

The situation in North America appears equally mixed, and in some respects even less promising than in South America and Europe with respect to representation in government institutions. In Canada, the situation is rather unimpressive. There exist a few networks, mainly scholarly in character, and some involvement in local activism, but in elected office ecosocialism remains insignificant.

The highest achievement in terms of the spread of ecosocialist ideas has been in Kalaallit Nunaat (Greenland), through the Inuit Ataqatigiit Party (Community of the People, founded in 1976; https://ia.gl/). This is a democratic socialist and pro-independence formation (Kalaallit Nunaat is still a Danish colony, even if somewhat self-governing). Their platform includes some ecosocialist policies (e.g., anti-mining, self-sufficiency). They won the 2021 elections, so there is potential for the introduction of explicitly ecosocialist policies, even if they are not an ecosocialist party[4] (see also Wall 2010).

Within the US, a settler colonial empire with military bases in Kalaallit Nunaat along with hundreds of other military bases worldwide, there are self-defined ecosocialist currents within Democratic Socialists of America (DSA) and Green Party USA, each of which has had modest successes in gaining a few municipal and state offices. The anti-democratic structure of US electoral laws, even according to bourgeois standards, ensures no leftist anti-capitalist representation at the federal level, so far. Hence, it might be necessary to resort to supporting policies or legislation that could align with aspects of ecosocialism, such as the Green New Deal. This is contingent on using such policies, or the profitability crisis that might ensue from their implementation, as an additional tool whereby to wage class struggles while trying to avert even worse effects of climate change (Albert 2023, 18–21; Levidow 2022; Schwartzman 2011; Sica 2020; Singh 2022). Even if not articulated as such, this is an ecosocialist

[4] Conflating part with whole, the Australian publication Green Left have misleadingly characterised the Ataqatigiit Party as ecosocialist (see: https://www.greenleft.org.au/content/greenland-ecosocialist-party-wins-election-opposing-australian-mining-project).

strategy applicable in situations like the US or the UK, where, unlike other bourgeois parliamentary systems, anti-capitalist forces are effectively excluded from government representation.

Even under such unfavourable conditions, the Green Party USA have officially inscribed ecosocialism as part of their platform in 2016, while ecosocialism is represented in merely one among many caucuses within the DSA. The Green Party USA participation in national elections is contingent on securing ballot access, which is actively stymied by Democratic or Republican Party interference through the introduction of more stringent criteria for eligibility. To circumvent the manoeuvre, Green Party activists, as with DSA, may resort to supporting Democratic Party candidates with environmentalist positions, among other ways of gaining indirect representation.

The DSA, founded as an anti-communist leftist organisation, are not a political party and are largely tied to the liberal wing of the very much capitalist Democratic Party. They reflect an alternative approach to having some form of government representation in the restrictive US electoral system. The DSA have recently moved leftwards to include openly communist caucuses, though retaining an exclusionary clause for any organisation based on democratic centralist principles. Since 2016, there has been an influx of thousands of especially younger people becoming members, as interest in socialism surged in large measure thanks to the repeatedly failed campaigns for Senator Bernie Sanders' Democratic Party presidential nomination. Through the Democratic Party, DSA activists have succeeded in securing elected posts at municipal and state levels in a few parts of the US. However, unlike in the Green Party USA, ecosocialism is not necessarily represented in policies by DSA-affiliated elected officials. It is decidedly negligible at the national level, even if one were to count as ecosocialist the failed effort at introducing a Green New Deal. Otherwise, sparse and uncoordinated ecosocialist projects have been established within certain cities, including Cooperation Jackson, discussed in Chapter 3.

The Fourth International

Of special significance, because of its presence in virtually all continents, is the Fourth International. It is a federation of political parties and in some ways illustrates a combination of centralism and decentralism. Founded by

Trotsky and associates in 1938, the international committee of the organisation drafted a resolution in 2008 redefining socialism as ecosocialism[5]. The document was introduced, discussed, and approved in 2010 at their 16th World Congress (Löwy and Tanuro 2022, 351–352). All member parties in different parts of the world presumably abide by the ecosocialist principles adopted and promote ecosocialism in their respective countries. Several ecosocialism meetings have been held, but only in Western European countries, until 2024, when the sixth meeting was held in Buenos Aires, Argentina.

Aside from the above-described Revolutionary Workers Party-Mindanao, examples of Fourth Internal parties are Amandla! (South Africa) and Socialist Resistance (UK). Other parties exist in South Asia and are involved in struggles over food sovereignty and land reform (Löwy and Tanuro 2022, 351–352). There is a sizable presence in Latin American countries. The largest and more influential party are the PSOL (Socialism and Liberty Party, founded in 2004), who split from the Brazilian Workers' Party (currently in government) and adopted ecosocialism in 2011. They have some parliamentary seats at federal and state levels. There is an internal faction of PSOL (Movimento Ezquerda Socialista) who seem to act independently. In Argentina, the Fourth International are represented by the Partido de los Trabajadores Socialistas, who have had a few representatives in government). In México, there is a strong ecosocialist content to the Partido Revolucionario de las y los Trabajadores (Comisión Ecosocialista). In Venezuela, Marea Socialista (founded in 2007 as syndicalist current) and the Aporrea group (founded in 2002) are further examples of political organisations affiliated with the Fourth International and taking an ecosocialist stance, though not necessarily as openly as with other member parties.

There are still other Trotskyist parties coordinated at the international level. In Argentina, there is the Movimiento Socialista de los Trabajadores, who are part of the International Socialist League. With Izquierda Socialista, they form the Frente de Izquierda y de Trabajadores Unida (established 2011). The International Socialist League (founded in 2019 in Barcelona, Spain) seems the larger of the other Trotskyist formations espousing ecosocialism. Overall, these Trotskyist parties and their respective international coordination organs at most hold minoritarian

[5] See https://climateandcapitalism.com/2008/03/07/fourth-international-redefine-socialism-as-ecosocialism/, accessed 12 April 2024.

opposition party status in a small number of countries, rarely at national levels. Their accomplishments are mainly in organising protests on the ground, but lasting institutions and projects of ecosocialist substance, if any exist, do not exceed what has been done in a small area of the Philippines. This in itself is important, including by providing models and concrete examples, but they fall short of shaping policies and environmental practices at the national scale. At the same time, all advances made by Trotskyist parties combined, regardless (and perhaps in part because) of mutual antagonisms, remain inadequate towards addressing the global problems they identify, as priority like climate change.

International Affinity Groups

Relative to what they set themselves as political objectives, decentralist international outfits unaffiliated with political parties have fared even less favourably than political parties. One historical example is the Ecosocialist International Network (EIN). It was founded in Paris in 2007 by more than 60 activists from 13 countries; however, it never overcame the combined problems of being composed overwhelmingly by whites from Northern Europe and North America and of male voices predominating in discussions. The degree of the network's inclusivity was delimited by an adherence to anti-capitalism, socialism (a version that excludes almost all socialists in socialist countries), and a vaguely ecocentric worldview. This was an elastic enough set of parameters to keep otherwise diverging currents together, which were mostly Trotskyist and "subsistence perspective" oriented, including Ecofeminist (Kovel 2008).

The EIN, in its time, served as a discussion forum mainly focused on presenting and diffusing ecosocialism at the World Social Forum (WSF). The Belém Declaration (Kovel 2009, 7–12; Wall 2010, Appendix 1) was one important product presented at the homonymous Brazilian location for the 2009 WSF, where the EIN held their second meeting. There, Hugo Blanco was among those bringing a First People's perspective to an EIN assembly of about 100 people. The EIN fizzled out shortly thereafter, its implosive last meeting held again in Paris in 2011.

Similarly, a so far more enduring Ecosocialist Horizons, officially a non-governmental organisation founded by Joel Kovel in 2011, features the diffusion of ecosocialist ideas, often by means of convergences of affine organisations and a greater emphasis on Pan-Africanism and New

Afrikan liberation.[6] Other such networks have arisen. Closely related to Ecosocialist Horizons and Cooperation Jackson is the First Ecosocialist International, founded in 2017 in Veroes, Venezuela (Saul 2023). It is a loosely connected group of individuals (and through them several communities), mainly in Venezuela and the US, who occasionally meet for discussions and updates on their respective progress towards reaching ecosocialist objectives (Saul 2018).[7] In part out of the EIN's ashes, the Global Ecosocialist Network was established in 2019, thanks to the efforts of the late John Molyneux. It functions mainly as an online discussion and information-sharing group.[8]

There are, finally, instances of international political organising that seem a direct fit with ecosocialist objectives, yet there is little evidence for the possibility of such instances becoming ecosocialist forces. La Vía Campesina is a major example of an international network that could have such potential. Certainly, ecosocialists have been drawn to it, at least for inspiration (Giacomini, 2021; Löwy 2014). Founded in 1993, the network encompass and coordinate 183 organisations in more than 80 countries. Ecosocialists may find in La Vía Campesina a way in which ecosocialism can be shown to be a workable alternative, combining aspects of centralism, in terms of international coordination, with a federated, decentralist structure. But La Vía Campesina do not espouse ecosocialism and there are no signs that they ever will. Perhaps ecosocialist ideas are redundant with what the network already do or the policies for which they advocate. Perhaps La Vía Campesina are too internally diverse politically to provide a coherent anti-capitalist framework. This could be a worthy area for further investigation.

Ecosocialism's Geographical Limits and Potentials

As far as can be discerned from the ecosocialist literature, ecosocialist movements or groupings are geographically delimited to parts of the Americas, Europe, as well as South Africa and Australia, with some presence in a few countries in Asia by way of the Fourth International.

[6] https://www.ecosocialisthorizons.com/, accessed 10 April 2024.

[7] The plans of action are described in: https://www.ecosocialisthorizons.com/posts/the-first-ecosocialist-international, accessed 7 April 2024.

[8] http://www.globalecosocialistnetwork.net/, accessed 10 April 2024.

Participation in and organising from within political parties has generated some tangible results, but mainly limited to having some representation in government institutions in relatively few countries in the lesser populated world regions. Much more is discussed about political strategies and achievements by ecosocialist formations in Chapter 9.

Representation of ecosocialist ideas seems to have been more often materialised at subnational administrative levels, but it is unclear whether any specifically ecosocialist policies have been introduced and implemented and whether lasting impacts have been generated. Grassroots or movement-oriented strategies have produced a few localised projects of marginal political consequence within capitalist contexts, principally in the Americas. For the most part, such strategies have led to the formation of loose networks whose main activities are an online presence, publications, public statements, conferences, and the like.

None of this suggests that so far nothing noteworthy has come out of the rise and spread of ecosocialism outside of the Venezuelan context. Political parties latching on to ecosocialist ideas, thanks to ecosocialists' involvement, have created more space for discussion and possibilities for ecosocialist policy consideration and implementation at state levels. Cooperation Jackson also provide an important, concrete model of what could be achievable, for example. Joining and organising protests for climate justice, lending a hand to organisations like the MST, spreading information on ecosocialist alternatives, canvassing for political parties espousing ecosocialism, and even academic publications, among other activities, are valuable contributions. Sometimes the effects are indirect as well. For instance, the 2001 Ecosocialist Manifesto, "a message in a bottle tossed into the ocean by two intellectuals" (Kovel 2009, 3), has reached and inspired many people, some of them to the extent of becoming interested in ecosocialism. Still, the problem is that the amount of writing and speeches on ecosocialism have been inversely proportional to the amount of practical or concrete ecosocial achievements so far. In part, this is due to rejecting anything that could resemble "first epoch" socialism or failing in the attempt to attain institutional prevalence.

One world-consequential exception is in Bolivarian Venezuela, where ecosocialism is enshrined in government institutions and in part supported through the comunas movement. Millions have benefited from wealth redistribution, including provisions of public housing and medical care to many for the first time, as well as the constitutional provisions to recognise (historically for the first time) and defend First

Peoples' and Afro-Venezuelan cultures. At the same time, efforts have been underway to reduce dependence on fossil fuel and mineral extraction while expanding protected areas for biodiversity. But all these efforts have been in part stymied and in part even rolled back because of the violent reaction of the Venezuelan capitalist and landlord classes supported by political interference, military threats, and economic sabotage from the US. Venezuela, to reiterate, is exceptional in terms of ecosocialist achievements. Moreover, the ecosocialist component of the Bolivarian movement, first led by the late Hugo Cháves Frías, came years following the takeover of much of the state through the 1999 Bolivarian Revolution. The accomplishments and setbacks of ecosocialism in Venezuela are tied to a strategy that many, if not most ecosocialists would find authoritarian or centralist. There has been much more critique than tangible support by ecosocialists in the rest of the world. Notably, there is no US ecosocialist movement attempting to bring an end to US aggression against ecosocialism-promoting Venezuela. This is most unfortunate, as Bolivarian Venezuela could be the springboard for expanding the geographical reach of ecosocialism.

References

Adaman, Fikret, and Pat Devine. 2022. "La coordination négociée, un modèle écosocialiste." *Nouveaux Cahiers du socialisme* 28: 41–46.

Adi, Hakim. 2018. *Pan-Africanism: A History*. London: Bloomsbury Academic.

Akuno, Kali, and Ajamu Nangwaya. 2023. "Toward Economic Democracy, Labor Self-Management, and Self-Determination." In *Jackson Rising Redux. Lessons on Building the Future in the Present*, edited by Kali Akuno and Matt Meyer, 56–90. Oakland: PM Press.

Akuno, Kali, and Sacajawea Hall. 2021. "Building Economic Democracy to Construct Eco-Socialism from Below." In *The New Systems Reader: Alternatives to a Failed Economy*, edited by James Gustav Speth, and Kathleen Courrier, 433–440. New York and London: Routledge.

Albert, Michael J. 2023. "Ecosocialism for Realists: Transitions, Trade-Offs, and Authoritarian Dangers." *Capitalism Nature Socialism* 34 (1): 11–30.

Albritton, Robert. 2019. *Eco-socialism for Now and the Future*. Geneva: Springer International Publishing.

Ali, Tariq. 2009. *The Idea of Communism*. London: Seagull Books.

Alvarado, Rómulo and José Daniel Romero Bastidas. 2022. "La Dinámica Agroecológica para la Construcción del Nuevo Estado Ecosocialista y Autogestionario en Venezuela. Perspectiva Evolutiva de la Política Agroecológica y su Relación con las Iniciativas Populares." In Agroecologías Insurgents en Venezuela. *Territorios, Luchas y Pedagogías en Revolución*, edited by Francisco F. Herrera and Olga Domené-Painenao, 251–282. Caracas: Fonacit.

Atkinson, Adrian. 1991. *Principles of Political Ecology*. London: Belhaven.

Baer, Hans A. 2016. "Toward Democratic Eco-Socialism as the Next World System." https://seed.uw.edu/wp-content/uploads/baer-eco-socialism-2016.pdf. Last accessed 4 September 2022.

Baer, Hans A. 2018. *Democratic Eco-Socialism as a Real Utopia: Transitioning to an Alternative World System*. New York: Berghahn Books.

Baer, Hans A. 2021. "Toward Democratic Eco-socialism as the Next World System: A Vision for the Future." In *The New Systems Reader: Alternatives to a Failed Economy*, edited by James Gustav Speth, and Kathleen Courrier, 127–144. New York and London: Routledge.

Baer, H.A. and M. Singer. 2022. "Can Ecological Modernisation Contain Climate Change: An Eco-Socialist Perspective." *Journal of Australian Political Economy* 90: 75–91.

Barca, Stefania. 2023. "Forces of Reproduction." In *A Map to Ecosocialism, Emancipations: A Journal of Critical Social Analysis*, edited by Marco Armiero, Erik Swyngedouw, Erik Massimo De Angelis, Stefania Barca, Sutapa Chattopadhyay, Mina Lorena Trujillo Navarro, Theordore Steinberg, David N. Pellow, and Salvatore Engel-DiMauro, vol. 2, no. 1, Article 8: 7–9.

Bartels, Dennis A., and Alice A. Bartles. 1995. *When the North Was Red. Aboriginal Education in Soviet Siberia*. Montréal: McGill-Queen's University Press.

Bond, Patrick. 2022. "The Case for Ecosocialism in the Face of the Worsening Climate Crisis." *Science & Society* 86 (4): 485–515.

Boykoff, Jules. 2013. *The Suppression of Dissent: How the State and Mass Media Squelch USAmerican Social Movements*. New York: Routledge.

Brain, Stephen. 2011. *Song of the Forest: Russian Forestry and Stalinist Environmentalism, 1905–1953*. University of Pittsburgh Press.

Burbach, Roger, and Camila Piñeiro. 2007. "Venezuela's Participatory Socialism." *Socialism and Democracy* 21 (3): 181–200.

Burden-Stelly, Charisse, and Gerald Horne. 2020. "From Pan-Africanism to Black Internationalism." In *Routledge Handbook of Pan-Africanism*, edited by Reiland Rabaka, 69–86. New York: Routledge.

Campbell, Susan. 1994. "'Black Bolsheviks' and Recognition of African-America's Right to Self-Determination by the Communist Party USA." *Science & Society*, 440–470 (1994)

Ciccariello-Maher, George. 2016. *Building the Commune: Radical Democracy in Venezuela*. New York: Verso Books.

Clarke, John Henrik. 1988. "Pan-Africanism: A Brief History of An Idea in the African World." *Présence Africaine* 145: 26–56.

Colle, Marijke. 2022. "Philippines. À Mindanao, les Paysan.ne.s Mettent l'Écosocialisme en Pratique." In *Luttes Écologiques et Sociales dans le Monde. Allier le Vert et le Rouge*, edited by Michael Löwy and Daniel Tanuro, 105–120. Paris: Les Éditions Textuel.

Devine, Pat. 2017. "Ecosocialism for a New Era." In *Varieties of Alternative Economic Systems. Practical Utopias for an Age of Global Crisis and Austerity*, edited by Richard Westra, Robert Albritton and Seongjin Jeong, 33–51. Oxon: Routledge.

Engel-Di Mauro, Salvatore, Danny Faber, and Christina Schlegel. 2023. "Social Struggles in Post-Bolsonaro Brazil: An Interview with Cassia Bechara of the Movimento dos Trabalhadores Rurais Sem Terra (MST)." *Capitalism Nature Socialism* 34 (2): 57–68. https://doi.org/10.1080/10455752.2023.2212513

Flanagan, E., and D. Raphael. 2023. "From Personal Responsibility to an Eco-Socialist State: Political Economy, Popular Discourses and the Climate Crisis." *Human Geography* 16 (3): 244–259.

Geiss, Immanuel. 1974. *The Pan-African Movement: A History of Pan-Africanism in America, Europe and Africa*. New York: Africana Publishing Company.

Giacomini, Terran. 2021. "Agroecology and Ecofeminist Activism." In *The Routledge Handbook on Ecosocialism*, edited by Leigh Brownhill, Salvatore Engel-Di Mauro, Terran Giacomini, Ana Isla, Michal Löwy, and Terissa Turner, 189–200. New York: Routledge.

Gilbert, Chris. 2023. *Commune Or Nothing!: Venezuela's Communal Movement and Its Socialist Project*. New York: Monthly Review Press.

Haro García, José Luis. 2022. "Ecosocialism and Workplace Democracy." In *The Routledge Handbook on Ecosocialism*, edited by Leigh Brownhill, Salvatore Engel-Di Mauro, Terran Giacomini, Ana Isla, Michal Löwy, and Terissa Turner, 214–222. New York: Routledge.

Hayward, Tim. 1990. "Ecosocialism—Utopian and Scientific." *Radical Philosophy* 56: 2–14.

Huan, Qingzhi. 2014. "Development of the Red–Green Environmental Movement in China: A Preliminary Analysis." *Capitalism Nature Socialism* 25 (3): 45–60.

James, Cyril Lionel Robert. 1969. *A History of Pan-African Revolt*. Washington: Drum and Spear Press.

Kabwe, Zitto. 2021. "The Arusha Declaration: The Case for Democratic Socialism 50 Years on." In *The New Systems Reader: Alternatives to a Failed Economy*, edited by James Gustav Speth, and Kathleen Courrier, 449–460. New York and London: Routledge.

Kovel, Joel. 2005. "The Ecofeminist Ground of Ecosocialism." *Capitalism Nature Socialism* 16 (2): 1–8.

Kovel, Joel. 2008. "The EIN." *Capitalism Nature Socialism* 19 (1): 1–4.
Kovel, Joel. 2009. "The EIN: Chapter Two What Is to Be Done?" *Capitalism Nature Socialism* 20 (2): 1–12
Levidow, Les. 2022. "Green New Deals: What Shapes Green and Deal?" *Capitalism Nature Socialism* 33 (3): 76–97.
Li, Minqi. 2009. *The Rise of China and the Demise of the Capitalist World Economy*. New York: Monthly Review Press.
Li, Minqi. 2016. *China and the 21st Century Crisis*. London: Pluto Press.
Li, Minqi. 2017. "Political Economy of China from Socialism to Capitalism, and to Eco-Socialism?" In *The Political Economy of Emerging Markets Varieties of BRICS in the Age of Global Crises and Austerity*, edited by Robert Westra, 79–100. London: Routledge.
Löwy, Michael. 2014. "Ecosocialism: Putting on the Brakes Before Going Over the Cliff." *New Politics* XIV (4). https://newpol.org/issue_post/ecosocialism-putting-brakes-going-over-cliff/. Accessed 10 April 2024.
Löwy, Michael. 2015. *Ecosocialism. A Radical Alternative to Capitalist Catastrophe*. Chicago: Haymarket Books.
Löwy, Michael, and Daniel Tanuro. 2022. "The Fourth International's Contribution to Ecosocialism." In *The Routledge Handbook on Ecosocialism*, edited by Leigh Brownhill, Salvatore Engel-Di Mauro, and Terran Giacomini, Ana Isla, Michael Löwy, and Turner Terisa, 346–356. Oxon and New York: Routledge.
Magdoff, Fred, and Chris Williams. 2017. *Creating an Ecological society: Toward a Revolutionary Transformation*. New York: Monthly Review Press.
Magdoff, Harry, and Fred Magdoff. 2005. "Approaching Socialism." *Monthly Review* 57 (3): 19–61.
Malisa, Mark, and Thelma Quardey Missedja. 2020. "The Origin and Evolution of Pan-Africanism." In *Routledge Handbook of Pan-Africanism*, edited by Reiland Rabaka, 35–47. New York: Routledge.
Malm, Andreas. 2021. *How to Blow Up a Pipeline*. New York: Verso Books.
Marquina, Cira Pascual, and Chris Gilbert. 2020. *Venezuela, the Present as Struggle: Voices from the Bolivarian Revolution*. New York: Monthly Review Press.
Marx, Karl. 1992 [1867]. *Capital: A Critical Analysis of Capitalist Production, Volume 1*. Translated by S. Moore and E. Aveling. New York: International Publishers.
Mendoza Suárez, Jesús. 2023. "Desarrollo de la economía turística comunal ecosocialismo en el Consejo Comunal la Concepción del "B"." *Miradas* 18 (1): 207–214.
Miliband, Ralph. 1994. *Socialism for a Sceptical Age*. Cambridge: Polity Press.
Mulberg, Jon. 1992. "Who Rules the Market? Green Versus Ecosocialist Economic Programmes: A Response to Eckersley." *Political Studies* 40 (2): 334–341.

Nelson, Anitra. 2022. "Ecosocialism from a Post-Development Perspective." In *Post-Capitalist Futures: Paradigms, Politics, and Prospects*, 31–40. Singapore: Springer Singapore.

Núñez, Miguel Ángel. 2021. *La Ciencia del Konuco y Su Visión Integral*. Caracas: Fonacit.

Núñez, Miguel Ángel. 2022. "Model Learning Spaces Towards Transitioning to Ecosocialism." In *The Routledge Handbook on Ecosocialism*, edited by Leigh Brownhill, Salvatore Engel-Di Mauro, Terran Giacomini, Ana Isla, Michal Löwy, and Terissa Turner, 301–309. New York: Routledge.

O'Connor, James R. 1998. *Natural Causes: Essays in Ecological Marxism*. New York: Guilford Press.

Ojomo, Philomena Aku. 2019. "Niger Delta Environmental Crises and the Limitations of Africanizing Aldo Leopold's Land Ethic: Towards an Earth-Eco-Socialist Model." In *African Environmental Ethics*, edited by Munamato Chemhuru, 291–308. The International Library of Environmental, Agricultural and Food Ethics, vol. 29. Cham: Springer.

Pepper, David 1984. *The Roots of Modern Environmentalism*. London: Croom Helm.

Pepper, David. 1993. *Eco-Socialism. From Deep Ecology to Social Justice*. London: Routledge.

Raymundo, Sarah. 2022. "International Solidarity Against US Counterinsurgency." In*Sanctions as War. Anti-Imperialist Perspectives on American Geo-Economic Strategy*, edited by Stuart Davis and Immanuel Ness, 333–344. Leiden: Brill.

Rabaka, Reiland. 2020. "Introduction. On the Intellectual Elasticity and Political Plurality of Pan-Africanism." In *Routledge Handbook of Pan-Africanism*, edited by Reiland Rabaka, 1–32. New York: Routledge.

Roelofs Joan. 1993. "Charles Fourier: Proto-Red Green." *Capitalism Nature Socialism* 4 (3): 69–88.

Ryle, Martin. 1988. *Ecology and Socialism*. London: Radius.

Salleh Ariel. 2008. "Eco-socialism and 'Ecological Civilization' in China." *Capitalism Nature Socialism* 19 (3): 123–129.

Satgar, Vishwas. 2018. "The Climate Crisis and Systemic Alternatives." In *The Climate Crisis: South African and Global Democratic Eco-Socialist Alternatives*, edited by Vishwas Satgar, 10–16. Johannesburg: Wits University Press.

Satgar, Vishwas, and Jacklyn Cock. 2022. "Ecosocialist Activism and Movements in South Africa." In *The Routledge Handbook on Ecosocialism*, edited by Leigh Brownhill, Salvatore Engel-Di Mauro, and Terran Giacomini, Ana Isla, Michael Löwy, and Turner Terisa, 179–188. Oxon and New York: Routledge.

Saul, Quincy. 2018. "Reflections on the Founding of the First Ecosocialist International." https://radicalecologicaldemocracy.org/reflections-on-the-founding-of-the-first-ecosocialist-international/. Accessed 7 April 2024.

Saul, Quincy. 2023. "Red, Black, and Green Destiny Weapon: Cooperation Jackson the Ecosocialist International." In *Jackson Rising Redux. Lessons on Building the Future in the Present*, edited by Kali Akuno and Matt Meyer, 464–479. Oakland: PM Press.

Schwartzman, David. 2011. "Green New Deal: An Ecosocialist Perspective." *Capitalism Nature Socialism* 22 (3): 49–56

Schwartzman, David. 2013. "4 Scenarios for 2050." *Capitalism Nature Socialism* 24 (1): 49–53.

Schwartzman, David, and Peter Schwartzman. 2018. *The Earth Is Not for Sale: A Path Out of Fossil Capitalism to the Other World That Is Still Possible.* Singapore: World Scientific Publishing.

SERA [The Socialist Environment and Resources Association]. 1980. *Eco-Socialism in a Nutshell.* London: Writers and Readers Publishing Cooperative, also available at https://climateandcapitalism.com/wp-content/uploads/2013/11/Ecosocialism-in-a-Nutshell.pdf. Last accessed 9 February 2024.

Sica, Carlo E. 2020. "For a Radical Green New Deal: Energy, the Means of Production, and the Capitalist State." *Capitalism Nature Socialism* 31 (4): 34–51.

Singh, Pritam. 2022. "Green Reforms and Individual Interventionsin the Green New Deal Transition to Ecosocialism." In *The Routledge Handbook on Ecosocialism*, edited by Leigh Brownhill, Salvatore Engel-Di Mauro, and Terran Giacomini, Ana Isla, Michael Löwy, and Turner Terisa, 235–241. Oxon and New York: Routledge.

Smith, Richard. 2021. "Six Theses on Saving the Planet." In *The New Systems Reader: Alternatives to a Failed Economy*, edited by James Gustav Speth, and Kathleen Courrier, 173–192. New York and London: Routledge.

Somerville, Peter. 2022. "Problems with Ecosocialism." *The Ecologist*, December 5. https://theecologist.org/2022/dec/05/problems-ecosocialism. Accessed 7 January 2022.

Sonderegger, Arno. 2020. "Ideas Matter: Framing Pan-Africanism, Its Concept and History." *Stichproben, Vienna Journal of African Studies* 38: 5–31.

Tanuro, Daniel. 2020. *Trop Tard our Être Pessimiste! Écosocialisme ou Efondrement*. Paris: Textuel.

Terreblanche, Christelle. 2018. "Ubuntu and the Struggle for an African Eco-Socialist Alternative." In *The Climate Crisis: South African and Global Democratic Eco-Socialist Alternatives*, edited by Vishwas Satgar, 168–189. Johannesburg: Wits University Press.

Trewhela, Paul. 1988. "George Padmore: A Critique. Pan Africanism or Marxism?" *Searchlight South Africa* 1 (1): 42–63.

Tunteng, P. Kiven. 1974. "George Padmore's Impact on Africa: A Critical Appraisal." *Phylon* 35 (1): 33–44.

Turner, Terisa E. 2022. "Ethnicized, Gendered Class Analysis: A Theoretical-Methodological Framework for Analyzing Ecofeminist, Ecosocialist Praxis." In *The Routledge Handbook on Ecosocialism*, edited by Leigh Brownhill, Salvatore Engel-Di Mauro, and Terran Giacomini, Ana Isla, Michael Löwy, and Turner Terisa, 39–48. Oxon and New York: Routledge.

Vettese, Troy, and Drew Pendergrass. 2022. *Half-earth Socialism: A Plan to Save the Future from Extinction, Climate Change and Pandemics*. New York: Verso Books.

Wall, Derek. 2010. *The Rise of the Green Left: A Global Introduction to Ecosocialism*. London: Pluto Press.

Wallis, Victor. 2001. "Toward Ecological Socialism." *Capitalism, Nature, Socialism* 12 (1): 127–145.

Wallis, Victor. 2018. *Red-Green Revolution: The Politics and Technology of Ecosocialism*, .ePUB edition. Toronto: Political Animal Press.

Wallis, Victor. 2020. *Socialist Practice: Histories and Theories*. London: Palgrave Macmillan.

Weston, Joe, ed. 1986. *Red and Green: A New Politics of the Environment*. London: Pluto Press.

White, Damian. 2020. "Just Transitions/Design for Transitions: Preliminary Notes on a Design Politics for a Green New Deal." *Capitalism Nature Socialism* 31 (2): 20–39.

Wilpert, Gregory. 2022. "The US War on Venezuela." In *Sanctions as War. Anti-Imperialist Perspectives on American Geo-Economic Strategy*, edited by Stuart Davis and Immanuel Ness, 273–289. Leiden: Brill.

Witherspoon, Alex, Yu Zhou, and Alle Fang. 2022. "The Korean Miracle's Rural Legacy." *Cosmonaut Magazine*, https://cosmonautmag.com/2022/11/the-korean-miracles-rural-legacy/. Accessed 19 February 2024.

Worthington, Richard. 1984. "Socialism and Ecology: An Overview." *New Political Science* 5 (1): 69–83.

Xu, Zhun. 2024. "Industrial Agriculture: Lessons from North Korea." *Monthly Review: An Independent Socialist Magazine* 75 (10): 30-47.

Xue, Jin. 2016. "Arguments for and Against Economic Growth." In *Crisis System. A Critical Realist and Environmental Critique of Economics and the Economy*, edited by Petter Næss and Leigh Price, 69–88. London and New York: Routledge.

Zhu, Liyuan, and Kevin Lo. 2022. "Eco-socialism and the Political Ecology of Forest Conservation in the Greater Khingan Range, China." *Political Geography* 93: 102533.

CHAPTER 5

Ecological Marxism

Environmentalism and Marxism

As outlined in Chapter 3, ecosocialism emerged by the 1980s from socialist currents combining environmental concerns with mainly Marxism-based socialism, elaborating in theoretical terms the kind of organising efforts that fused trade unionism and/or party politics with environmentalism in liberal democracies. This would be socialism in a relatively broad sense, ranging from anti-capitalist egalitarianism-seeking currents within the left, including unions, religious organisations, and labour parties, to self-identified socialist, democratic socialist, and communist formations (post-1950s social democracy excluded, as it ceased to be anti-capitalist). The 1980s Alliance of Forest Peoples in the Amazon furnishes one example of merging environmentalism with an anti-capitalist leftist outlook, as discussed in the previous chapter.

The story goes back decades further, though. Affected by a rising and predominantly bourgeois environmental movement, some Marxists in industrialised liberal democracies started by the 1960s to question aspects of Marxism itself and develop an alternative or modified Marxist framework to explain cases of environmental degradation, if not the environmental crisis. To a minor degree, a similar phenomenon occurred in the guise of dissident intellectuals in some of the industrialised socialist countries, like the USSR and the German Democratic Republic (Amberger 2023; Weiner 1999). Contemporaneously, the same or other socialists

S. Engel-Di Mauro, *Ecosocialism*, Marx, Engels, and Marxisms,
https://doi.org/10.1007/978-3-031-74288-0_5

were dissatisfied with environmentalists' tendency to ignore relations of power. Certainly, much of mainstream environmentalism tended to be remiss (and still is, though less so) on class differentiation, but also on issues of racism, patriarchy, homophobia, and much else related to social inequalities (Barca 2019; Löwy 2015; O'Connor 1998; Tello 2016). In the eyes of Marxists or leftists living in capitalist countries like the US, there had to be a way of uniting commitment to socialism with environmental issues, including for the sake of political relevance in practice (Faber and O'Connor 1988; Worthington 1984).

One way that Marxists addressed environmental issues in core capitalist countries was by showing, often indirectly, that using a Marxist methodology makes it possible to reveal the intrinsic contradiction between "the accumulative nature of capitalism and the finiteness of earth resources and delicacy of the environmental envelope" (Peet 1979, 168; see also Commoner 1971; Enzensberger 1974; Fitzsimmons 1989; Gorz 1974; Katz and Kirby 1991, 263; Parsons 1978; Sheehan 1985; Smith 1984; Smith and O'Keefe 1980). This included, in a small minority of cases, applying Marxist methodology directly to some of the technical and biophysical sciences, mainly biology and scientific ecology (Haila and Levins 1992; Levins and Lewontin 1985, 2007). These efforts in part laid the ground for those seeking to develop what would be called Ecological Marxism or Eco-Marxism (e.g., Burkett 1999; Foster 2000; O'Connor 1988; see also Layfield 2008).

According to James O'Connor, an early introduction of the term was in a 1987 book by Ben Agger (*Western Marxism, an Introduction*). He held that capitalism's main challenge had become the ecological crisis, which had supplanted the economic one (see also Foster-Carter 1974, 93–94). O'Connor rejected such a thesis, arguing for a combined set of contradictions inclusive of the ecological crisis (O'Connor 1998, 174, fn 11; see below). In any case, parallel Marxist perspectives on environmental degradation continued to be developed throughout the 1980s and 1990s, especially in academic geography and, to waning degrees, in academic political ecology (e.g., Castree 2000; Harvey 1996; Hughes 2000; see also Moore 2017). These have not been attempts to build an eco-Marxist current, much less any form of ecosocialism. This is why they are treated here as scholarly endeavours peripheral to eco-Marxism.

Such perceived need to demonstrate the environmentalist credentials of Marxism, though, would have been redundant had it not been for historical shifts among leftists in liberal democracies away from, if not against

Marxism. By the late 1980s and early 1990s, the debate over Marxism and environment was centred mostly about the former's relevance. At most, Marxism was and continues to be viewed as an insufficient or inadequate framework to address environmental issues. Often, this opinion rests on allegations that socialist countries' environmental records have been poor or worse than those of capitalist countries (e.g., Henderson 2016, 289). At times, the problem is traced to Marx's and Engels' very ideas. Environmentalist thinkers often deem such ideas to be tinged with "Prometheanism"[1] and to be founded on a productivist[2] notion of the economy, where there is no consideration for ecological limits (e.g., Benton 1989; Eckersley 1992; see also Pepper 1993, 60–61) and a "domination of nature" ideology is promoted (e.g., Lee 1989; Schmidt 1971; but see Harvey 1974; Grundmann 1991; Hughes 2000).

A rebuttal to such charges came about not by examining the environmental record of socialist movements and states (socialism in practice), but by embarking on close readings and reinterpretations of Marx's and sometimes Engels' original writings to recover or underscore their ecological underpinnings (Dickens 1992; O'Connor 1988; Foster 2000). These are alleged to have been lost through violent suppression in socialist countries (Foster 2000; Tello 2016; see also Gare 1993b; Saito 2022, 7), where instances of disastrous environmental impacts are interpreted as showing the inadequacy of "conventional Marxism and socialism" (Dickens 1996, 4; see also Boxley 2022, 304–305; Pepper 1984). To a comparatively miniscule degree, the textual revisitation exercise to "green" Marxism has been extended to other socialist thinkers, such as William Morris, Rosa Luxemburg, Anatoly Lunacharsky, Nikolai Bukharin, and Otto Neurath (Buhle 2013; Franco 2021; Gare 1993b; Tello 2016; Wall 2010).

By the late 1990s, attention was increasingly turned towards the degree of ecological sensibility inhering the philosophical edifice developed by

[1] This alludes to the mythological figure of Prometheus, who stole fire from the Olympian gods to give it to humanity in the form of knowledge, including technological know-how. For this, he was condemned to an excruciatingly painful death. The term Promethean has been reformulated among environmentalist thinkers to signify the anthropocentric push to exploit resources for human benefit without minding the wider destructive effects.

[2] Productivism refers to production for the sake of production instead of satisfying human needs. This pejorative epithet has also been a way to equate capitalism and socialism as both ecologically destructive.

Marx and Engels, and then, to a still meagre extent, to those of Marxist (or even non-Marxist socialist) thinkers. It was thus that Marxism in liberal democracies was made to return, in the guise of Eco-Marxism, as the way to explain the socio-ecological crisis and to develop ideas to address it as part of a joined socialist-environmentalist ("Red-Green") politics. Camilla Royle has succinctly described this version of Marxism as being "defined by an appreciation of how differing relationships between humans necessitate different ways of relating to the rest of the living and non-living world" (Royle 2021, 444). Put with a different inflection, Eco-Marxism is an application of Marxist or Marx-Engels theories to explain the ecologically destructive tendencies of capitalism, arguing that the contradictions intrinsic to capitalism lead to its ecologically destructive tendencies.

More accurately, it could be said that Eco-Marxism is primarily concerned with applying "Marxian analysis" (Wallis 2020, 16), in some contrast to other forms of Marxism, to matters of ecological concern. This is because it is a matter of methodological application.

> More generally, what is needed is to continually develop and apply the type of critique pioneered by Marx himself. If anything within the Marxist tradition deserves to be rejected, it would be certain types of strategic choice made by activists or power-holders that have led to political failures. Whether or not those choices should be blamed on Marxism is, in my view, not a worthwhile debate, because it does not affect the usefulness of Marxian analysis for understanding the broader framework within which all political activity unfolds. (Wallis 2020, 17)

This usually unstated difference between Marxian and Marxist may seem at first to be insignificant. However, the political content and project become evident in Eco-Marxists' nearly exclusive attention and reliance on Marx's texts and rarely on those of Engels. Only Marxists assumed to have been truer followers of Marx's original formulations are deemed worthy of attention. Whatever has not conformed to whatever was "envisioned by Marx" (Wallis 2020, 18) is to be discarded, expunged, rejected.[3] Thus, Eco-Marxism is excised of any form of Marxism found

[3] At times, Eco-Marxians will go beyond Marx and draw from Lenin when his work becomes useful, as in the notion of the vanguard party, while ignoring or disparaging Marxists who built on Marx's and Lenin's ideas to develop political practices but are deemed unworthy of such inheritance (e.g., Wallis 2020, 34–35).

to be unappealing or inconvenient, most often the varieties associated with socialist states. The level of Eco-Marxists' judgementalism tends to be inversely proportional to the degree of self-reflection regarding their own political accomplishments. For the sake of simplicity and in conformity with prevailing uses of terms, Eco-Marxism will stand for what is essentially an Eco-Marxian position.

More constructively, in the main, Eco-Marxists argue that endless capital accumulation necessarily begets ecosystem and environmental destruction and the current socio-ecological crisis. As powerful an indictment and as persuasive an argument as this may be, the environmental credentials of Marxism have been staked on the results of analyses of ideas and theoretical development. Eco-Marxism thus revolves about the realm of ecological philosophy with a smattering of social science studies on the environment, but remains virtually irrelevant to scientific ecology and, more egregiously, to policy formation and environmental practices in socialist countries.

Ways of Linking Marxism and Philosophical Ecology

As adumbrated above, three general avenues have been undertaken within capitalist countries to relate Marxism to environmental questions (cf. Benton 2023, 680). These three ways of linking Marxism and ecology overlap historically and are not mutually exclusive. This is especially the case among those who view any problems regarding ecology in the works of Marx and Engels as easily correctable by extending the application a Marxist methodology to those very works.

One has been an attempt at general reconciliation, an endeavour inchoate in the early to middle 1970s (Ehara 2023, 121; Enzensberger 1974; Schmidt 1971) and maturing by the late 1980s (Benton 1996; Grundmann 1991; O'Connor 1988; Vlachou 1994). This has at times involved allegations of inadequacies in Marxism itself or stemming from the theories developed by Marx and Engels that compel the adoption of complementary frameworks founded on ecological principles (Bahro 1978; Benton 1989; Boxley 2022; Breen 2014; Gare 1995; Leff 1993; Redclift 1984; Tanuro 2010).

Another approach, still underdeveloped, is to examine the environmental concerns expressed in Marxist thought and politics in different socialist or communist formations or countries (Barca 2019; Ehara 2023;

Gare 1993b; Tello 2016). This way of thinking may include the study of academic currents within capitalist countries. Recently, by way of illustration, Saito (2017a, 105–119) points to the Japanese Marxist Kuruma school, who had, already by the 1950s, shown that Marx understood his abstract labour category as underpinned by physiological or material conditions (e.g., the biological limits of all workers' efforts combined), rather than just as an outcome of labour-power commodification (the exchange-value or wage received from the process of selling of one's ability to work).[4]

A third way to combine Marxism and environmentalism has been to look for and underscore the ecological aspects already existing in the writings of Marx and Engels and, often, to offer reinterpretations accordingly (Benton 1993; Dickens 1992, 1996, 20; Foster 1999a; Hughes 2000; Lamy 2016; Parsons 1978; Pepper 1993; Vogel 1988). Related to this textual reinterpretation approach, Layfield (2008, 82–85) distinguishes between Marxist ecology and ecological Marxists. The former search for ecocentric principles in Marx's original writings (Marx as "proto-ecologist") to construct a different Green ontology and ethics, while the latter draw from Marx's framework to explain the socio-ecological crisis and formulate alternatives. Marxist ecology is a way of fitting Marx into philosophical ecology (Green thought) and cannot explain capitalism's ecologically destructive tendencies. Ecological Marxism is instead a way of using Marx's theories about social relations and insights about how capitalism works to demonstrate the specifically capitalist causes of the current ecosocial crisis. The Eco-Marxist works discussed actually blur this distinction, so Layfield's categories will not be utilised. A clearer yet not mutually exclusive difference is between Eco-Marxian and Eco-Marxist viewpoints, which depart from each other in terms of main sources used and objectives. This is further explained below.

[4] Had Marx's concept of abstract labour not been tethered expressly to material conditions, it would be difficult to explain Marx's insistence on the struggle to shorten the working-day, which would undermine or restrict the extent to which capitalist processes can render labour-power into an abstraction (say, wages) severed from people's daily needs (Saito 2017a, 129), as exchange-value hardly satisfies the necessities of life or human species being.

Ecological Understandings in the Writings of Marx and Engels

An early and still largely ignored case of this third avenue was the systematic elaboration of the late Marxist philosopher Howard Parsons (1978). In that work, he methodically and incisively refuted mainstream environmentalist criticisms of Marxism, including the charges of "Prometheanism," the denial of the value of "external nature," the overemphasis on conflictual relations, and the downplaying of "basic human values" (Parsons 1978, 35–67; see also Hughes 2000). Parsons saw dialectical materialism as an elaboration of the understandings developing in the biophysical sciences under industrial capitalism, where nature was starting to be perceived "as a unitary and dynamic system of systems" (Parsons 1978, 7). But that was a "mechanical materialism" characterised by reductionistic tendencies (e.g., everything and everyone being explainable by physical laws). As ecofeminists like Carolyn Merchant (1980) would show several years later, the mechanical materialism fed into a wider society-nature dichotomisation, where nature and anyone deemed close to nature would be treated as exploitable inert material. The dialectical view offered by Marx and Engels was and remains a crucial corrective to conventional science in advancing a "grasp of the dialectical effects of social practice upon the world of external nature" (Parsons 1978, 8). That is, what happens in society shapes what people do to the rest of nature and in turn what happens in the rest of nature, including as a consequence of human impacts, lead to changes in society (see also Levins and Lewontin 1985).

Parsons collated numerous and extensive quotations from the works of both Marx and Engels to demonstrate what would now be regarded as ecological underpinnings to their philosophy. This is not to give any impression, unlike some Eco-Marxists do, that Marx and Engels offered any extensive treatment of the contradictory relations between society and the rest of nature (Parsons 1978, 17). The concern was to move away from both Hegelian idealism and reductionistic or vulgar materialism to develop a "dialectical, ecological standpoint"[5] (ibid., 10; see

[5] Such a standpoint, however, was based on a surpassed notion of ecological balance (or balance of nature). This is a still widespread belief, including within Eco-Marxism, that finds little support in scientific ecology. Parsons understandably invoked the view of the balance of nature, given that such a presumption would not start to be disputed in

also page 28–29). Importantly, relative to Eco-Marxism, Parsons recognised Marx's use of "metabolism" as a way to expand on contemporary scientific notions, riddled with vulgar materialism, to express the mutually transforming processes involved in the relationship of humanity and other parts of Nature (ibid. 11). This interpretation of the works of Marx and Engels contrasts with Foster's more recent metabolic rift theory (discussed below), where the same concept of metabolism is regarded as central to Marx's philosophical edifice, instead of as illustrative of applying dialectical and historical materialism to conceptualisations in the biophysical sciences (see also ibid., 21–22). In other words, Parsons understood that Marx used universalising, mainstream science concepts to show how they can be put in the service of dialectical and historical materialism, not to make of such materialist concepts the basis of Marxist method. Eco-Marxist thought in part missed and selectively built on such prior Marxist work that debunked anti-Marxist environmentalism. Parsons' work was at most acknowledged in lists of references and hardly studied.[6]

Social Context and Marxist Understandings of the Environmental Question

The above three kinds of perspective on combining Marxism and environmentalism mainly apply to the development of Marxism in liberal democracies. Elsewhere, several schools of thought had long-established linkages between Marxist method and the environmental degradation question by the middle of the 1980s in, for example, the People's Republic of China (PRC) (Li 2021; Zhang 1986) and USSR (Frolov 1986; Gare 1993a, 92–99; Mattley 1966). In fact, in the PRC, some researchers have long taken note of Ecological Marxism and ecosocialism (Wang et al. 2014). One interpretation sees it as linked but separate from ecosocialism, the former associated with North American Marxist thinkers and the latter with aspects of environmentalist organisations mainly in

the mainstream of scientific ecology until the 1980s (see, e.g., the numerous case studies in the edited volume by Zimmerer and Young 1998).

[6] Those Marxist works from the 1970s, for example, feature in neither O'Connor (1998), nor Foster (1999a, 2000). Foster (1999a) and Foster and Burkett (2000) mention Parsons, but do not discuss the content of his work.

Europe (Huan 2010, 2022). Some have drawn inspiration from Eco-Marxist works to insist on the ecological nature of Marx's framework (Zhang 2022).

There has even been an appreciation for Eco-Marxist work in furthering a global agenda that can be aligned with Xi Jinping Thought (Chen 2015; Huan 2022; Ren and Lin 2019). Regrettably, the attentiveness Marxist scholars in socialist countries have given to Eco-Marxist and ecosocialist thought has yet to be commensurably reciprocated by their counterparts in the liberal democratic imperialist countries.

Some of the issues, especially in combining Marxism with environmental protection, had been taken up within Bolshevik ranks even while the Russian Civil War was raging following the 1917 Revolution (Gare 1993b) and continued in different ways thereafter (Weiner 1999). This should not be surprising. In socialist countries, environmental matters are part of government policy and everyday political life, at least in those countries where there had been relative successes in fending off capitalist military aggression and attaining relative domestic stability (as opposed to civil war). In such social contexts, the issue of Marxism and environment is of direct consequence to government policy and intervention and is therefore part of mainstream institutional discourse and practices.

Marxists in capitalist countries have had no such institutional influence, or societal reach, and instead must contend with hostile institutions and well-funded reactionary movements. This specificity and its ramifications are addressed towards the end of the chapter. Suffice it to state here that the debates that eventually led to the emergence of Eco-Marxism were in part a defensive response to most environmentalists' charge that Marxism or socialism is anti-environmental. Many environmentalist intellectuals and activists, including of the leftist variety, chide Marx and Engels for being "Promethean" thinkers. They claim that Marx and Engels' ideas are centred on raising economic productivity (productivism) in ways that are thoughtless about the environmental consequences and thereby little different from mainstream capitalist thinkers and their philosophical foundations. Ecology, according to such environmentalist positions, is alien to the thought of Marx and Engels. Thus, capitalism and socialism are alleged to be at least equally destructive of the environment, with a tendency for detractors to believe that socialism is worse (e.g., Barrillon 2013; Boggs 2020; Breen 2014; Eckersley 1992; see also Foster 2000; Saito 2017a, 10). Marxists in liberal democracies have therefore devoted much ink to countering this widespread uninformed posture, if not disinformation campaign.

To return to the issue of combining Marxism with environmentalism from within liberal democracies, the principal early response to alleged ecological failings in Marxism was to demonstrate how Marxism explains the socio-ecological crisis itself and offers a way out, while conventional theories do not. This in some ways clashed with much of the environmental movement and their demands, which did not necessarily cohere with prevailing understandings of working-class struggles in socialist movements and parties. Among at least some Marxists and certainly within communist parties, there was some distrust about the politics of such movements as well, given as they were (and often still are) prone to pretending away class differences and to be easily coopted by or subordinated to capitalist forces or politics (e.g., Enzensberger 1974; Faber and O'Connor 1988; Gorz 1974; Weston 1986). Some would find in Marx's own writing a lack to be overcome by rethinking the Marxian framework. One way would be in considering "natural conditions" in addition to the relations and forces of production, as part of class struggle (Skirbekk 1973; see also Gorz 1975). Regardless, many Marxists would increasingly feel compelled to make themselves relevant to environmentalism, within a wider attempt to resonate with widely influential social movements emerging by the late 1960s (or earlier, like the civil rights movement).

Another kind of or parallel Marxist response, like that of ecosocialists more broadly (see Chapter 3), again confined mainly to liberal democratic countries, was to continue to or to start to disavow socialist states by claiming that such states have never been or had ceased to be socialist and/or have been unreflective of or contradictory to Marx's ideas. In this way, many Marxists could claim that any environmental degradation in such countries has nothing to do with Marxism, or at least the sort that they espouse (Benton 1989; Dickens 1992; Foster 2000; Williams 2010, 171–173).

This stance became especially important by the late 1980s as a result of pervasive propaganda tracing the Chernobyl accident, for instance, to a presumed anti-ecological nature of Marxism, or what would be otherwise called "orthodox" Marxism. Of course, the same environmentalist critics or anti-communists, including many leftists, will purposely fail to associate democracy and all their related liberal philosophers with Bhopal—the worst industrial massacre in human history relative to lives claimed—as well as tens of nuclear accidents at places like Idaho Falls and Three Mile Island in the US (not to mention intentional radioactive releases at Hanford), or more recently in Fukushima, Japan.

In other words, there could have been a different discursive strategy that would acknowledge environmental problems in socialist countries by considering the global capitalist context and siege conditions of those countries and by highlighting the overwhelming evidence for the much more devastating ecological effects of capitalist societies. A few Eco-Marxist ecosocialists recognise those objective conditions and decisive pressures shaping environmental practices within socialist countries, yet insist on a new, ecological form of socialism as if such conditions and pressures no longer exist.

A partial exception can be found in O'Connnor's writings (1998, 260–264), where the issue of environmental degradation in socialist countries is more constructively interpreted in terms of resource constraints under capitalist pressures. This notably approximates the view promulgated by Parsons roughly two decades earlier:

> Since 1917 ... socialist states, starting from a poor and underdeveloped economic base, have been continuously subverted and encircled, and repeatedly attacked, by capitalist states. The emphasis on action and mastery of natural and social problems bearing on food, raw materials, industry, defense and the like has reflected these conditions of economy, subversion, encirclement, and attack. Even so, the love and enjoyment of nature appear abundantly in artistic expressions in socialist countries, while ecological concerns and esthetic mode of perceiving nature have not until the last decade or so figured prominently in the philosophical literature. (Parsons 1978, 68)

Such sensitivity was no lofty longing for that which did not yet exist. It was grounded in the USSR's institutional framework.

> ... ecological care in the Soviet Union is a central concern at state and local levels, and the Soviet Union has renewed its call for an all-European conference on environmental protection, transportation, and energy. (Parsons 1978, 92)

Policies and practices consisted of much greater achievements than sympathetic Eco-Marxists seem to have grasped. What is more, ecologically destructive impacts, such as there have been under socialism, could be and have been addressed in swifter and more effective ways, compared to capitalist countries, because of the socialist state ownership of the means of production, despite military encirclement and varying levels of economic blockade. Even pro-capitalist specialists on the topic had to concede that the socialist countries had made substantive strides in many areas of ecological concern, nationally and internationally. These strides often surpassed those alleged to have occurred in the (genocide and war-based) capital-endowed, (stolen) resource-abundant countries, like the US. Examples included the amount and treatment of waste, public transport, urban planning, heating infrastructure, and much else (Goldman 1972; Parsons 1978, 90–101; Pryde 1972).

Eco-Marxism, like much of its oft-called Western Marxist predecessor, is comparatively ill equipped to counter bourgeois propaganda on socialist states. This seems due to the circumstances and peculiar Marxist lineage out of which Eco-Marxism emerged. It was from a historical unfolding, within liberal democracies, of widespread social discontent, waning working-class economic leverage, declining socialist or communist parties' political influence, intensifying environmental degradation in much of the world, and effective liberal democratic prevarication about socialist countries (internalised by many leftists to this day). Marxist currents attempting to find ways of becoming politically engaged with and relevant to the general social tumult through which they were living included the ever more politically salient question of the environmental crisis. Eco-Marxism, relating Marxism to ecological and other quality-of-life concerns spreading across major segments of society, was taken up within countries with a long-established, well-entrenched form of capitalism. The objective of articulating philosophical ecology and Marxism has been in response to environmental movements that have been shaped by the most historically effective anti-socialist forces, which have succeeded in popularising anti-socialism even among those most oppressed. Viewed this way, striving to relate to environmentalists imbued with anti-socialist ideology and demarcating oneself from socialist states, partially or totally, can be understandable.

Combining Marxist and Ecological Perspectives: The Second Contradiction Thesis

One attempt to accomplish the objective of combining Marxism and environmentalism took shape with a socialist university group formed to address this problem and led by Marxist sociologist James Richard O'Connor. By means of seminars, the group debated how Marxism can explain ecological and social crises together. What the group concluded was that Marx had important insights about the environment and social problems, but not enough to address the questions posed by the new social movements, including the ecological crisis. Something more had to be developed, building from Marx's theories. Both academic and more widely engaged work would start to be published to contribute to developing this line of thinking. The main outlet for this became the journal *Capitalism Nature Socialism*, founded in 1988 by O'Connor with the crucial assistance of Barbara Laurence, as well as others in the group, such as Ariel Salleh and Danny Faber. The journal has since then been dedicated to combining Marxism and issues of social justice with environmentalism, becoming a main Anglophone forum for Ecosocialist thought (Engel-Di Mauro 2024).

Drawing from the debates and discussions in the seminars and the early issues of *Capitalism Nature Socialism*, O'Connor formulated and refined a way to mesh Marxism with environmentalism by considering and theorising about the conditions of production that prevail under capitalist systems. In this understanding, the socio-ecological crisis is really a general social crisis in the conditions of production. Capitalist relations undermine both the ecological conditions of life (ecological crisis) and the social conditions of production (both social and individual). This process takes the form of two principal contradictions. The first one is about the forces and relations of production. The second is related to the mode (forces, relations) and conditions of production (O'Connor 1988, 1998, 167–171).

The contradiction between the forces and relations of production is manifested through tendencies for crises of over-production. This is when capitalists have so much power over workers that the rate of exploitation is too high and not enough commodities can be sold to make profits. This problem is deferred (never solved) by developing credit systems, technological innovation, marketing, intense competition, state subsidies, and other means used to stave off crises and dump them on others somewhere else or on the working classes generally or on the next generation. Characteristics of the first contradiction is that it affects capital on the demand side. When capitalists try to lower costs to raise profit rates, they unintentionally reduce demand for their commodities because fewer people have enough wages to afford to buy the commodities. This generates a crisis of realisation of value (too many commodities go unsold). There is no problem in producing surplus value, that is, making commodities for sale, which is by extracting labour from the working class over and above what is necessary for workers as a whole to reproduce themselves.

O'Connor thought that this contradiction between the relations and forces of production is insufficient to explain the destructive character of capitalism. It misses the impact on the conditions of production. These include the quantity and quality of resources needed for production and how the use of those resources affects the degree of surplus value extraction. There are, in other words, wider (external) effects to capitalist production that are recognised even by bourgeois economists (they call them "negative externalities").

This second contradiction, between the mode and conditions of production, unfolds in two ways. For one, it affects capitalists on the cost (or supply) side. The process is not directly about exploitation (surplus labour extraction and surplus value appropriation). It is a crisis of "liquidity"—a problem of producing surplus value. The second characteristic of the second contradiction is about raising the costs of production. When capitalists try to lower costs to raise profit rates, they externalise costs onto the conditions of production for wider society in general. This can take the form of undermining the health of workers (reducing the availability of labour-power), reducing soil fertility, unsettling or destroying urban infrastructure, upending individual psychological health, impoverishing public water quality, and other social and ecological conditions that provide the unpaid and mostly unpayable backbone of production (unpayable when it comes to the biophysical processes, since one, for instance, does not pay trees for carbon dioxide absorption). Such overall negative impacts on the conditions of production increase costs for other capitalists who have workers producing other kinds of commodities.

The second contradiction is about the socio-political and economic problems caused by capitalism. For instance, high rent affects people's ability to have shelter or small businesses' ability to run a shop. Struggles over urban planning are behind urban traffic and air pollution. A struggle between environmentalists and capitalists underlies the cost of water, such as over who will pay for sanitation and water pollution. As O'Connor articulated the matter, "[T]he basic cause of the second contradiction is capitalism's economically self-destructive appropriation and use of laborpower, urban infrastructure and space, and external nature or environment" (O'Connor 1991, 108). The reason for the existence and persistence of the second contradiction is because capitalists keep dumping the costs of production on the rest of society and on the rest of nature. It is a destructive tendency that undermines the ability of capitalists to keep having workers producing commodities.

Capitalism is thus characterised by the combination of first and second forms of contradiction. For instance, individual businesses continue to try to lower costs, but by so doing they tend to raise the costs of business as a whole (second contradiction), as well as threatening their own markets (first contradiction). The implication is that the ecosocially destructive tendencies of capitalism lead to rising costs (of inputs, infrastructure, reproduction of labour-power) and weak market demand. As these costs of the conditions of production rise and wages keep being squeezed, capitalist societies are increasingly exploding with crises and social revolts. This latter aspect is how O'Connor understood the social movements rising during the 1960s. These kinds of social movements represent the pressures created by the double contradiction intrinsic to capitalism. Hence, they are not movements concerned primarily with raising wages, having more benefits, and related demands (the traditional concerns expressed by labour movements). They are about defending the conditions and quality of life, lived areas (e.g., cities), health, the environment, and the like.

The second contradiction thesis furnished an innovative and effective way of applying Marxist insights not only to environmental questions, but also to social justice issues in the manner they have been articulated since the 1960s in liberal democracies. O'Connor accomplished even more than this relative to Marxist theory. He showed the key role Marxism can play in explaining the socio-ecological crisis in all its dimensions and thereby develop strategies to draw environmentalists and socialists together (Faber and O'Connor 1988).

Socio-Ecological Crisis, Philosophy, Science, and Marxism

O'Connor seems to have proposed his second contradiction thesis as if in isolation from previous Anglophone work on the subject area.[7] Drawing from original texts, others had already demonstrated how the ideas of Marx and Engels help explain people-environment relations generally, bourgeois ideologies on nature and science, as well as the socio-ecological crisis by means of dialectical and historical materialism (e.g., Harvey 1974, 266–267; Parsons 1978, 24; Sheehan 1985).

The early 1990s work of British sociologist Peter Dickens, who ignored works such as those of Parsons (1978) and has in turn been largely overlooked in much of Eco-Marxism, represents another systematic formulation of an Eco-Marxist framework by drawing directly from the writings or Marx and Engels. In some respects, it was like reinventing an already well-trodden explication. Still, Dickens' analysis of the ecological content in the ideas of Marx and Engels remains among the more thorough examples of how to develop linkages between Marxism and ecology (see also Wallis 2020, 73–85).

He starts by finding points of convergence between Marxism and environmentalism. This is done by dissecting notions about nature in capitalist societies, which deform science, particularly in the core capitalist countries or Global North (Dickens 1992, 60–88; see also Levins and Lewontin 1985, 2007; Smith and O'Keefe 1980, 36–37). The reason to focus on science, and with that the philosophy of science, is that it is the primary knowledge production system that has made the socio-ecological crisis visible institutionally and to the public in general. A fundamental obstacle to addressing the socio-ecological crisis in capitalist countries is

[7] For example, Redclift (1984, 6) had underlined that development prospects themselves (as outcomes of forces and relations of production) are undermined by environmental destruction, the magnitude of which could not have been conceived by Marx or Engels during their lifetimes. Far from rejecting Marxism, however, Redclift advocated for its modification to encompass what O'Connor would later call the conditions of production. Redclift's understanding of Marxism leaves much to be desired. For instance, he imputes Marxists with regarding superstructure simply as impediment to developing the productive forces (ibid.) and clings to the liberal environmentalist stereotype (common also to some proponents of degrowth) that Marx's idea of socialism was premised on a growth imperative in continuity with capitalism (ibid., 122). Nevertheless, Redclift's embrace of the thesis of development being contradicted by ecological crisis was not too far off from the one espoused in the second contradiction thesis.

thus identifiable in the canon of Western European philosophies, which underlies much of conventional scientific work. Since the 1500s (Bacon, Descartes), such philosophies developed a predominant view of nature that conformed to the emerging ideologies of capitalism. They are ideas that arose to justify the destruction of nature and of many societies as well as multiple modes of gendered and racialised oppression that underpin the making of profits (Merchant 1980).

Under capitalism, four main ideas about nature have come to predominate. These are that nature is passive (malleable), inert (static), and separate from as well as inferior to humanity. Part of the ideological armoury developed by ruling classes to legitimise their imperialism, colonialism, slavery systems, and genocides includes the notion of inferior peoples or social groups who are to be deemed inferior because they are closer to or part of nature. The obverse argument, often still found among environmentalists, is to put certain communities on a pedestal precisely on account of their being closer to nature, implying that such environmentalists are somehow outside of nature (possibly supernatural or extraterrestrial, one might think). Whether inferior or superior, nature and oppressed communities are viewed through a society-nature dichotomy that contradicts an ecological worldview and denies the full humanity of the oppressed. When the dichotomy is based on an ideology of inferiority and superiority, it is more appropriately called a society-nature dualism, as the late Val Plumwood (1993) articulated it. These prevailing understandings of nature remain powerful influences in the sciences and society more widely by means of two broad tendencies, namely objectivism and the society-nature dichotomy. These continue to undermine a scientific understanding of the socio-ecological crisis and policies to confront it (Dickens 1992, 61–63).

Objectivism (unlike objectivity) is the view that science is outside of that which is being observed or studied. It is a claim to having a capacity of viewing matters from above, in a detached manner, as if from nowhere (Haraway 1988). This is how scientists can delude themselves into thinking that research or experimentation, for example, is independent of or unaffected by the prevailing sets of ideas or beliefs held by the society in which scientists exist. Objectivist ideology stifles the critical self-reflection needed to detect how one's worldview and social position affects the kinds of research questions asked, the topics and methodologies preferred, and much else in scientific practice and theory. In other words, objectivism undermines objectivity, which is a result of processes like understanding the effects of one's social position or political commitments or material circumstances on one's ways of doing science, developing openness to multiple perspectives, and the back and forth between theories and methodologies and research findings.

The other tendency in capitalism-subordinated science is to view society as separate or separable from nature. Institutionally, the ideological dichotomy is mapped into the sciences as the separation between the social and natural sciences. Marx and Engels identified and critiqued this segmentation, as it is underlain by the assumption that people are not part of nature (Dickens 1996, 19), and called for and developed a unitary historical and dialectically materialist science (or, as related terms, Scientific Socialism, Marxist method).

The repercussions of the society-nature dichotomy with respect to ecological thought are rather damaging because it implies that the relationship between people and ecosystems is of lesser importance in the sciences, as Dickens has pointed out. Even on sheer empirical grounds, the nearly ubiquitous imprint of human impact on the planet's surface and atmosphere takes some effort to miss, as Marx and Engels underlined more than a century and half ago (Dickens 1996, 18–19). However, the entry of environmentalist sensibility in the mainstream of science since the 1970s has not overcome the anti-ecological implication of the nature-society dichotomy. In fact, because mainstream environmentalism is underpinned by capitalist principles and practices, the result of a still limited environmentalist re-orientation in the sciences has been to reproduce the same dichotomy in the form of separating environmental science from environmental studies.

Conventional science in core capitalist countries has nevertheless been challenged to its foundations by the socio-ecological crisis, even if this is seldom recognised. For one thing, the political effects of the development and spread of environmental movements, materially prompted by the socio-ecological crisis (O'Connor 1991), have contributed to reshaping much of scientific work and institutions (think of the emergence of environmental science by the 1970s as a scientific field with its own university departments). But the socio-ecological crisis has had a more profound influence than that on scientists, forcing them to question both objectivism and the society-nature dichotomy. There have been repeated calls and attempts to construct a new science that is relevant to and capable of meeting the many challenges presented by the socio-ecological crisis. The new science often invoked is to focus on the society-environment relationship, which is an indirect way of refuting the separation between the natural and social sciences. This is evident in calls for inter- or trans-disciplinarity, among other initiatives to overcome the divisions. The continuing institutional separation of the sciences and the difficulties faced by scientists in achieving such integration speak to a contradiction between actually objective science and capitalist ideology and practices, based as they are on separating people from each other and from the rest of nature, materially and discursively (Dickens 1992, 62–63).

The Promise and Limits of Systems Theory

With the widespread discontent and revolts of the 1960s, there has been a search for an alternative paradigm that stops thinking of only humans as active agents and avoids this society-nature dualism, within which lurks objectivism (since it is a belief that relies on thinking of oneself as detached). This has been called a Systems perspective or Systems Theory, which has its roots in major respects in the 1930s work of the biologist von Bertalanffy. His lectures were not widely known until they were published in 1968 (Dickens 1992, 81–82; Levins and Lewontin 2007, 104–105).

The perspective was initially used to model organisms and it was not applied to studying society-nature relations until the 1960s. In the mainstream of European and Euro-settler philosophy, there were also some critiques of reductionism identified in the sciences that eventuated into a Systems understanding as well. By the 1940s, a similar view emerged in engineering through the study of cybernetics or servomechanism (self-regulating mechanisms). At that point Systems Theory expanded mainly into military and business management applications. Much more constructive versions and applications aimed at social advancement and building socialism had emerged in the USSR by the 1950s (with earlier precursors in the 1920s, represented by the likes of Bogdanov's Tektology) and in the Allende administration of Chile (1970–1973) until capitalist forces gained the upper hand, through a US-backed military dictatorship in the latter case.

The sort of Systems perspective that developed mainly within capitalist countries can be regarded as being characterised by four main elements. There is an emphasis on connections and on the realisation that all phenomena being ultimately interrelated and interdependent. A third aspects is the conceptualisation of nature in terms of balance, self-repair, and self-regulation. Finally, interdisciplinarity is viewed as essential, that is, the combining of different natural and social sciences to solve problems. It may at first glance seem as if such a paradigm should be most apt in confronting the socio-ecological crisis, but there are certain fatal problems.

In a Systems worldview, everything has a function within a system, which explains nothing. For example, climate change is a function of earth evolution and the evolution of the planet, as a system, involves climate change as among the functional processes that enable planetary change. This is a logical flaw called functionalism, where any change in the system can only be explained by some outside force and the presumption is that change is an aberration (a sort of political conservatism sublimated into a view of nature as static, unchanging; see also Levins and Lewontin 2007, 108–109, 112–121).

A second problem, especially pertinent to environmental thought, is the assumption of a pristine nature outside of society (Dickens 1992, 83–84). This defies the basic understanding of humans as part of nature and makes such Systems viewpoint already contradictory with basic scientific notions. This underlies a third flaw in that humans are still viewed as separate from nature, as outsiders doing damage or causing imbalance in nature. In this manner, a Systems view of nature reinforces a society-nature dichotomy. In terms of environmental policy, the repercussions would be an inability to think in terms of how to be part of ecosystems constructively instead of thinking in terms of mass expulsions and separation to preserve nature.

Finally, changes to nature are understood in Systems thinking as disturbance events, and so human impact is reduced to a perturbation that will be resolved by nature's self-repair or self-regulation. There is no room for an understanding of systems (wholes) being parts of larger systems or of systems and their constitutive parts being mutually transforming (Levins and Lewontin 2007, 109–112). For example, human impact is exogenous to the atmosphere, which is supposed to be balanced. Eventually, nature will bring self-regulation, called climate change, to return to a balance. By means of this fetishism of nature (see further below regarding fetishism), this view amounts to denying the capitalist causes of the recent form of climate change and, relatedly, of the importance of political struggles. The causal effects of capitalist relations of power in society on atmospheric chemistry (or nature generally) are erased from view or swept under a generic human cause.

In fact, scientists adopting a Systems perspective must cling to objectivism. They could scarcely do otherwise. This is because to recognise themselves as part of the same system within which their object of study exists is to clash with the functionalism inhering Systems perspectives. If scientists and their objects of study are part of the same system (or cogs, perhaps a more apt analogy), scientific findings or discoveries are merely functional to the stability and reproduction of the overall system. The Systems scientist must therefore pretend to exist outside of the system within which lies the object of study to be able to explain its characteristics, dynamics, etc.

This pretence may seem persuasive for some forms of research, such as in astronomy or geology, but astronomers and geologists still live in a system that includes social relations of power resulting in differing levels of access to the instrumentation or technology needed to carry out research. They may be studying systems ostensibly external to the one that they inhabit, but by ignoring the system of which they are part they systematically fail to understand the processes that enable or constrain their research, and this will affect the results obtained about an external system and the theories developed about such an external system (Levins and Lewontin 2007, 108). For a more direct example, the climate scientist, according to a Systems viewpoint, is supposed to be outside of or unaffected by climate change (as object of study) even as climate change, as part of planetary dynamics (the wider biophysical system), is the source of data for the research on climate change and often impacts the environments where the climate scientist lives.

At bottom, Systems perspectives cannot offer the sort of understanding of interlinkages and the critical self-reflection required to overcome objectivism and its close relative, the nature-society dichotomy. Thereby, their contributions, though useful in terms of tracing connections (as well as in providing empirical evidence), fall short of explaining the socio-ecological crisis and are even less helpful in developing any political guidance to address it. In fact, proponents of Systems thinking and scientists employing Systems theory are often aligned with the prevailing capitalist system.

Marxism as Basis for a New Paradigm

It is precisely this incapacity of knowledge systems and paradigms under capitalist regimes to change course that could or should open the door for Marxist methodology to form the basis for a new ecological paradigm worthy of the planetary challenges at hand. Even prior to Charles Darwin's evolutionary theory, Marx had already begun, by the 1840s, to connect social relations and nature in ways that preclude objectivism and a society-nature dichotomy. The holding back of evolutionary theory's potentials to overcome anti-ecological attitudes in science exemplifies how a capitalism-subordinated science (relations of scientific production) impairs the development of science generally (as a force of production). This contradiction between the relations and forces of scientific production (including paradigms) can be addressed by adopting Marx and Engels' framework because it helps understand and explain the linkages of society and nature. There are four components to this framework.

First, Marx's method, developed in consultation mainly with Engels, centred on reciprocal connections between organisms (humans included) and environments (Dickens 1992, 63–64). Marx and Engels, as part of what they viewed as scientific socialism, developed a dialectical materialist understanding of reality whereby it is understood that beings change and are changed by environmental processes and other beings. Dialectical materialism means things or organisms have latent potential that can only become manifested in relation to other things or organisms. As Dickens aptly summarised the issue, "an organism's being resides not just within the organism itself but within its relationship with the rest of the ...world" (Dickens, 1992, 64). Marx understood objects, organisms as sets of relations (dialectically related to each other), not as isolated or separable entities. A reciprocal connection between organisms and environments means that it is not possible for scientists to be outside of what they research. It is similarly impossible for society to be separate from nature because they are interrelated. Given that the relationship between beings and environment or society and nature is characterised by mutual determination and transformation processes, it follows that any notion of society-nature dichotomy is untenable (see also Vlachou 1994).

This is, though, the least that could be said on the matter. One should go further in pointing out that the notion of humans as natural beings, as part of nature, is a given even in the mainstream of science. Therefore, the dichotomy, unless humans are to be regarded as supernatural beings, is unjustifiable even in conventional terms. Nevertheless, many scientists continue to uphold such an ideological dichotomy, engaging thereby in incoherent thought without necessarily being aware of their incoherence. This situation should not be surprising, since scientists are just as susceptible to ruling ideologies as the rest of society. Moreover, by and large, scientists are like other workers. They depend on wages for their livelihood and their highly unevenly distributed privileges, which are contingent on ruling classes' profitability prerogatives or even whims. There are, in other words, material conditions behind such scientific incoherence.

The second component of Marxist science is inextricably bound to the first. It is an ontological theory of humans as natural and species-beings. The first aspect, humans as natural beings, accords with the mainstream of science. The latter aspect on species-being, further discussed below, often diverges from most scientists' understanding, which typically defaults to liberal ideology of humans as "naturally" individualistic or egoistic and with a view of nature as fixed or unchanging ("human nature"), free of any historical change, despite and flatly contradicting prevailing evolutionary perspectives on humans and all other organisms, on physical environments, and on ecosystems. It is as if most scientists were obdurately holding on to flat earth theory while taking for granted a spherical earth.

The first two components of Marxist method should be easily adopted even within conventional science, since there is much overlap and Marxist method, relative to biophysical processes and ideas about nature, draws from scientific work, evidence, and theorisation, including theories of evolution. Where there is the most evident departure from the scientific mainstream is in the third and fourth component of Marxist method. One is a critical appraisal of capitalism, and any class-based society, as restricting and distorting of human nature. The other, which follows from this, is the argument for the possibility of recovering human species-being under capitalism, a thesis developed mainly by Engels (Dickens 1992, 64–71; Henderson 2016, 268–269).

Marx identified specifically capitalist constraints and distortions of human nature that affect the formulation of ideas, the production of knowledge, and much else, including within the sciences. One could extend this to the more recent environmental movements, who also draw from scientific ideas and data. At a most basic level, our dependence on the rest of nature even for our very existence already refutes the capitalist ideology that sees people as separate from nature (and with that any objectivistic attitude). This is one way that capitalism deforms our human nature, by obfuscating the obvious, the reality of our being part of nature. The society-nature dichotomy is a reinforcement of a historically imposed disconnection between an increasing number of people and the rest of nature (Dickens 1996). This is why bourgeois forms of environmentalism quickly become self-contradictory as well as racist and colonialist, exemplified in preservation policies involving land appropriations and expulsions of peoples whose activities are part of making those ecosystems to be preserved. By extension, the ever more widespread and frequent destructive impacts resulting from capitalist relations undermine human health and even the survival of many communities, especially racialised communities. Some Eco-Marxists would even say that capitalist relations eventually lead to self-destruction, as discussed below.

Our dependence on the rest of nature for survival presents one way of seeing through capitalist ideology and grasping the environmental consequences of capitalist relations. Marx, however, went beyond this basic fact to insist that we are part of nature (nature as humans' "inorganic body," Dickens 1992, 64). It is precisely because we are natural beings (organisms) that what we do in the environment will impact us and/or other people. As in the case of any organism, people's abilities and capacities, our innate though differential potentials, depend on the relationship with the rest of nature. But we are also as a species different from other species, according to Marx and Engels, in being able to develop new abilities or powers (Dickens 1992, 65–69; 1996, 57–58; Vogel 1988).

Marx in particular distinguished between natural and species-being in terms of specifically human capacities (powers) and needs. That is, we are characterised by both natural and species-being. Natural being includes powers and needs related to survival and reproduction as well as other needs (e.g., to explore, to be with others of the same species, to move about, etc.). Species-being is, on the other hand, what distinguishes the human species in terms of having additional powers and needs. These include spiritual, intellectual, and aesthetic potentials. Human activities (including production) can span all of nature (i.e., the entire planet's surface and beyond), can be on behalf of all humanity and even other species, and are beyond immediate needs. Put differently, people must relate directly with the rest of nature to survive and reproduce (natural being), but also to realise their intellectual, spiritual, and aesthetic potentials (species-being).

In capitalism, people become estranged (alienated) from their *natural* as well as their *species*-being needs and powers (see also Parsons 1978, 18–19). Capitalist relations, as they evolve, result in the distancing or *alienation* of people from (1) their own labour (from the products of their labour); (2) other people (asociality); (3) themselves (lack of individual sovereignty); and (4) the rest of nature (the ideology of humans as "unnatural" beings, the nature-society dichotomy).

People work on (appropriate) bits of the rest of nature to make commodities and nature is divided into private property. Workers lose control over the products and over the understanding (meaning) of their labour. The direct relationship to nature is eventually denied to at least most people in capitalist societies, in some communities more than others.

Obligatory, mutually exclusive, destructive forms of competition among people deny or undermine the natural and species-being need for sociality (association with members of the same species). People become alien to each other because they are separated as single individuals, under legal frameworks, in everyday work experience, etc. Individualism is both an ideological and practical consequence of the full development of capitalist relations. Methodological individualism is its sublimated scientific and philosophical incarnation.

The alienation from one's labour and from others extends to one's very self. An example is of thinking of one's body as a marketable commodity or rationalising being sacked as due solely or mainly to one's foibles or inadequacies. Another is the belief in solving problems by using pills, applying facial creams, taking up wellness programmes, building up muscles, etc. The obverse is when individuals are blamed for any characteristic considered deviant, or departures from a presumed norm. The diffusion and popularity of self-help manuals speaks to this process of individualising faults that are social in character. Among other means, such as ubiquitous exposure to advertisements, they feed into a concealment of structural causes and often degenerate into self-harmful behaviour.

Environmentalists should find Marx's insights especially resonant with their concerns when it comes to the concept of alienation from nature, or, rather, the rest of nature (see also Parsons 1978, 63–65). For capitalist relations not only generate a distancing people from their work, each other, and even themselves, as part of nature, but also from all other beings and from physical environments. Organisms, for instance, are often thought about as things, not forms of life. Forests become timber or logs instead of interrelated ecosystems with many kinds of organisms, interacting with each other. Insects are reduced to the category of pests, to be destroyed, instead of essential parts of ecosystems that sustain our lives. These are the sorts of tendencies promoted by capitalist relations that should leave any environmentalist aghast. But environmentalists in capitalist societies are not spared the processes of alienation and they are often complicit with them. Cases of environmentalists embracing "free-market democracy," the basis of what they allegedly fight against, are paradoxical conditions created by internalising capitalist ideology and the alienation processes on which it is founded.

These processes contribute to distorting people's species-being needs. Their senses of the world (being) are reduced to possessing or wanting ever more things (having) or, the obverse, denying themselves in acts of individualist asceticism that elide the systemic causes of the socio-ecological crisis. Overall, however, capitalist relations tend to create a majority attitude of possessing nature becoming more important than appreciating nature (Dickens 1992, 69–73; see also Saito 2017a, 29–32).

There are possible antidotes to such a tendency that are built into capitalist relations themselves as they evolve. In other words, from a perspective of dialectical and historical materialism, capitalism, as any mode of production, contains contradictory tendencies out of which there exist potentials to recover human species-being. This is what was alluded to above as Engels' contribution to this issue. Capitalism may split people from their work, each other, themselves, and the rest of nature, but it also brings them together in different, even if alienated ways. This occurs through, for example, market exchange (profit-oriented, money-based), workplaces (offices, factories, mines, plantations, etc.), and resource consumption (including tourism).

To bring more clarity into the contradictory forces unleashed within capitalism that can lead to the overcoming of capitalism itself, a brief discussion of the concepts of commodity and commodity fetishism is useful. Because capitalist relations split people from each other, from self, and the rest of nature, the products of labour and the wealth of nature are thought about as external things. They are mainly related to us as commodities for sale, with no social or ecological context or history. This is among the central insights Marx contributed especially in the first and only completed volume of *Capital*. Commodities just appear, unrelated to the labour processes that created them and emptied of the ecological relations that enabled their existence. In this way, commodities become external objects with a life of their own (fetishes).

Because of this, our relationship to nature is more difficult to understand. It is distorted. In capitalist societies, many people think of markets as people or that, for instance, milk comes from the shop instead of cows or that technology itself causes bad things. As a more extreme expression, in the US, corporations have the same legal standing as people. The belief that what people produce (commodities, including firms) are independent of people, have a life of their own, or have causal properties is a result of capitalist alienation processes.

Engels developed a complementary view of the development of alienation and how to overcome it relative to nature, to our relationship to the rest of nature. To paraphrase, humans have evolved ever-greater capacities through labour (appropriation of "nature" and communication capacities). This includes the mental capacities of abstraction that can enable people to think that they are separate from nature. Both Marx and Engels thought that as human capacities develop further, including through capitalism, nature will be mastered, that is, understood, more and more. This is a main avenue leading to the potential for the overcoming of the alienation from nature, undoing the restricting and distorting effects of capitalism on people's natural and species-being. To clarify further, according to Engels, the contradiction between idealism (or beliefs in society-nature dichotomy) and our material relations in nature eventually leads to new mental capacities (greater understandings of nature) to overcome the alienation from nature and to realise that we are part of nature.

This would mean the emergence of a new kind of science (not objectivist, with no society-nature dichotomy) that is unified, founded on a dialectical materialist approach. In essence, Engels had been calling for this sort of science, like Marx, and had anticipated much of Systems thinking by decades. A systems view of phenomena's interdependence and interrelationship is redundant with Engels' concept of science of interconnections as he was developing in *Dialectics of Nature* during the 1880s. In some respects, this dynamic evolutionary view has been vindicated, as dialectical principles are increasingly, even if very unevenly, internalised in the biophysical sciences as an ever-greater understanding of nature proceeds (Levins and Lewontin 2007, 102). Perhaps progress towards a unified science has been in the works, but, so far, the process is stymied by capitalist relations.

In a current ecosocialist reading, such a unified science, coupled with Marx's theory of alienation, would be better equipped to address the socio-ecological crisis than Systems perspectives, conventional ecology, and predominant forms of environmentalism, and certainly any capitalist frameworks. Caused as it has been through capitalist relations, the crisis provides the material conditions from which people in capitalist societies are coming increasingly to understand that we are part of nature. This contributes to the overcoming of capitalist processes of alienation.

Engels and Marx, then, had long ago built the theoretical groundwork to help make sense of the ecological, not just social dimensions of capitalist relations and their consequences. This is not to claim that the Marx-Engels framework is spotless or ready to use regardless of historical or geographical context. But it is a germinal advance over capitalist perspectives, especially when it comes to providing self-correctives. For example, one potential problem some Marxists have identified is a residual dualism regarding the human and other species (Benton 1989; Schmidt 1971). There are passages in Marx's work that suggest humans as intrinsically different or even superior to all other beings, including in what Marx and Engels advocated for as the need for the "mastery of nature" (Dickens 1992, 86–88; Smith and O'Keefe 1980, 34). This is to miss the dialectical materialist understanding laid out by Marx and Engels, who wrote that people are a mutually constitutive part of nature and hence mastery is about comprehension, not control (Parsons 1978, 69–71). Leaving aside detractors' wilful misinterpretations, even if such faults did exist in some of Marx's or Engels' writings, they do not extend to the entirety of the methodological framework. As Ted Benton (1988; reported in Dickens 1992, 87) has explained, Marx's notion of species being, for example, can be extended to all beings, recognising all species' peculiarities but also overlapping characteristics, without resorting to any dualistic notions. In other words, the Marxist method itself provides the sort of self-corrective process that capitalism-subordinated science cannot, wedded as it is to a view of nature as separate from humanity and, in the more extremist form, of human nature as innately capitalist.

To recapitulate, the socio-ecological crisis has led to the questioning of one of the foundational aspects of capitalist ideology, which is the nature-society dichotomy, as well as its scientistic corollary, objectivism. Ecosocialism, as a movement mainly in core capitalist countries, has come to be mainly informed by Ecological Marxism precisely because the original works of Marx and Engels offer both explanations and principles helpful in identifying causes and formulating alternatives. At the same time, the writings of Marx and especially Engels help build a new science that overcomes the limitations of Systems thinking, its functionalism, fetishistic ideas about nature, nature-society dichotomy, and uncritical approach to capitalism (the foremost cause of the socio-ecological crisis). In contrast, Marxism (at least in this reading of it) is not functionalistic but dialectical materialist. That is, it accounts for the mutual influences and transformations within and between totalities or systems. There is no assumption of external and pristine nature and no teleology. What is more, as an important alternative to much environmentalism, social relations of power are addressed and considered as the primary processes subtending the socio-ecological crisis.

Red-Green Fusion: Intrinsic Value and Ego

Dickens had demonstrated that Marxism is not only relevant to grasping and addressing the socio-ecological crisis, but essential to develop and put into practice the sort of politics, research, and policies necessary to build an ecologically sustainable society. Neither environmentalism, nor science can do without Marxism, if the objective is to take remedial action relative to the socio-ecological crisis. Joel Kovel, drawing from O'Connor and apparently unaware of developments among Marxist intellectuals in the UK, would add that Marxism is in a unique position to explain capitalism as the efficient cause of ecological crisis (Kovel 1995; further elaborated in Kovel 2007). Kovel put the matter thus: "Caught in the cycle of 'grow or die' that characterizes accumulation under the terms of relentless competition, capitalism is revealed as the destroyer of nature as well as of society" (Kovel 1995, 32).

This could also be aptly described in terms of a first and a second contradiction, as understood by O'Connor, or even as Foster's later notion of metabolic rift. However, Kovel went further by characterising the socio-ecological crisis in existential terms, expressed in dramatic fashion. But Kovel did not find O'Connor's interpretation satisfactory and likely would have found Dickens' just as wanting. Kovel claimed that the Marxist critique of capitalist relations is too centred on the first contradiction, which includes concern over "the lived, suffering experience of human beings" (Kovel 1995, 33).

On the one hand, a point Michael Löwy would repeatedly make as well, much of ecological thought pays no attention to social justice, and this is why it resonates poorly with Marxism (or socialism more broadly). This is hardly helped by a residual or reinforced anti-communism and latent reactionary politics (e.g., biological conservation through mass expulsions). There is a lack of understanding of capitalism in relation to socio-ecological crisis. Instead, there are mystifications like blaming the crisis on population growth (see also Foster 2000).

On the other hand, there is a chronic difficulty for Marxism to resonate with ecological thought. According to Kovel (1995), Marxism only provides an underdeveloped framework or language to address the *second* contradiction: the destruction of nature and people's care for nature. This motivates them to struggle for nature protection, for example. At bottom, there is a problem with privileging the human species over the rest of nature. This can explain the challenge for Marxists to become relevant to environmentalist movements.

It is not that Marx offered nothing about the society-nature relationship. He did provide a useful conceptual framework. It is just that the relationship was tackled at too abstract a level. For Kovel, ultimately, as well as for the likes of Ted Benton (1988, 7),[8] the problem is to be found in an inadequate attention to the intrinsic value of nature, a concept that Kovel would continue to refine in later writings (see Kovel 2007, 2014). The concept of intrinsic value is instead expressed in radical green movements (mainly in bioregionalism, some forms of Ecofeminism, and deep ecology) that unfortunately tend to offer little to no critique of capitalism.

Put differently, while critical or radical green perspectives are inadequate on capitalism, Marxist value theory, where only humanity-centred use and exchange-values are recognised, provides little guidance in valuing nature as such. O'Connor had devised a way to fit socio-ecological crises, including the movements thereby generated, within the ambit of Marxism, by way of identifying first and second contradictions and combinations thereof. Dickens, similarly, recommended that environmentalists and scientists adopt Marxism if they wish to understand the causes of the socio-ecological crisis and do something about preventing further damage and reducing its effects.

Kovel's solution was rather different. In the first place, he implicitly treated Marxism and environmentalism (green thought) as equally valid perspectives. This is what led him to seek a reconciliation of the two

[8] That Marxist work on intrinsic value would be in part subsequently acknowledged in Kovel (2007).

worldviews instead of thinking in terms of how they could be combined. Kovel's quest was more like a fusion than a linkage of worldviews. This is why he elected for a dialectical approach. For him, in agreement with many other Marxists, the

> dialectic refuses the either/or; it is no more a close relative of chaos theory than it is of systems theory. This is because there is a directedness within becoming, and a dialectic between determinacy and indeterminacy. (Kovel 1995, 40)

To proceed in this analytical direction, he found it useful to specify basic ecological principles because they contradict capitalism to its core. These are reciprocity, mutual recognition, and interconnection. He also found it just as important to identify problems in Marx's view of nature and humans' relationship to it, a view that remains crucial but incomplete. Marx understood society as part of nature. However, he considered nature as humanity's "inorganic body," which to Kovel suggested a split between humanity and nature. In such an ontology, nature is still regarded as passive and inert. The ecological interconnection is acknowledged, but not the reciprocity and mutual recognition (for a rebuttal, see Dickens 1992, 64; Foster and Burkett 2016).

The issue, then, is how to overcome a persisting dualism pervading Marxian thought. One way is to rethink nature and humans in terms of mutual recognition. Humans come out of nature, where there is at least proto-consciousness. But humans differ in having self-consciousness. Both Marx and Engels had arrived at this conclusion, according to Kovel, when thinking of matter as having some sort of agency, or inner activating qualities. For this, they referred to the concept of *qual*, following the philosopher Jacob Böhme. In this understanding, there are two aspects of nature. One (first nature) is nature that exists without us. Another (second nature) is nature that exists with us. Humanity is thus a product of nature and human labour. Kovel expressed the issue this way:

> The power of ecological Marxism may add an equivalent grasp of the origins of human being as nature is transformed. It is Marxism's project to see nature and labor as fully joined in the production of human being, and human being as second nature made self-conscious through social labor. (Kovel 1995, 38)

In other words, Kovel called for a return to the dialectical roots of Marxism because this allows us to see the world as composed of many processes that make nature as a whole and humanity, in particular.

Dialectics involves processes of differentiation (humanity-nature) and the overcoming of differentiation (unity and interpenetration of opposites). There is a unity of opposites within a whole, but also mutual determination and the negation of the negation (or transformation). It is *not* the oft quoted formula of "thesis-antithesis-synthesis" because dialectics is not a formula. There is no ultimate end point. Even more importantly to Kovel, dialectical thinking is an affirmation of life or being (Kovel 1995, 41). It implies subjectivity (self-consciousness) and the interactions of many different subjects. Subjectivity is also intersubjectivity in that individuals are always also products of social relations. Inspired by the work of Merleau-Ponty, Kovel sought an "ecologisation" of dialectics to help fuse Marxism and environmentalism, that is, build Ecological Marxism. By ecologisation, Merleau-Ponty meant that consciousness and subjectivity are also made by the rest of nature.

Furthermore, there is a spiritual sense of being part of nature that is constitutive of species-being. It is thus that Ecological Marxism simultaneously entails overcoming the Ego produced by capitalism, an Ego that tends to be ecologically destructive under capitalist regimes. It is not about effacing or doing away with the Ego since the Ego is part of being human. However, under capitalism, the Ego is taken to exaggerated levels and is based on destructive relations with nature. The destructive tendency is due to five main processes and characteristics evolving under capitalist conditions: the splitting of self from others; the limiting of recognition primarily or only to oneself (pridefulness, rationalisation, isolation, seeking power over others); productivism (involving efficiency, self-discipline, obeying orders, calculation, all for the sake of capital accumulation); the internalisation of commodity fetishism; and consumerism. The subjective basis of capitalism is founded on hyper-production and hyper-consumption, with socially and ecologically detrimental consequences. The Ego is the main way that people become subjects in capitalist society. Capitalist subjectivity is defined by Ego and egoistic relations in that "capitalism builds itself on egoic relations and reproduces egoic relations on an expanding scale" (Kovel 1995, 46). The Ego provides a psychological structure of capitalism that underlies the ecological crisis.

At the same time, environmental destruction creates possibilities to overcome capitalism. The second contradiction of capitalism (the undermining of the conditions of production) shakes up the Ego (egoic being) and creates an opening to transform society. This results by drawing people together to fight for better conditions of life, to defend human life, and to preserve "nature" (mainly as second nature). But a society that overcomes the capitalist Ego requires the expansion of interconnectedness within humanity (i.e., socialism) and between humanity and nature (i.e., ecological understanding). This is why, to Kovel, Red-Green transformative politics are needed to build the social structures or institutions to expand interconnectedness, overcome capitalism (egoic relations), and realise an ecologically sustainable society.

Transformative politics of this sort, that is a red-green or ecosocialist politics, are informed by several guiding principles. One is an ecological mode of production (that must be socialist by definition), where ecological principles are applied to relations and forces of production. Another is radical democracy, comprised of a network of places and freely associated small producers. A third is a deep subjectivity fostered by ecological rationality to connect second and first nature. Importantly, for Kovel, this third guiding principle includes spirituality, understood as awareness of being part of something much greater than oneself and thereby acting accordingly (Kovel 1991, 1995). All three principles involve the application of the concept of intrinsic value, which, like the consideration of the capitalist production of the Ego (egoic being), must be introduced into traditional Marxism if the second contradiction identified by O'Connor is to generate a potential for people to overcome capitalism. As he would phrase it in later writing, "The precondition of an ecologically rational attitude toward nature is the recognition that nature surpasses us and has its own *intrinsic value*, irreducible to our practice" (Kovel 2007, 151; italics in original).

Environmentalism and Spirituality

Spirituality is among the controversial sides of Kovel's transformative politics. Some Eco-Marxists might baulk at the notion and find the inclusion of spiritual matters troubling. Foster (2000), for instance, points out that environmentalism teems with ideas often predicated on unscientific and/or irrational "spiritual" beliefs. Hence, Eco-Marxism (or red-green perspectives) will founder if attempting to draw from ecological or green thought uncritically. Obscurantist notions promoted in some mainstream ecological thought ultimately undermine efforts to develop environmentally constructive practices and policies, since beliefs in supernatural or in processes beyond social control effectively deny capitalist causes and the need for fundamental social change.

This critique of spirituality in environmental thought is something that Joel Kovel would find to be a confusion of irrationality or religion with spirituality. This is if spirituality is understood as a sense of being part of a larger whole, such as nature. This aspect of human existence, as part of species-being, is something Kovel thought to have been dear to Marx (Kovel 1991, 1995, 2001). Kovel also finds spirituality to be another way through which the notion of intrinsic value can be integrated with Marxism as a complementary analytical category to use- and exchange-value (Kovel 2007, 212–213; 2014, 19). This interpretation coheres with Ted Benton's "transvaluation of values," based on William Morris' insights and a re-reading of Marx's early works where metabolism with the rest of nature entails "spiritual nourishment," not just the fulfilment of "bodily needs." It is an understanding of value that goes beyond "necessity for human survival and wellbeing" (Benton 2023, 684–685).

Nevertheless, there is potential trouble with an emphasis on spirituality as a basis of politics. While irrationality, religion, belief in transcendent being(s), and spirituality are distinct social processes, as well as phenomena at different levels of abstraction (e.g., religion is but one manifestation of spirituality), they often do coincide and in socially reactionary and devastating ways. This Kovel readily admitted and seized as a reason to struggle for egalitarian forms of spirituality as a main aspect of ecosocialism (Kovel 2007).

Relative to Eco-Marxist concerns, theistic religious spirituality has served to obfuscate relations of power, and with that, as mentioned above, the capitalist causes of the current socio-ecological crisis. Hence, Foster's critique of at least some forms of environmentalism, those replete with what could be termed spiritualistic decoys, is well founded. Furthermore, though for Marx spirituality is intrinsic to human species-being, he was sharply critical of religion (at least the sort prevailing in European societies). Marx may have been sympathetic to people's need for religion to cope with the horrors of capitalism, but he did not locate in spirituality per se a potential for overcoming capitalist relations. For this to occur, the concretely grounded processes of capitalist alienation must be overcome

(Brien and Bove 2020; Brentlinger 2010; Dickens 1992; Foster 2000, 73).

A debate over the role of spirituality in bridging Marxism and environmentalism has yet to develop fully among Eco-Marxists and is beyond the scope of the present work; however, in practice, movements like the Brazilian Landless Workers and Movement and Liberation Theology have for decades already combined spirituality with Marxism (Betto and Löwy 2010; Dussel 2020; Hammond 2014). These exemplify constructive ways of mobilising people's spiritual side towards socialist ends and, by extension, demonstrate the feasibility of an Eco-Marxism that adopts aspects of spirituality. In some ways, this vindicates Kovel's position but also converges, even if coming from a different direction, with Foster's own view that "we must understand how spiritual conceptions, including our spiritual connections to the earth, are related to our material, earthly conditions" (Foster 2000, 11). Spirituality is but one aspect of a larger relationship between humanity and the rest of nature. Foster's reinterpretation of Marx's metabolism concept would presumably encompass this, although the matter has not, to my knowledge, been explicitly linked in this way. A metabolism involves spiritual aspects, though whether in the universal or social metabolism or both could be a source of debate (the social seems more apt). It is, in any event, by means of concrete actions, like reappropriating land for communal farming, that, for example, the Brazilian Landless Workers Movement have been building a spirituality worthy of socialism.

Problems Intrinsic to Intrinsic Value Theory

An insistence on recuperating dialectical thought is among Kovel's valuable contributions to Eco-Marxism, where it is at times not evident what the role of dialectical historical materialism is if it is considered at all. Vlachou (1994, 122), for example, understands a Marxist approach to people-environment relations as the application of the "concepts of overdetermination, contradiction, and class," which seems a limiting rendition of dialectical historical materialism. However, there are several difficulties with the concept of intrinsic value aside from those raised in other work (saed 2019; see also Foster and Burkett 2016, 34–56; Saito 2017a, 107). This concept merits particular scrutiny because it underlies and constitutes the basis of Kovel's theorisation of transformative politics as well as his red-green political platform.

One problem is the conflation of a general with a specific category. Intrinsic value pertains to nature, while use/exchange-value is a social process that lies within a tiny part of nature. For instance, a table could be thought of having intrinsic value, but that is a much more abstract than saying that people devoted much work to creating the table as something socially useful or that the table was made to be sold. Both the social usefulness and the sale potential of a table are natural processes that are part of the intrinsic value of tables, which includes all the human activities that go into making and exchanging them.

Another problem is a lack of clarity about the valuation process (including any notion of what gets valued when and why) or criteria to establish intrinsic value, such as the underlying theorisation in the Marxist categories of use- and exchange-value (or surplus value, detractors notwithstanding). Without such clarification, everything and all beings are equally important at all times. All life forms, all ecosystems, all atoms, etc., can be said to have intrinsic value. At the same time, the clarifying process itself should make it evident that intrinsic value is not determined by reference to what is outside of humanity, but by means of humans deliberating among themselves over the term. How intrinsic value is determined is itself a political struggle that has nothing to do with the rest of nature. It has to do with what happens within some societies that are part of the human species.

More troubling is the unsubstantiated assertion (or, rather, leap) that attaining the realisation that nature as larger than humanity must entail a belief in intrinsic value. This is equivalent to claiming that spirituality is somehow hobbled or absent without buying into intrinsic value. But people's sense of wonder or curiosity about what exists beyond humanity is an important part of use-value, like aesthetic pleasures. It is socially useful to appreciate forests in many areas of the world for their own sake, for without forests in such places, life may become degraded for many people as well as many other beings. The notion of intrinsic value is redundant with what is already covered by the concept of use-value. In fact, positing intrinsic value as alternative to use-value distracts from the political struggles ensconced in claims about valuing nature, which is often done in the abstract precisely to evade the issue of concrete human needs and political struggle (see also Marx 1978 [1844], 124–125).

Relatedly, the notion of intrinsic value has little to do with the rest of nature because it conceals a humanity-centred reckoning of time and place. Over thousands of years, for example, forest decline, as part of gradual evolutionary processes, can be enhancing to some life forms, say, those thriving in grassland or savanna ecosystems. Given that all ecosystems change, emerge, and disappear, within what time scale should one view an ecosystem as valuable? What is more, cultivated land has often been at the expense of forest ecosystems, yet agroecosystems, depending on the type of farming practised, can be very biodiverse and be beneficial to many life forms, aside from human beings, who may find forests less hospitable or uninhabitable. Should one regard an ecologically sustainable agroecosystem less intrinsically valuable than, say, an old growth forest? How is one to decide? These types of questions indicate that it is the temporal range of human life spans and geographical contexts of human beings that delimit the meaning of intrinsic value. By pretending to include what is outside society, proponents of intrinsic value mystify the social nature and socially derived meaningfulness of value.

Much of the reasoning behind Kovel's recourse to the concept of intrinsic value and egoic being rests in any case on a mischaracterisation of Marx's notion of people's "inorganic nature," used as proof of an enduring form of dualistic thinking, with humanity on one side and nature on the other. Recall that to overcome this dualism, Kovel deems a dialectical approach necessary to arrive at the realisation of being part of nature and to overcome egoic being. The spiritual aspect of species-being is to be harnessed towards making this realisation or recognition happen. As Foster and Burkett (2016) have pointed out, this is a tendentious reading of Marx's text that is widespread among environmentalist intellectuals and segments of the left, including influential Ecofeminists (see the next chapter).

One need not go into pedantic philosophical quibbles to demonstrate that the notion of people's "inorganic nature" is, according to Marx, a manifestation of a dialectical relationship within nature in general, where humanity is expressly included. One can readily understand this by reading from the (translated) original (minding the arcane gendered language):

> Nature is man's *inorganic body*—nature, that is in so far as it is not itself the human body. Man *lives* on nature—means that nature is his *body*, with which he must remain in continuous intercourse if he is not to die. That man's physical and spiritual life is linked to nature means simply that nature is linked to itself, for man is a part of nature. (Marx 1978 [1844], 75)

Thus, Marx had pithily described a way of thinking about nature and society that showed exactly the opposite of any dualism or dichotomy. It is abundantly clear that Marx understood nature as a differentiated unity based on dialectical relations. As some Marxists had already shown several years prior to Kovel's writing on intrinsic value and egoic theory (see Dickens 1992, 64).

The transformative politics advocated for by Kovel are thereby on shaky philosophical grounds. Environmentalist thought, pervaded as it is by bourgeois dogma (e.g., abstract individual freedom, moralisation of environmental problems) and fetishism (e.g., population growth, harmony in nature), is largely unhelpful towards achieving ecological sustainability because of a denial of specifically capitalist causes, which often leads to a refusal to consider socialist alternatives. More specifically, the notion of intrinsic value distracts away from the realisation that we are intrinsically part of nature precisely because of our very existence. This is something that ecological science, rather than philosophy, affirms even while often contradicting that understanding (Haila and Levins 1992; see also above on Systems Theory).

Use-value, far from wresting our consciousness away from recognising some intrinsic value to nonhuman nature, involves an appreciation of nonhuman nature in its specific, concrete manifestations in relation to actually existing human life. It is by finding a tallgrass prairie socially useful or, in the case of communities directly dependent on it, essential to human life that people really understand the importance of that ecosystem and will fight to preserve it as if their very lives were at stake. No environmental movements struggle to safeguard montmorillonite deposits or polioviruses for their own sake, for example, but there are plenty of struggles over aspects of nonhuman nature that impinge directly on human life, such as climate change and water quality.

Positing intrinsic value as alternative to use or exchange-value amounts to displacing the product of social relations (such as values) onto processes outside of humanity, and therefore outside of social struggle and politics. This leads to a dubious political strategy that would include embracing movements where politics are deferred to the prerogatives of some supernatural being. Prosaically stated, value is a multifarious process ordained by the small human fraction of nature. One can understand that we belong to a much larger nature while also recognising that values about nature are socially created. In this manner, one can engage in political struggles that include changing the actual source of values promoting harm to people and other beings, and on developing and spreading values about nature consistent with, for instance, ecosocialism.

Metabolic Rift

Several Marxist scholars had shown that the ideas of Marx and Engels, contrary to liberal or "green thought" distortions, contain the basis for truly ecological philosophy, in contrast to mainstream environmentalism. The works of Parsons, O'Connor, Dickens, and Kovel exemplify differing perspectives on how to grapple with and propose ways out of such environmentalist obduracy. Marx and Engels, for example, had long understood the wealth and value of the rest of nature and that ecological conditions broadly affect production and human existence.

But it has not been enough to refute common misconceptions even within socialist circles, nor to avoid sharp disagreements among Marxists regarding the relationship between ecological thought and Marxism. These contradictions, in a context of recurring politically reactionary distortions of the ideas of Marx and Engels, propelled the intensification of Marxological debate by the middle of the 1990s. That is, Eco-Marxists in liberal democracies have tended to focus on the degree of ecological content in the works of mostly Marx (in terms of philosophy more than science) more than developing political strategies combining Marxism with environmental struggles, studying the relationships between Marxist currents and ecological philosophies, or contributing to building scientific socialism.[9]

For the most part, there has been relatively little interest in Kovel's ideas among Marxists in core capitalist countries concerned with combining Marxism and environmentalism. This is likely due to Kovel's claim that some of Marxism's foundational principles must be modified or overcome for a viable Eco-Marxism to be developed. The work of Dickens, however, did not garner the level of engagement one might expect, given his arguments in favour of keeping and building on the insights and framework of Marx and Engels. This may be due to his claims that Marx had not entirely rid his philosophy of a nature-humanity

[9] To date, the development of a scientific ecology based on Marxist principles or a Marxist methodology for the ecological or environmental sciences is largely beyond Eco-Marxism, mainly due to the prevailing social science and philosophy backgrounds of the overwhelming majority of Eco-Marxists.

dualism and had neglected the subject matter in subsequent writings (Dickens 1992, 62). Regardless, it was the second contradiction thesis that received attention, especially from Paul Burkett (1999, 194–197) and John Bellamy Foster (2000, 300, fn 73), and then paradoxically faded from view. In the eyes of the Burkett-Foster school of thought, O'Connor's argument gives the impression that Marx's framework is limited to considerations of the first contradiction and therefore inadequate or insufficient towards addressing the socio-ecological crisis.

According to Burkett and Foster, Marx had already developed an ecological perspective (a "dialectical naturalism"; Foster 2000, 228), one that was central to his entire work. This, it turns out, is a repetition of Parsons' understanding that Marx and Engels had developed a "dialectical, ecological standpoint" (Parsons 1978, 10). The term ecology, it should be noted, was coined in 1866 and not by Marx. Scientific ecology came about only in the 1900s, and in ways that took on a bourgeois perspective on nature (see Dickens 1992). Marx, in other words, had developed philosophically an "ecological" framework prior to ecology. This is backed up by samples from Marx's various writings that Parsons, as explained above, had already indicated. Marx drew from agricultural chemistry (especially the work of Justus von Liebig) to propose a general critique of capitalist exploitation that includes not just workers but also the means of production appropriated from the rest of nature. A main and oft-cited illustration of Marx's ecological understanding, though it never materialised as a detailed study, was the degradation of both soil (*Erde*, which can also mean earth) and workers by means of capitalist agriculture (see also Parsons 1978, 21–23).

Underlying this precocious insight is the concept of metabolism, a concept that gained scientific currency by 1815. It came to be used in agricultural chemistry by the 1840s to describe the material exchanges between organism and environment. Eventually, though some decades after Marx passed away, metabolism became an important concept in the science of ecology to include the exchange and flux of energy through ecosystems. The concept of metabolism (*Stoffwechsel*), in Foster's view, was a key aspect of the framework developed by Marx. Marx understood that humanity relates to (appropriates the materials of) nature through labour (work). This labour-based relationship is captured by the notion of *Stoffwechsel*, which refers to the exchanges of materials (and presumably energy) that are involved in the relationship between humans and the rest of nature. In other words, labour mediates the metabolism between humans and the rest of nature (Foster 2000, 155–170; see also Saito 2017a, 62–64, 68–79).

As in the case of natural and species-being (see above), Marx used the concept of *Stoffwechsel* in two ways, as universal (general) and social (specific to humanity). Universal metabolism is that occurring between humanity and nature, mediated by human labour. Social metabolism (that is, *within* society) refers to dynamic and interdependent sets of needs and relations created socially, which under capitalism become alienated. This is expressed particularly in the town and country contradiction, which Marx identified as part of the development of capitalist relations with respect to large-scale industry. As more people concentrate in cities and industrial centres, the countryside is depopulated. This is the *social metabolism* aspect of the contradictory process. At the same time, increasing quantities of materials are taken to industrial centres and cities from the countryside, which is thereby depleted of resources. This is the *universal metabolism* aspect of the contradictory process. Capitalist farming displays this double contradiction, in the form of interrupted metabolic cycles. Nutrients are extracted from soils (mainly by means of crop harvests) and not returned. This is how in capitalism agriculture develops in such a way as to no longer being able to rely on locally available resources. As a parallel problem, sewage from cities is also not cycled back into soils as fertiliser. Thus, the metabolic cycle (in this case, nutrient cycling) is interrupted. While cities get polluted with sewage, soils are depleted of nutrients.

According to Foster, Marx had shown that capitalism causes a rift in both social and universal metabolism. Just as dramatically as Kovel, Foster and others in the metabolic rift school have described the current planetary situation as an "irreparable rift" between "society and nature" (Clark and Foster 2021, 89), which is one way to interpret how Marx articulated the issue.[10] That is, it is claimed that Marx theorised that capitalism causes a general metabolic rift. By rift, Marx meant an interruption in

[10] Clark and Foster are quoting from the unfinished third volume of *Capital*, where Marx, drawing from Liebig, states:

> … large landed property reduces the agricultural population to a constantly falling minimum, and confronts it with a constantly growing industrial population crowded together in large cities. It thereby creates conditions which cause an irreparable break in the coherence of social interchange prescribed by the natural laws of life. (Marx 1998 [1883], 799)

the metabolic cycle that sustains life or society. In this way, Marx's ideas can be understood as precursors to ecological thought, if not ipso facto constituting an ecological perspective (see also Saito 2017a, 64).

Foster goes further and contends, controversially, that metabolism and rift are essential concepts in Marx's approach. That is to say, Marx's ideas were precocious and, one could argue, more profound than ecological thought because he understood that the capitalist mode of production leads to a *rift* in the *metabolic* relationship between society and the rest of nature (earth). This is how Marx's writings cannot be charged with

The problem is framed in terms of whether environmental practices, not "society", are consistent with "natural laws of life," not "nature." This is not an inconsequential slippage. Crucially, the natural laws to which Marx referred are drawn not from Marx's own work, but from that of the biophysical sciences of his time. Presumably, with new discoveries about natural laws, Marxists should modify their people-environment theories accordingly. So, for instance, findings about threshold and non-equilibrium conditions as well as theories of ecosystem resilience should inform if not reshape Eco-Marxist theories. Instead, there persists a concept of natural balance (as if biophysical processes amounted to equalising inputs and outputs, as if time or spatial scale did not matter) and a non-relational view of physical processes (like soils) as static or passive. These problematic views are reinforced by insisting on metabolic rift theory (at times confused with method) as the centre piece of Marx's understandings of nature and people-environment relations (e.g., Foster and Burkett 2016, 233) instead of dialectical historical materialist method (ibid., 235). More importantly, metabolic rift theory provides no guidance over which biophysical science theories are more valid. Eco-Marxists act as if unaware of theoretical debates in the biophysical sciences or unaware of the ramifications of such debates for Eco-Marxist theories (the tendency is to assume that if something is in the scientific mainstream, it is valid, instead of studying scientific findings, theories, and debates). Marx, after much studying of and note-taking on the scientific literature, elected to follow Liebig instead of Lavergne, for example, regarding soil fertility and plant nutrition, as Foster and Burkett show (2016, 28–30). Marx's rejection of Lavergne's theory illustrates a mode of discernment that escapes much of Eco-Marxism, where criteria or methods for such discernment are undeveloped. Regardless, if there really is an "irreparable rift," as claimed in metabolic rift theory, then such a "rift" cannot be undone. Yet the claim is often otherwise, that the rift can be "healed" (e.g., Clausen 2007; Foster 2015b; Foster and Burkett 2016, 239–241; Gunderson 2011; Schneider and McMichael 2010) instead of thinking in dialectical materialist terms, such as mutually transforming activities with ecological relations in ways that enable better living conditions and raise biodiversity, among other wider ecological benefits (none of which are guaranteed, given the openness intrinsic to dialectical relationships). Moreover, it can be as well the case that Marx's reliance on the science of his day precluded a more dynamic understanding of the physical environment and of ecosystems, such that there can be reversible and also irreversible processes rather than just irreparable rifts. Marx's method is what should be essential to Eco-Marxists, not historically contingent theories and observations, which can be incomplete, outdated, or even incorrect (Engel-Di Mauro 2014).

any anti-ecological content. If anything, Marx's ideas, as well as those of Engels, were well ahead of his time (one might wish to add that this was so at least in Europe and European settler colonies). Marx's framework, interpreted by Foster as metabolic rift theory, in some ways prefigured a socialist environmentalism and the concept of ecological sustainability, among other related issues and notions.

Meta-Industrial Work and Metabolic Value

Whether one adopts this perspective or not about Marx's writings, the work that goes into universal and social metabolism includes that of social reproduction, which in capitalist regimes is largely unacknowledged in a direct manner. As Salleh has repeatedly argued, "most women are reproductive labor in the sexual, economic, cultural, and ecological senses of that word. They are associated with—but are certainly much more than—'conditions of production'" (Salleh 2005, 10; see also Salleh 1995). Similarly, the metabolic rift thesis does not capture the deeper gendered contradiction of what for the past centuries is to be described as capitalist patriarchy. That is, according to such an ecofeminist view, capitalism is the most recent iteration of male domination (the subsequent chapter is devoted to Ecofeminism in relation to ecosocialism).

Eco-Marxist theories that do not consider the biological aspects or sources of social oppression are missing the socially reproductive mediating step in the metabolism process that is largely done by women worldwide and mainly to the benefit of men. Salleh's nature-woman-labour nexus theory is discussed in the following chapter, but here it is important to mention because it underpins the thesis that the socio-ecological crisis is caused by the primary contradiction induced by capitalist patriarchy, not capitalism. In fact, women's labour is but one major component of the deeper contradiction. Capitalist patriarchy involves the manifold, unremunerated, and often coerced contributions of

> meta-industrial labor—mothers, small farmers, hunter-gatherers on the fringes of capital— who know most intimately the meanings of economic justice, social equity, cultural autonomy, and ecological sustainability. (Salleh 2005, 12–13; italics in original)

Elsewhere, the category is defined as "householders, peasants, and indegenes" (Salleh 2009, 6), and still more recently it includes "carers" and "parents" (Salleh 2010, 211–212). The category keeps expanding in analytically confusing and politically counterproductive ways (for instance, capitalists can also be parents). Regardless, these social categories, to Salleh, constitute a class whose regenerative activities make the global economy possible and dampen the effects of the metabolic rift from industrial capitalism. This broad class is the main source of social change necessary to overcome capitalist patriarchy and end the scourge of the socio-ecological crisis.

Importantly, the meta-industrial workers are also geographically differentiated according to broadly defined divides in what has been called Global South and Global North.

> Where 'development' has not consumed local resources, the labour of peasants or indigenous gatherers in the South demonstrates a good metabolic fit between human needs and biological growth. It preserves and generates metabolic value. In the North, the metaindustrial labour of mothers and other human care givers is also attuned to natural cycles. These reproductive labour forms give rise to a distinct set of economic skills and values – at once material and ethical. (Salleh 2009, 6)

In other words, whoever does meta-industrial work preserves harmony with the rest of nature by "reproducing nature's metabolic cycles" (Salleh 2010, 7). The outcome of such labour is metabolic value, which is "the value sustained and enhanced by this kind of worker in supporting ecological integrity and the social metabolism" (ibid., 212). Metabolic value signifies the "recursive web of self-regulating matter/energy flows" (ibid. 216). It could be useful to know whether meta-industrial labour is redundant if flows are self-regulating. That could save people a lot of chores. However, predation (including of humans by other species), volcanic eruptions, and earthquakes are also processes of self-regulating flows of matter and energy, so it seems more information is needed to determine what metabolic value signifies ecologically and socially.

Be that as it may, the social composition of the meta-industrial workforce is not confined to women. Included is also a specific category of men. They are those who are

> 'outside of' capital, such as small farmers and forest dwellers, [and] undertake regenerative or metaindustrial labour. Unlike factory work, or academic work, the labour of these socially diverse groupings oversees biological flows and sustains matter/energy exchanges in nature. (Salleh 2009, 7)

Elsewhere, however, one learns that "In every society, women have been accorded the meta-industrial labor of mediating the humanity–nature interface" (Salleh 2010, 207). Whether this extends to all men "outside of capital" is unclear. It is also unclear whether business-owning males doing parental work or capitalist females recycling plastic wares for their local community belong to the same class of meta-industrial labourers. One could resolve the difficulty through what amounts to a call for cross-class collaboration, recognising that "there are masses of people, who strictly speaking may not be part of the working class, but who for various reasons are becoming politicized and radicalized in anti-capitalist directions" (Albritton 2019, 6). However, one would have to address contradictions related to matters of access to or control over the means of production as well as to property relations, wage differentials, and other processes reproducing multiple levels of class differentiation and undergirding class conflict.

Still, Salleh provides an important corrective to Eco-Marxist analysis in identifying unremunerated labour as producing metabolic value. This theorisation contrasts with the notion of intrinsic value, where the meta-industrial labour underlying it or making it possible is effaced.[11] The notion of conditions of production is much more refined as well when it is understood that such conditions also result from meta-industrial labour. Relative to metabolic rift theory, there is additional specificity in that the extraction and deterioration of metabolic value contributes to metabolic rift. One may add that metabolic rift is too limiting a way of understanding the socio-ecological crisis by missing the key element of metabolic value, which if enhanced will reduce or repair the, if one adheres to a "balance of nature" viewpoint. In fact, the notion and analysis of meta-industrial labour, as the source of metabolic value, could be applied towards evaluating the extent to which the effects of the metabolic

[11] Like Salleh, Kovel regarded gender relations of power as a primary contradiction, but it is unclear whether he understood this gendered problematic of intrinsic value.

rift are or can be countered by the production of metabolic value. This would be a promising avenue of investigation in quantitative terms as well.

There are some challenging dimensions to this view in that male meta-industrial labourers benefiting from patriarchal systems nevertheless contribute metabolic value, that is, to positive environmental effects. This suggests that the primary contradiction subtending the socio-ecological crisis is not necessarily gendered. There is anyway no specification of multiple forms and variable histories of patriarchy and the ecological significance or repercussions of such historical and spatial variability. Thus, it is not possible to explain any form of environmental degradation from, say, peasant communities over time or of different kinds of peasant communities. They are all patriarchal, yet regenerative, provided they are outside of capital.

It is furthermore unclear whether all men outside of capital have been accorded meta-industrial tasks or how they exploit women in non-capitalist societies (given thousands of years of patriarchy). There is no room for considering societies that have been recognised as matriarchal or that are gender egalitarian, due to the tendency to generalise from capitalist patriarchy (the examples given of, for instance, equating nature with woman in pejorative ways is not universal).

As for the notion that meta-industrial labour exists outside of capital and what it means in practice, that would be an additional issue to investigate. However, it would not take long to notice that virtually all societies (save for self-isolated ones in parts of the Amazon and South Andaman Islands, for instance) are either formally or really subsumed under capitalist relations. There would need to be greater precision regarding what being outside of capital entails in practice and relative to ecological relations. Without criteria being spelled out, it is difficult to verify whether the beneficial effects of meta-industrial labour are the result of being part of the circuit of capital or outside of it.

More worrisome is the insistence that "Human attempts to redesign nature, by say, dam construction or nanotechnology, invariably risk entropic disintegration" (Salleh 2010, 210). Assuming "entropic disintegration" refers to something like environmental degradation, this assertion could be applicable to any farming system, for example, as it is people designing what to cultivate and for what and, by doing so, often permanently altering ecosystems. One might also wonder about beaver dams or termite mounds as not being also ways that nonhuman beings redesign "nature." Perhaps beavers and termites also destroy ecosystems by redesigning them, to return to the debate about the problem of human exceptionalism discussed above. Making abstract universalisations of this sort is not only unhelpful towards identifying and addressing the socio-ecological crisis, but also in describing ecological relations and what counts as destructive. This lack of ecological understanding in the guise of ecologically sounding terminology results from a continuing reliance on the myth of natural balance or harmony, where nature is reduced to a scale-unspecified equalising of inputs and outputs in a perennial cycle.

Prefigurative Revolutionary Subjects and Class Differentiation

For decades, in capitalist countries, one of the challenges that Marxists have faced is what to make of the class composition of environmental movements, given their tendency to involve people from different class positions, not just the working class (e.g., Enzensberger 1974; Gorz 1975; O'Connor 1988). This is of importance towards political strategy in terms of identifying the most promising social groups for revolutionary social change. If the socio-ecological crisis is the most important contradiction to confront, then it stands to reason that environmental struggles are central to bringing about an ecologically sustainable society based—in a red-green or ecosocialist understanding—on social equality or the end of class divisions and exploitation.

Part of the difficulty lies in a traditional and relatively narrow notion of the working class as composed of industrial workers, in the more restricted version, or of employees, in a more encompassing conceptualisation. Those social categories miss major portions of society involved in environmental struggles or alternative environmental practices, especially in the Global South and among Indigenous Peoples (Ofreneo 2022; Salleh 2009). Eco-Marxists, but also Marxists more broadly, have accordingly expanded the range of people to consider, those radical enough to bring about the social change needed to overcome capitalism and end the socio-ecological crisis (Brownhill and Turner 2020; Löwy 2015; Napoletano et al. 2023; Schwartzman and Engel-Di Mauro 2022; White et al. 2016). Ecosocialists' assertions regarding Indigenous Peoples, discussed in the subsequent chapter, can here serve as critical qualification.

Löwy and Tanuro (2021, 32–33) identify six characteristics shared by movements combining ecological and social concerns, or ecosocial movements: (1) the central role or leadership of women; (2) the occupation or blockage of a critical area or route; (3) a local community struggle against an environmentally destructive project; (4) a democratic structure; (5) non-violent struggle; (6) reliance on spiritual, cultural, and/or religious reference points; and (7) a scepticism of and refusal to compromise with institutions. These define at the same time a prefigurative process, providing the practical contours of an ecosocialist future. By their own admission, however, such movements have failed to stop what they deem as "the infernal productivist machine" (ibid., 29; translation by author). They locate the main causes in state repression and the difficulty of gaining trade unions and institutional Left support (ibid., 33). It is left to the reader's imagination to explain such recurring challenges.

Movements as Class: The Environmental Proletariat

Salleh's concept of meta-industrial labour is a way to unify such disparate social groups analytically as well, but more ambitiously as a class. For the most part, in Salleh's theory, the diversity of classes involved in environmental movements is not directly confronted. Presumably, it is understood as part of a contradictory gendered, racialised, sexualised processes generated through capitalist relations and their related imperialist and colonial spawns. But the problem of conflicting material interests among classes in an inter-class alliance remains in abeyance.

Neither her metabolic value nor meta-industrial labour theory has found favour in much of Eco-Marxism, though. Within the metabolic rift school, there has instead been a re-conceptualisation of the rise and diffusion of environmental movements as a coalescence of a revolutionary subject named the environmental proletariat. This is a variegated and disparate group of social forces, mainly in the Global South, who represent "a new struggle for ecological hegemony." They have been emerging "to challenge both the ecological degradation and social exploitation associated with capitalist expansion" (Foster et al. 2011, 47).

By "ecological hegemony,"[12] the authors mean an ideology of ecological sustainability that insists on social equality (ibid., 397–398). This is to be achieved by building alliances (a historical bloc in Gramsci's sense)

[12] One could alternatively refer to attaining leadership or dominance in the environmentalist world or just making socialist environmentalism the dominant form of environmentalism (since anti-capitalist environmentalism is not necessarily socialist).

with various social groups who "would depend fundamentally … on a wider *environmental proletariat*" (italics in original), instead of the usual understanding of (industrial) working class. The environmental proletariat are those who increasingly have nothing to lose because their living conditions—including the environments where they live or from which they derive a living—are being threatened or have been undermined. They are thus the new revolutionary class, one that struggles against capitalist depredations on people and environment and for "ecological" hegemony, especially in areas of the world most affected by environmental degradation (ibid., 439–440).

This bears a close resemblance to O'Connor's above-described theory that new social movements have arisen as a result of the ruination of conditions of production (and one could add reproduction) through the effects of capitalism. And in similar fashion, environmental proletariat theory may founder in drawing attention away from the task of undoing the capitalist mode of production, the ultimate cause of the socio-ecological crisis, especially in Global North countries where revolutionary potential continues to be slim (Stoner 2021). It is anyway a formidable endeavour to form such a historical bloc as indicated by the concept of environmental proletariat. Even to call it "proletariat" is of doubtful analytical rigour, as it represents such diverse range of often contradictory material interests as those of Indigenous communities (who are internally divided and could be in direct conflict with any working-class fraction descended from settler communities), lumpen proletarians, unwaged houseworkers, nationalist anti-colonial and anti-racist groups (which may be class collaborationist), small property owners, office workers, landless farm workers, and industrial workers, among others.

Debating Metabolic Rift Theory

Those concerns should be extended to meta-industrial labour and metabolic value theories. Salleh's ideas nevertheless await a full reckoning with respect to defining the scope of the environmental proletariat category and to widening the social composition of the causal processes of the metabolic rift. Nevertheless, to refocus attention on Foster's metabolic rift theory, it has the merit of demonstrating how essential Marx's work is to understand why capitalism is necessarily destructive, both socially and environmentally.

Some, however, find exaggerated or tenuous the claim that the concept of metabolic rift was central to Marx's thought (e.g., Baer 2018, 26). First, the term metabolic rift hardly ever appeared in Marx's writings, so it is rather Foster's theory about Marx's use of metabolism. Second, Marx used the term metabolism sporadically and in such a historically different context that make Marx's ideas at most inchoate and in part superseded by subsequent scientific findings on ecological dynamics (see the 2001 symposium on Foster's *Marx's Ecology* in *Capitalism Nature Socialism*).

These absences or limitations, however, do not necessarily vitiate the legitimacy of Foster's claim, seeing how, for instance, the concept of totality or the term dialectics, central to Marx's methodology, also figure rarely in Marx's writings (Harvey 1974, 264–265). To some extent, the issue was long before prefigured and addressed by, for example, Parsons (1978) and Smith and O'Keefe (1980, 32–33), who recognised the lack of systematicity yet an implicit conceptualisation of nature in the work of Marx. On the other hand, the problem of relying on archaic scientific conceptualisations and theories can also be rectified by updating metabolic rift theory according to the most recent scientific findings, especially in scientific ecology (for instance, non-equilibrium conditions instead of balances). This remains to be done and in accordance with dialectical and historical materialism.

Others argue that even if the Marxian framework could be shown to be ecological, there is no way that Marx could have foreseen the magnitude and scale of the effects of capitalism and could not provide the conceptual instruments to address what did not exist during his time. Simply stated, capitalism may be so self-destructive as to preclude any socialist transformation (Altvater 1989; Skirbekk 1994; see also the discussions in the above-cited 2001 symposium in *Capitalism Nature Socialism*). Yet rethinking Marxian analytical categories through the inclusion of "natural conditions" seems redundant with metabolic rift theory, an explanatory framework to understand, in broad terms, how capitalism necessarily brings about an ecological crisis. Metabolic rift theory is not about identifying processes necessary for the development of socialism or about socialist strategies. Rather, it is a generic description of how capitalism necessarily leads to widespread environmental destruction. In this, the theory adds little to what Parsons (1978) had already compiled to show the ecological nature of Marxist method.

In contrast, Jason Moore (2011) critiques metabolic rift theory for reinforcing a society-nature dichotomy ("binary"). Huber (2013, xvii–xviii) levels a similar critique of "undialectical dualism" at both O'Connor and Foster. All this ironically resembles a charge Foster and Burkett have levelled at O'Connor, as pointed out above, but the issue is treating the environmental crisis as the outcome of capitalism instead of being constitutive of capitalism. Moore offers a theory of "capitalist world-ecology" as alternative explanation of the environmental crisis. At bottom, the contradiction is between exchange-value ("economic equivalence") and the concrete differences between commodities ("natural distinctiveness"). Importantly, the contradiction is expressed historically as cyclical crises, leading to long-term world-ecological shifts (Moore 2015). Moore recounts arriving at this from an intellectual journey that eventuated in building on and drifting away from metabolic rift theory (Moore 2017, 298–299).

It is unfortunate that the second contradiction thesis has yet to find any space in Moore's intellectual travelogue. His tortuous navigational efforts have landed him on what amounts to a reformulation of the second contradiction thesis, but also a restatement of Barry Commoner's exposition of the contradiction between capitalist economies and the "four laws of ecology" that Foster himself had earlier adopted (Foster 1999b). In other words, when the matter is viewed more broadly than Moore's understanding, exchange-value is one of the major pillars of the capitalist mode of production that contradicts the conditions of production (within which is a commodity's "natural distinctiveness"). Moore narrows the focus, compared to O'Connor's thesis, to certain subsets of the conditions (biophysical commodity differences) and mode of production (exchange-value), while O'Connor's thesis, like Foster's metabolic rift and Moore's "capitalist world-ecology," rests on a truncated understanding of ecology that considers only people-environment interactions, rather than entire ecosystems always undergoing transformations (Engel-Di Mauro 2014). In any case, Moore's application of a world-systems perspective constructively complements both metabolic rift and second contradiction theories, as he well demonstrated in an earlier work (Moore 2000).

This is likely an unappetising interpretative resolution to Saito (2022, 3), who instead insists that Marx held dear a methodological "dualism" predicated on the concept of metabolism. Moore's capitalist world-ecology is jettisoned, lumped with the work of conservative anti-communist Latour, as a form of "monism" typical of "critical Marxism," where the natural and social fuse or "hybridise" into one another. In fact, all Marxists who dissent from (or, more often, just ignore) the metabolic rift thesis are consigned to the dustbin of monism or social constructivism. The issue is presented as a stark choice between planet defenders (metabolic rift school) and crisis denialists (the rest) (Foster 2016, 398).

Löwy (2017, 19), proposing a "third position," prefers to view the works of Marx and Engels as "incomplete and dated" regarding ecological issues, but they nevertheless "have real relevance and methodological significance today." This would be one way to obviate the reduction of Eco-Marxist currents to monists against dualists. Eco-Marxists White et al. (2017) would nevertheless demur, countering that mutually antagonistic worldviews and politics are being lumped together into categories like "first stage" ecosocialism and "crude" or "abstract" "monism" (Foster 2016, 402). In contrast to such a dichotomous treatment of what are in fact multiple debates, White et al. point to a "critical hybridity." By this they mean the discernment of substantively divergent worldviews and the acknowledgement of the legitimacy of different Marxist positions, all equally supportable by reference to Marx's original texts, to enable dialogue and to find commonalities in ecosocialist struggle. Such equanimity, though, could be disputable as well, as some Eco-Marxist contentions rest on contrived interpretations and self-serving textual selections, as discussed below.

Controversies over the Ecological Substance of Marx's and Engels' Ideas

While the above-described Eco-Marxist positions consist of allegedly divergent interpretations or reinterpretations, other viewpoints, as in the above-discussed case of Kovel, express dissatisfaction with what are found to be crucial flaws in the thought of Marx and Engels. Salleh had made the point in 1997 and recently reiterated that, in spite of Marx and Engels being quite ahead of their time in terms of ecological thought, their

> ecological understanding … was undermined by their own Enlightenment conviction that reason with technology might shape the 'forward march' of History. Marx's vision of human dominion over the natural world spoke a linear notion of progress – an idea reinforced by his contemporary Darwin's evolutionary schema. (Salleh 2017, 100)

Furthermore, Marx contradicted his "monistic naturalism" when distinguishing between human and other species, something Salleh considers to be an "evolutionist oppressive distinction" (ibid., 102).

Though sharply at odds with the view of Marx held by the metabolic rift theorists, Salleh's contention has yet to be directly addressed by Eco-Marxists of any stripe. This may be due to the fact that Marx's ideas of progress and also that of communist luminaries like Kollontai, Lenin, and Gramsci were anything but linear, based as they are on dialectical and historical materialist reasoning. Similarly, distinguishing human animal characteristics from those of other species is not in itself a rejection of understanding ourselves as part of nature. It is to understand the specificity of how we are part of nature. Distinction is elementary to any perspective, including Salleh's, where distinctions between women and men are made paramount. Distinguishing one process or phenomenon from another is not equatable to dualism, which implies dichotomisation and inferiorisation (Plumwood 1993).

In contrast, Daniel Tanuro (2010) identifies two major flaws in the aspects of Marx's writings that relate to environmental issues much more directly. To him, they represent conceptual limitations that need to be recognised and overcome. For Tanuro, absent is a systematic vision for an ecologically sustainable transition to socialism and a differentiation between renewable and non-renewable forms of energy. These problems fatally undermine Marx's otherwise major contributions to the development of ecological thought. Another major and related difficulty stems from a lack of clarity about and elaboration on the finiteness of resources.

Before getting to the flaws, Tanuro explains that Marx furnishes important insights about capitalism as ecologically destructive, calling for the rational metabolism (material exchange) between humanity and environment. Using Marx's concept of metabolic rift, according to Tanuro and in agreement with the Foster-Burkett school, is therefore effective in describing the disruption of global carbon (as well as nitrogen) cycles by burning fossil fuels, which results in global atmospheric radiative forcing (climate change).

Additionally, and something under-appreciated by other Eco-Marxists, Tanuro finds that Marx's theory of capitalist ground rent (particularly in Volume 3 of *Capital*) implicitly recognises natural limits. Marx showed that landowners are able to appropriate surplus-value from society in the form of rent (and even super-profits) by monopolising natural resources that are finite and/or concentrated in particular places, like hydropower, soil, mineral, old growth forests, wetlands, etc. These kinds of monopolised finite and concentrated resources include fossil fuels, which are since the nineteenth century the engine of world capitalism.

As an obverse reading of the issue, Ehara (2023) finds that Marx's analysis of ground rent lays the basis for a critical grasp of capitalist mechanisms involved in natural resource exploitation. It is especially when capitalists successfully depress absolute rent (the wealth derived by means of sheer ownership, rather than from differences in land or natural resource characteristics or differential rent) that the landlord class intensify land or resource exploitation. Furthermore, rent-seeking in general not only clashes with the total surplus value extracted globally (and reduces capitalist profit rates), but also, if unchecked, infringes on planetary limits.

Tanuro, however, considers the specific characteristics of natural resources, rather than natural resources in the abstract. Because of ground rent, when it comes to shifting to renewable resources, capitalists tend to prefer industrially produced biofuels (including ethanol) since using biofuels guarantees them capital accumulation by means of ground rent. But to mitigate climate change, there must be massive decrease in energy consumption, not just a replacement of fossil fuels with renewable energy sources. Marx (and Engels, and almost all Marxists until recently) never offered a roadmap for an ecologically sustainable socialism, only some fragmentary comments, generic ideas, and incomplete analyses scattered in his work.

More importantly, Marx did not distinguish renewable (stochastic) from finite (stored) energy sources. This leads to a self-contradictory approach. The rational management of natural cycles is supposed to happen along with a utilitarian and linear concept of resources that is not very different from classical bourgeois economics. Marx and Engels ignored the ecological and social implications of the form of energy used and, in particular, of the switch from renewable to non-renewable energy that has been foundational to industrial capitalism. About this, there are ecological and social aspects to consider.

Ecologically speaking, fossil fuels cannot be rationally managed over the long term. This is because they "are not only limited but exhaustible, non-recyclable and irreplaceable historically, if not geologically" (Tanuro, 2010, 94). In terms of social relations, the switch from biomass to coal meant concentrating workers in single locations. This contributed to creating the conditions for workers to organise and struggle to change the conditions of work and even fight for socialism (see also Malm 2016).

There are substantive implications for Eco-Marxism. Different sources imply different technologies that are not neutral. Relying on fossil fuels leads to a linear view of resource management (resource use, production, waste) that is contrary to Marx's insistence on the rational management of natural cycles (or natural metabolism, see above). This means that developing the forces of production there must be very close attention given to the qualitative aspects of energy resources and the technology needed to make use of specific kinds of energy resources.

With respect to ecosocialist struggles, there are also important considerations. As stated above, to mitigate climate change, switching to renewable energy is not enough; energy consumption must decline. But this means material production must also decrease. And it must increase for the world's majority because they have too little energy access and low material well-being. Therefore, global wealth redistribution is necessary to apply Marx's principle of a rational management of natural cycles, which would mitigate climate change. As Tanuro concludes, "a decline in material production is compatible with an improvement in well-being, and in the richness and quality of life of the immense majority of humanity" but only through global wealth redistribution (Tanuro 2010, 101).

Löwy concurs with Tanuro that ecological concerns were not central in the approach of Marx and Engels. This is because, when they lived, the ecological crisis was just starting.

> Marx and Engels' discussion on ecological issues is incomplete and dated, but despite these shortcomings, it does have real relevance and methodological significance today. (Löwy 2017, 19)

Put differently, Marx and Engels contributed arguments and concepts that explain the capitalist causes of the ecological crisis and provide an ecologically sound social alternative.

Löwy surmises that they did so in three main ways. One was by developing a naturalist philosophy. Marx and Engels thought of humans as natural beings, as part of nature. The communism of Marx and Engels was a form of humanism that complements naturalism, as explicated as well by Parsons (1978, 104). In other words, for them naturalism and humanism mutually complete each other. This idea contributes to creating an ecologically constructive alternative to capitalism. Secondly, Marx thought that real wealth is from both human labour and nature. Real wealth is use-value (satisfaction of human needs), not exchange-value (for endless capital accumulation). But use-value is produced through human labour as well as from nature. Finally, the forces of production are to be developed to enable people to fulfil their needs and full potentials (as species-beings). The development of the forces of production is about being, not about having. It is not for infinite economic growth and endless consumption of material goods.

However, there are tensions springing from the concept of developing the forces of production. Marx and Engels praised the great technological achievements and development of forces of production in capitalism. But they also criticised capitalist degradation of soils and forests (as well as workers), or capitalism's metabolic rift (see Foster 2000). They thought socialism will bring rational (even unlimited) development of the productive forces and rational use of resources and technology. To reconcile the tension between developing the forces of production and environmental degradation thereby produced, Marx clarified that capitalism's metabolic rift will be overcome through socialism. That is, it will be a new kind of harmony (metabolism) between society and nature achieved by means of rational and planned use of resources. For Marx and Engels, it is not nature that must be dominated, but the society-nature metabolism that must be made rational, controlled, mastered (cf. Salleh 2017, 104–108).

As Tanuro has pointed out, there is still a problem that Marx and Engels viewed technology as if it were neutral. Löwy sees this understanding formulated by Marx and Engels as in need of amendment. One way is to relate technology to health and environment. Technology under capitalism is for exchange-value and capital accumulation, so working classes must break with the ways technology is developed and used to redirect it to improving people's health and to raise the levels of environmental protection. The whole productive system must be transformed to achieve this.

It is in this manner that Marx's and Engels' ideas help to develop Ecosocialism. They provide a materialist understanding of capitalism's destructiveness, a radical critique of the capitalist destruction of the environment (the ecological crisis), and a perspective of socialism that respects the "inalienable conditions" of life on Earth (see the above discussion on O'Connor's second contradiction thesis). To Löwy, Ecosocialism is "the social appropriation of the main means of production and a democratic planning of the economy which takes into account the preservation of the ecological equilibrium" (Löwy 2017, 20). It is socialism without productivism and with ecological sustainability.

The general objectives promoted by Tanuro and Löwy are likely shared by most if not all Eco-Marxists, but their critiques of Marx and Engels are off the mark. As Parsons had pointed out decades before when analysing the energy crisis of the day:

> Marx's and Engels' description of capitalism was an account of how the production and distribution of energy are frustrated by a system of private property relations and exchange, so for them such a crisis [the energy crisis] would be credible and predictable. (Parsons 1978, 74)

Rather than a flaw in Marx's understanding of energy or "natural limits," the issue is the private ownership of energy sources, including fossil fuels. Socialist control of fossil fuels would avert energy crises that are, rather, matters of social maldistribution (the episodic to chronic deficiency in energy availability imposed on much of society), about which Tanuro and Löwy would concur. Socialist ownership of fossil fuels would eventually reduce the extent of deleterious ecological repercussions, while at first leading to intensive use and harmful impacts (crises due to pollution), due to external and internal capitalist pressures and, thus far, historically, to initial social conditions. The potential for terminal resource depletion is not an issue for socialist countries because policies must prioritise social spending and ecological health, which undercuts any prospect for an economy based on endless economic growth. This was becoming evident in socialist countries, including in the USSR, and currently socialist countries like Cuba and the People's Republic of China. Tanuro and Löwy mistake contingency with permanence and survival strategies with systemic properties (see also Parsons 1978, 83–84, 90–93; Suing 2018).

From an Eco-Marxist point of view, Foster and Burkett (2016, 18–23) have pointed out that Marx and Engels actually did distinguish between forms of energy and technology and demonstrated the political nature in the development and spread of energy and technological complexes (see also Wallis 2020, 17).[13] Tanuro and Löwy are thus reiterating conventional disinformation about Marxism common among

[13] Regrettably, even while usefully demolishing this fiction, Foster and Burkett reproduce another one by claiming that "...Marx and Engels recognised that industrial capitalism originated with waterpower, which they took careful note of and compared to steam power." Marx and Engels, much more importantly, "took careful note of" the genocides, slavery system, and plunder out of which industrial capitalism originated.

environmentalist thinkers, who confuse concepts like forces of production with just one component thereof, technology. One could similarly argue that allegations—made also by the likes of Tanuro and Löwy—of historical Marxist failures to address ecological concerns are equally unsupported and merely present a leftist rehash of prevailing bourgeois environmentalism. This mistaken view is recently repeated by the likes of Roos and Hornborg (2024), who displace their frustrations with solar power technologies under a capitalist world economy onto a phantomatic "Promethean Marxism."

Marx as Ecosocialist, Marx as Degrowth Communist

In contrast to all the above Eco-Marxists, Kohei Saito concludes that Marx had already fully expressed what are now recognised as Ecosocialist principles, at least by the late 1860s. In other words, Marx's work was not just ecological, but Ecosocialist.

> ... Marx's ecological critique possesses a *systematic character* and constitutes an *essential moment* within the totality of his project of *Capital.* Ecology does not simply exist in Marx's thought—my thesis is a stronger one. I maintain that *it not possible to comprehend the full scope of his critique of political economy if one ignores its ecological dimension.* (Saito 2017a, 13–14; italics in original)

Saito bases this claim on a close reading of Marx's unpublished 1865–1868 research notebooks, part of the *Marx-Engels Gesamtausgabe [Complete Edition]* (*MEGA*). He finds Marx was studying the natural sciences intensely and, by 1865, had progressed from what could be deemed an occasional productivist to an entirely ecological outlook. Elements of this transition exists in other notes Marx had taken in 1844, which were not published until 1932 (Marx 1978 [1844]) and are discussed above relative to the concept of alienation.

In this reading, there is continuity, an evolution of thought, where the *MEGA* plays an incisive role as explanatory clues in the development of theories or transitions from one to another kind of research emphasis (e.g., Saito 2017a, 32–33, 44–51). The crucial turning point is 1868, after which Marx engages in more intensive studies of some of the biophysical sciences in his analysis of capitalist processes. This is evinced as well in notebooks linked to content that was to be included in the second and third volumes of *Capital*. In those notes, it is clear that Marx understood how profit rates would be brought down by destructive impacts on environments and workers' health. Rather than leading to ecologically sustainable production, the tendency is for capitalists to search for other exploitable areas or types of resources. Thus, capitalist forms of exploitation leading to "decreasing natural conditions of productivity" will only spread and/or intensify deleterious ecological impacts (ibid., 134).

For Saito, then, the main purpose of Marx's rising interest in the biophysical sciences over time—namely agricultural chemistry, geology, and botany—was "to analyze the contradictions of the material world as a result of its modifications by capital" (ibid., 136). Furthermore,

> … Marx examined in a detailed manner how the mediation of the social and natural interaction between humans and nature by the logic of capital's valorization organizes social production and circulation in such a way that their metabolic interchange necessarily gets disrupted. While the capitalist mode of production structures a particular human metabolism with nature on a national and global scale, the forces of nature are, though elastic, always limited in various ways, resulting in eco-crises in multiple spheres. (Saito 2017a, 120)

But this is not all. Saito further claims that "Marx's seminal contribution in the field of ecology lies in his detailed examination of the relationship between humans and nature in capitalism" (Saito 2017a, 15). It helps little that the term ecology is left undefined, so it is difficult to understand to what sort of field Marx really contributed.[14] More decisively, evidence of such contribution is not presented. And it cannot be, given

[14] The only time he defines ecology is indirectly via Haeckel:

> All living creatures must go through constant interaction with their environment if they are to live upon this planet. The totality of these incessant processes creates not a static but an open-ended dynamic process of nature. Before Ernst Haeckel called this economy of nature "oecology," this organic whole that consists of plants, animals, and humans was often analyzed with a concept of "metabolism" (*Stoffwechsel*). (Saito 2017a, 63)

that Marx carried out no detailed ecological study of human environmental impacts (for that, one is better served by reading the works of von

Recall, as mentioned above, that Saito claims that Marx studied certain biophysical sciences to see about capitalist impacts, not to get a grasp of the dynamics of ecosystems in their specificity, as relations betwen organisms and their physical environments that generate ecological change. Metabolism is one aspect of ecological dynamics and a generic one at that. It says little about interspecific relations, ecological functions, biogeochemical cycling, co-evolutionarry processes, and much else of relevance to scientific ecology in ways that are more than metaphorical. As a more concrete example, metabolism applies to any interaction among species, ranging from mutualism to parasitism to resource partitioning to predation. Such interactions become indistinguishable under a metabolism lens. Otherwise put, using the concept of metabolism to study ecosystems may help track matter and energy exchanges and construct foodwebs, but it does not help, for instance, to distinguish between the kinds of relations among organisms or to find out about the co-evolution of different species' roles in making ecosystems function or change. Yet these are decisive aspects of ecological relations to know about if one is interested in how an ecosystem is structured and develops so as to devise preservation policies and ecologically sustainable environmental practices. This absence of competence in any biophysical science is painfully evinced in the following remark, in reference to Liebig's erstwhile manufactured manure theory: "Since nature can slowly supplement mineral substances through weathering of the soil, it is absolutely necessary to add manure" (Saito 2017a, 193). It is minerals that are weathered, as soil is composed of primary and secondary minerals, along with other inorganic and organic substances. This may seem pedantic until it is realised that this wording is like confusing the mineral weathering process common to all soils (which only gradually releases nutrients, and at widely different rates) with weathered, low-nutrient content soils often found in high-rainfall Tropical areas (or in areas that used to have such environmental characteristics). When soils are weathered, manure may hardly suffice to maintain fertility compared to the introduction of biochar and/or periodic burning of forest patches, as practised by many non-capitalist and capitalism-subsumed communities in the Amazon and West Africa (WinklerPrins 2014). It is difficult to see how Eco-Marxists can claim to have any ecological credibility when they are incapable of making such elementary ecological distinctions.

Regardless, it has since Liebig's time been understood that most plant nutrients in soil are stored and released through soil organic matter, especially humus, which largely results from microbial metabolite. It would be more apt and akin to Marx's practice to find out about current soil science and agronomy research on plant nutrition ("the recent and very latest in agriculture," as Marx noted, as reported in Saito 2017a, 234), rather than limit oneself to the nineteenth century, and see how Marx's framework on capitalist farming should be modified. For example, capitalist farming in humid temperate regions has been shown to lead in many instances to soil nutrient overloading and yet, through accelerated acidification processes, nutrient deficiencies in crops (Engel-Di Mauro 2014). This is clearly not a matter of nutrient replenishment problems, as Saito would have it by following nineteenth century notions via Marx, but one of hampered crop access to abundant nutrient levels. One could obdurately argue that it is a problem of nutrient cycling disruption (metabolic rift, if one prefers), but this would be a useless observation in terms of explaining the ecological issue (e.g., what are the causes of nutrient deficiency)

Humboldt or Reclus). Aside from lacking the scientific background, he had not the personal means (Marx and his family lived in poverty), nor the institutional support to conduct an agronomic, botanical, or geological investigation. Nor does Marx's work find much if any citation or acknowledgement in scientific ecology.[15] Not even in philosophical ecology—if this is what Saito means by ecology—has Marx had much of an impact. As Saito himself laments (and I share in that disappointment), Marx has been largely met with dismissiveness in that kind of field (Saito 2017a, 10).

It is anyway not much of a revelation that Marxian analysis is effective, if not indispensable in uncovering the social relations of power that are behind destructive environmental impacts (e.g., Saito 2017a, 213). Many

and in terms of practical applications like remediation (e.g., using materials to raise soil pH levels while reducing input for certain fertilisers). Such gross insensitivity to substantive ecological differences in Eco-Marxism results from a reliance on inappropriate levels of abstraction and a manifest disinterest in any biophysical science. This continues to escape the likes of Saito (2022, 39, fn 7), even as he concurs that "that today's critique of capitalism cannot be grounded upon the outdated science of the 19th century."

[15] For a similar reason, the claim of Marxian influence in the development of ecology, made in Foster (2020b), is unconvincing. For instance, in their scientific publications, neither Tansley nor Lankester ever cited any work or theoretical concepts by Marx or Engels. Bernal referred to Marxism only when discussing the philosophy of science, not his primary fields of biology and crystallography. The influence of Marx and Engels would instead be found in scientific publications in socialist countries, though inconsistently. Otherwise, in capitalist countries, the application of Marxist methodology in the biophysical and technical sciences has been rare and short-lived. It is rather in the philosophy of science and in the realm of science policy that the works of Marx and Engels have made their more direct impact and have exerted some influence, but overwhelmingly confined to socialist countries. Arguably, this should be viewed as much more important than any contribution to any biophysical or technical science in itself, which could nevertheless be enlightening in terms of guidelines (as I myself have found for the study of soils). The issue for Marx and Engels was ultimately about wresting science from capitalist influence and steering it towards building socialism (Levins and Lewontin 2007; Suing 2018). But from this a Marxian ecology does not follow except in terms of philosohical and political guidance. Foster, rather than showing how Marx's work influenced ecology, has pointed to the exact opposite. As he has posited, "Ecology as we know it today … represented the triumph of a materialist systems theory. … ecology today focuses on the human disruption of ecosystems from the local to the global" (Foster 2015c, 3). First, ecology focuses on the interactions of multiple species and their physical environments, and secondarily on "human disruption." Second, the very notion of "human disruption" is unhelpful. The matter is one of transformations from within ecosystems, not of ecosystems that would otherwise harmonious. This is precisely the reinforcement of bourgeois ideology about nature (the society-nature dichotomy) discussed above, quite far from the Marxist principles that Foster himself champions and with which I agree.

Marxists have demonstrated this already for decades, as discussed above. Similarly, Saito's findings that Marx's framework harbour principles that point to what today would be called ecological sustainability or ecological philosophy corroborate or reinforce what has been demonstrated even without perusing the *MEGA*, as also discussed above (e.g., Dickens 1992; Foster 2000). Those claiming otherwise (e.g., Löwy, Tanuro) should construct more convincing counterarguments. In this, Saito's work provides another detailed rebuttal but with copious supporting evidence that had not yet been considered.

Substantive rebuttals to Marxists arguing that Marx retained some form of "Prometheanism" are in themselves useful. But the main importance of Saito's contribution lies elsewhere. His meticulous appraisal of Marx's notebooks in relation to Marx's published manuscript has the great merit of demonstrating the processes informing the changes in Marx's theoretical emphases and formulations over time. A noteworthy example is how Marx's view of capitalist farming turned critical as new theories on plant nutrition emerged in the 1860s within agricultural chemistry, mainly by way of Liebig, where it was argued that soil "exhaustion" is due to a lack of nutrient replenishment, rather than any law of diminishing returns. It is a problem of disrupting the cycling of nutrients (Saito 2017a, 216).

Marx's attentiveness to and acceptance of Liebig's newer theories are especially interesting in light of the fact that Liebig's material interests lay in the manufactured fertiliser business. This made it less likely that Liebig would arrive at rejecting capitalist farming. It was instead someone like Marx, who had developed a critique of political economy, who could conclude from Liebig's findings that the issue of soil fertility decline is one of profit-maximising farming, not farming generally, nor industrial farming in particular (Saito 2017a, 196–202). Similar demonstrative illustrations emerge with respect to the debate between Liebig and Fraas, where Marx found the latter's findings and arguments more compelling. This suggests that theories about capitalism and environment should follow from scientists' interpretations of their findings, even if such scientists subscribe to capitalist ideology, but always in relation to a Marxian perspective on political economy, or perhaps better stated as historical and dialectical materialism (see also Parsons 1978, 83–84). This, arguably, should find broad agreement with Foster and Burkett, as discussed above regarding criteria to decide on scientific theories. The difficulty lies in what one makes of the framework of Marx and Engels, how one understands dialectical and historical materialism, whether metabolic rift or even dialectics is central to Marxian analysis, what one includes in terms like relations and forces of production, and much else pertaining to ongoing debates within Marxism.

By now, it should be clear that, at least for some Eco-Marxists, Marx merits several epithets that place him constantly at the cusp of historical developments relative to environmental concerns. Environmentalists have no idea just how far behind Marx they have been. Nor have Marxists themselves, apparently. First, Marx is discovered to be an ecologist, then, through a "process of self-critique" later in life (Saito 2022, 8), Marx converts to ecosocialism.[16] Of course, Marx "never coined the word 'ecosocialism'" (Albritton 2019, 5), but his prescience knows no bounds. His notebooks erupt with innumerable surprises. He is an ecologist decades avant la lettre. He is an ecosocialist about a century before anyone even dreamed up that term. Now, under the Anthropocene's full spell, Marx incarnates "Degrowth Communism." This time, however, his third and final, near-death conversion anticipates a concept and movement that has yet to come. As Saito explains, Marx's "last vision of post-capitalism in the 1880s went *beyond* ecosocialism, and it can be more adequately characterized as *degrowth communism*" (Saito 2022, 6; italics in original). This is defined as a "non-productivist and non-Eurocentric view of the future society" (ibid., 8). Whether a desire in the present for a better future needs to be justified by telescoping that desire into Marx's writings is an open question.

Saito's point, to be sure, is not just to reinvent Marx once again to make him conform to the latest intellectual fashion, but to "put forward a wholly new Marxian vision of post-scarcity society adequate to the Anthropocene... a revival of Marx's ecological vision" defining a "new ecosocialist project" (Saito 2022, 4). There is a practical purpose to this project in that "different approaches to the ecological crisis will provide different solutions to it" (ibid., 5). The purpose is to fight against "post-Marxist" relational monism, which posits that nature is produced and that everything is a nature-society hybrid. This is the latest weapon pointed

[16] Interestingly, Tanuro, though in a footnote, converges with this view by claiming that Marx arrived at a sort of anti-productivism (Tanuro 2020a, 241–264). It is not the same terminology, but this may be one way to reconcile the two positions, where Saito's "ecosocialism" is a way of saying anti-productivist in Tanuro's terms.

against metabolic rift theory and the "ontological dualism" it is falsely deemed to reflect (see also Bergamo 2023).

According to Saito, Marx developed a methodological "dualism" (the analytical distinction between society and nature), not an ontological one. This is misunderstood by "critical" Marxists like Moore, Smith, Harvey, and Castree, and the rest, who opt for a monistic view and fear any notion of natural limits as attempts to obscure politics (Saito 2022, 112–113). By doing so, they are reviving Prometheanism, and thus converge with "capitalist realism" (e.g., ecomodernism, green capitalism, etc.) by denying that there is a separate or independent aspect of nature. The same sort of accusation about misunderstanding Marx's method is levelled at "utopian Marxists" like Bastani, who conflate the development of the forces of production with unlimited resource consumption by means of technological innovation (see also Hughes 2000).

The consequence is an environmentalism where it becomes decisive to intervene and refashion nature in such ways as to be able to survive in it. Saito calls this "geo-constructivism." Ecosocialism will not suffice to avoid this grave mistake.

> Only by going back to Marx's own texts is it possible to offer a positive vision of a future society for the Anthropocene. Such a radical transformation must be the new beginning of history as the age of degrowth communism. (Saito 2022, 6)

Holding us back is the inheritance of Engels' distortion of Marx's intended word, and out of this has come about the general misunderstanding of Marx's method.[17] Saito takes Rosa Luxemburg as an

[17] Saito is here following a tired minoritarian script of old vintage and refuted numerous times. To some extent this opinion is traceable to a handful of hagiographical zealots following Marx's passing in the late 1800s. The diffamation of Engels returned more vociferously by 1956 with a few self-styled Marxist in liberal democracies engaging in tendentious readings of Lukács' early work (Blackledge 2020). This is among other dissembling interpretative manouvres motivated by an obsession with dissociating Marx from Stalin, as if the written word were in itself a causal agent of historical change. If so it were, the writings of such "Marxist" detractors of Engels should be held responsible for the historical defeats of the left in their repsective countries. It is difficult to distinguish such "Marxism" from the typical bourgeois view of history, where ideas, not people, change society. Most Ecosocialists of Marxian persuasion do not subscribed to such attempts to tear Marx and Engels asunder by discrediting the latter (e.g., Albritton 2019, 30).

illustration of the long-term repercussions of Engels' textual meddling. She had got things right about the exploitation of environment and peoples in the colonies. However, she was critical of Marx on the grounds of his limited understanding of capital accumulation relative to questions of colonial expansion (except that, I would add, this concern for colonialism is clearly articulated in the first volume of *Capital*; Marx 1990 [1867], 914–926; is this also Engels' fault?). She applied the concept of metabolism as part of her critique, but because she was misled by Engels' naughty editing, she could not grasp the full sense of that term relative to its environmental implications (of capitalism inevitably inducing a metabolic rift). Thankfully, Saito, having revealed Engels' mischief, is earnestly embarking on correcting Marxism's hitherto faulty historical trajectory.

Hence, one must go back to Marx's own texts, though only according to Saito's preferred textual interpretations. Or, rather, as one rapidly learns on the subsequent page (Saito 2022, 7), one must also go back to Lukács, "the founder of Western Marxism"[18] to develop this positive vision. However, and just like the case of Marx, to accomplish this it is necessary to read what Lukács never published and that dates to the 1920s (it appears that, unlike Marx, Lukács did not experience any late-life conversions). And this needs to be supplemented by the close readings of the theory of social metabolism (see above) as developed by István Mészáros, a student of Lukács who fully grasped the environmental repercussions of capital's inherent inability to observe natural limits (Saito 2022, 17–23). Only then is it possible to understand that he was following Marx's methodological dualism regarding nature and emphatically on the basis of the concept of metabolism. The reason that we have not known about this is because "his unique insight was suppressed by both orthodox Marxism and Western Marxism, leading to the marginalization of Marxian ecology throughout the twentieth century" (Saito 2022, 7). First Engels, then orthodox and Western Marxists have conspired to keep the truth from us. By overcoming this recurrent suppression of Marx's real viewpoints,

[18] Some would take issue with this characterisation, as the intellectual movement known as "Western Marxism" involved multiple people and diverse, sometimes mutually antagonistic schools of thought, ranging from the Institute for Social Research (the "Frankfurt School"), bankrolled by the US secret services by the 1930s, and portions of Bolshevik parties in Central and Western Europe drawing from the likes of Korsch, Gramsci, and Lukács (Anderson 1976; Losurdo 2017; Rockhill).

by bringing to light unpublished archival materials, it will be possible to draw together proponents of degrowth and Marxists. This comprises a major motivation subtending the reinterpretation of Marx's last writings as a manifestation of degrowth communism. Notably, Saito's assertions not only depart from most Marxist perspectives worldwide, but also flatly contradict Foster's own view about Engels' contributions (Foster 2017, 2023).

Debate over Relabelling Marx

Whether the evidence presented by Saito suffices to make of Marx an Ecosocialist or Degrowth Communist is rather disputable. The elasticity of the label Ecosocialist appears anyway to be quite capacious. For instance, William Morris (1834–1896) also enters the Ecosocialist club (Saito 2022, 209). At this point, one might find slave-revolt leader Spartacus (103–71 BCE) to be a communist because of his egalitarian principles and perhaps one may find him to have been an ecological thinker as well, one never knows. More obviously, perhaps, is the Ecosocialist status of the Epicurean Lucretius (99–55 BCE). Not only did he challenge the notion of human exceptionalism, but he also exhibited sensitivity to nonhuman beings. As Boggs (2020, 67) has pointed out, given the overwhelming lack of environmental concerns where they lived, Marx would have achieved nothing short of a miraculous feat, one that is beyond mere mortals. Relabelling Marx as ecologist (why not an ethnologist?), ecosocialist, or degrowth communist is to conflate form with time and to bid farewell to historical materialism.[19]

Saito's understanding of Marx the Ecosocialist and Degrowth Communist rests on dismissing other forms of Marxism as misinformed on Marx's

[19] In the case of making Marx into an ecologist, it is not only to reduce the scope and substance of ecology to tracing human impacts in one mode of production, unless one means ecologically inspired philosophy, but it is also to confuse studying the results of science with doing science itself. But there is no reason to stop with ecology. The same should be said about Marx's Ethnology, something Saito avoids, but someone else has considered (Smith, D., 2002). One could go much further. Given that Marx also studied maths and maths were central to his analysis of capitalism (especially evident in Volume 2 of *Capital*), there should surely be a volume dedicated to showing how Marx was a mathematician. In this kind of Marxology, all that Marx studied becomes Marx's areas of professional expertise.

method. This *ipso facto* excludes non-conforming Marxists from Ecosocialism and from what he surmises as the best version of Communism. Moore, for one, would doubtless object, finding the metabolic rift school redolent of Cartesian dualism and thereby incorrect compared to his own interpretation of Marx's work (Moore 2015). This is an evidently irksome position to those in the metabolic rift school (Foster and Burkett 2016; Saito 2022, 118–119).

Much of Saito's argument also involves accusing other Marxists with insensitivity to "the objective biophysical limits of the Earth," which he deems to constitute a slippery slope leading to "the problematic endorsement of further technological intervention in the form of genetic engineering, geo-engineering and nuclear fusion" (Saito 2022, 113). Thus, Saito offers an unfortunate parallel countercharge to what he finds to be a rightfully unhelpful accusation of apocalypticism or millenarianism, to which, one could add, Kovel has also been subjected. Yet those deemed to be "monists," mostly Marxist geographers, also have legitimate reasons for their different interpretation of Marx's work. They are suspicious of narratives or theories that invoke causes external to society and therefore to politics.

Each faction accuses the other of misinterpretation or self-serving selectivity of Marx's texts. At a minimum, at least, they all converge on the necessity to start from Marx to explain the socio-ecological crisis and arrive at the same conclusions that capitalism is inherently awful socially and ecologically. A possible way to reconcile these Marxist tendencies is to recognise that they share a commitment to socialism and to the Marxian ontological principle that we are part of nature. Methodologically and epistemologically, there can be diverging emphases appropriate to different contexts or research objectives, rather than correct or incorrect lines. If one is to try to motivate political struggle more directly in NATO countries, then something like making of Marx a "degrowth communist" could make sense. If the issue is to take apart environmentalist platforms that conceal capitalism-friendly policies and rest on a bourgeois nature-society dichotomy (ontological "dualism"), then a "socio-nature" or "monist" perspective could be more useful.

In any case, whether Marx was a precocious "Ecosocialist" also depends on which Ecosocialist current one considers, something Saito does not specify except indirectly by terms like "orthodox," "critical," "Western," and "utopian" Marxism. This specification is important because there are other Ecosocialists, living in the present, who do not identify with Saito's reinterpretation of Marx. Fourth Internationalist Ecosocialist Tanuro (2020b), for example, finds the claim exaggerated. That is, Marx's "ecological" understanding was incomplete; it is an unfinished project that Ecosocialists can now develop further. Marx's work is something to build on, but, Tanuro avers, what must be avoided are his occasional scientism (disdain for non-scientific knowledge; see also Boggs 2020, 78–82; Salleh 2017, 107), a lack of attentiveness to patriarchy and women's oppression, and a dismissiveness towards peasants (who often play key anti-capitalist roles, as vanguards in socialist struggle).

These charges are countered by Foster (2020a), who shows that Marx's alleged scientism is a decontextualised imputation derived from misconstruing Marx's argument, which was aimed at the pro-capitalist politics behind a particular scientist's take on plant nutrition. Marx otherwise showed an appreciation of peasant knowledge in other writings, especially with regard to Ireland. As for patriarchy, Foster agrees, though it is also untrue that Marx ignored domestic labour as part of the processes of oppression in capitalism. Furthermore, such a flaw has been amended, including through Marxist Feminist work, precisely by using Marxist method.[20] However, Foster fails to address Tanuro's main argument that Ecosocialism is about the further development of Marx's incomplete work on the ecological front.

The issue of incompleteness is echoed by Boggs (2020, 70–71). He points out that Saito, along with all the metabolic rift school, provides insufficient evidence to support such bold assertions. In fact, even the manner in which the proof is presented is in terms of hints or intentions ("discursive minimalism"), and in what amounts to "truisms" and "abstract generalities," rather than any systematic ecological study of capitalism. Moreover, Saito's allegation of Engels' editorial distortions of Marx's framework is inconsistent with the fact that Marx approved of Engels' *Anti-Dühring*, parts of which were published already by 1877, at least a decade after Marx's alleged "ecosocialist" conversion. Marx even wrote an introduction to Engels' *Socialism: Utopian and Scientific*, where the method of dialectical and historical materialism is explicated.

[20] One could add that Tanuro omits the fact that Engels made of patriarchal gender relations a major pillar of his critique of capitalism. And, as in the case of the development of dialectical materialism, it can be argued that Marx and Engels had a division of labour on the issue of social reproduction as well.

Boggs' critique of Saito's ideas veers off into speciousness by conflating the notion of mastery expressed by Marx and Engels with the idea of control over nature and by accusing them of speciesism. As shown above (Benton 1992; Dickens 1992; Hughes 2000), even if one believed such were true, it is a matter easily resolved by applying a Marxist framework. Boggs falls for the same ill-conceived and tendentious interpretations that Foster and Saito, among others, have shown to be erroneous. He resorts to creating, and without much evidence, the impression of an apocalyptic age during Marx's time, when "the very notion of harmony in natural relations had, in fact, already been torn asunder, fractured under the onslaught of industrialization, urbanization, and warfare" (Boggs 2020, 78). It should then be understood as nothing short of miraculous that any of us are alive more than a century since Marx's passing.

Going further than Boggs (and his zero-strategy anti-communist vitriol; see especially Boggs 2020, 84–85), one should note a much more important issue, one that pertains to Saito's work directly. Engels' latter work of the 1870s and early 1880s, read, commented, and approved by Marx, contains virtually nothing of what Saito claims that Marx had become by then, neither in "ecosocialist" or "degrowth communist" form. Even if one wants to read Engels' "revenge of nature" argument as a metabolic rift thesis, which would contradict Saito's contention about Engels (since he is supposed to have fundamentally diverged from Marx over this), it is a tangential and general statement that Engels applied to multiple modes of production, not capitalism per se. Moreover, the point is expressed in terms of understanding how humanity is dialectically related to the rest of nature, something that the concept of metabolic rift does not convey (Engels 2010 [1876], 461–462). Yet the lengthy passage about the "revenge of nature" is often used by proponents of metabolic rift theory to demonstrate the ecological aspects of Marxism, which Saito would now deny by claiming a rift between Marx and Engels over ecological principles. It is difficult to resist the sense that the works of Marx and Engels are being reduced by such Eco-Marxists to whatever is deemed worthy of their current preoccupations.

One could argue that there are more fundamental flaws with Saito's thesis. One is that it is predicated on treating a scholar's notebooks as equivalent to publications or public interventions. Had Marx considered "ecological critique" essential to his overall project, reasons must be furnished to explain the rarity of such "ecological critique" in whatever was completed or published of *Capital*. Even more so, if "ecology" was central to Marx, it must be explained why Marx never mentioned such central environmental concerns in official speeches, newspaper articles, or in any other such fora. Furthermore, it must be explained why Marx never carried out a project relating the effects of different forms of farming on the properties of soils, earthworm populations, or plant composition, to name a few examples. He contributed no such systematic study on any ecological issue covered in any published or unpublished writings. It seems evident that such concerns were of a lesser priority to Marx compared to what he stated publicly or published (e.g., for First International meetings) and to his political organising. Relatedly, if "ecology" was so foundational to Marx's objectives, it needs to be explained why Marx elected to transition towards political economy (Saito 2017a, 61), rather than a field in the biophysical sciences that he was studying closely. An ecological outlook or philosophy is not equipped to generate the systematic studies and primary data on ecosystems crucial to backing up arguments about the inherent destructiveness of capitalism.

The shaky foundations of the Ecosocialist (and by extension the Degrowth Communist) Marx thesis is readily admitted by Saito himself when he remarks that Marx

> hardly wrote on this theme [ecological critique of capitalism] after the publication of volume 1 of *Capital*. Yet one cannot imagine that, after this intensive study on natural limits against the formal logic of capitalist form determination, he suddenly abandoned his further research on ecological questions. (Saito 2017a, 216)

The trouble is that one can imagine exactly that. There is no evidence that Marx would have pursued the subject matter further, as would be expected of an Ecosocialist, or a Degrowth proponent. Instances of environmental impact analysis are too sporadic and brief (little more than a paragraph or two) in Volume 1 of *Capital* to sustain the Ecosocialist Marx argument. It seems there should be little lost by calling Marx a proto-Ecosocialist or a precursor to Ecosocialism if one must insist. However, Saito does not explain what is to be gained by this relabelling exercise.

The philology on which Saito embarks evinces some signs of desperation when he contends that "Marx studied the rich field of nineteenth-century ecological theory and integrated new insights into his own dissection of capitalist society" (Saito 2017a, 13). While it is true that Marx (and even more so Engels) gave much importance to the scientific findings in the biophysical sciences of their day, there was no ecology as an established scientific field until some decades following Marx's passing, much less any "ecological theory." If anything, it could be said that Marx was ahead of his time in making connections, even if sporadic, between capitalism and environmental degradation in generic terms. Such anachronism (similar to the wishful thinking about Marx's influence on the development of the biological and ecological sciences in Foster 2020b) undermines an otherwise valid argument.

The contrivance of treating informal or marginal intellectual endeavours as if representative of principal objectives might make for interesting reading but is an unnecessary manoeuvre. Marx had precociously understood at least the contours of capitalist tendencies for environmental destruction and had provided a framework whereby to analyse such tendencies, though at a level of abstraction that cannot address specific biophysical phenomena or ecological relations beyond those involving society. Saito, as also Foster and Burkett, have shown this useful yet limited proto-ecological framework in Marx's published work regardless of whether the *MEGA* materials are considered. One could more fruitfully return to Engels' writings, some unfinished and fragmented, to understand how dialectical and historical materialism provides principles and guidelines for scientific ecological studies and the development of practicable alternatives to existing destructive people-environmental relations under capitalist regimes (see, for example, Levins and Lewontin 1985). No more than this need be claimed to disprove conventional notions about Marx or Marxism as lacking in environmental or ecological sensibility (Hughes 2000). One can disprove this without recourse to exegetical acrobatics.

Saito's exhortation to return to Marx to discover degrowth communism is just as flawed, if not also troubling. It is founded on two explicit assertions and one implicit assertion, all of which are questionable. One is that classical Marxism and with that socialist states were productivist. This has been shown to be empirically false (Foster 2015a; Suing 2018). Hence, the incompleteness of the material bequeathed from Marx to the generation of socialists did not impinge negatively on the development of Marxism relative to ecological concerns. Another assertion is that Engels diverged from Marx over the concept of metabolism and, through posthumous editing, distorted Marx's words. This is an old motif that has been used by some Marxists to legitimise their own fanciful interpretation of Marx's text and to suit idiosyncratic political convictions often divorced from social struggles. After all, Engels' framework constituted a central foundation of scientific socialism and ecological sensibility that Marx shared, as amply and repeatedly demonstrated (see Foster 2023).

The third assertion bears the insidious assumption that succeeding generations of Marxists are unable on their own to develop theories adequate to the ecological or social challenges of their day. In this sort of narrative, Marxism only evolves by recourse to textual interpretation and by chance occurrences of what texts are available at what historical period. Marx's work (shorn of Engels' editing, to some) arrives ready-made for any eventuality, if only the appropriate texts are discovered and correctly interpreted. Unknown, obscure, or forgotten writings from the past are to dictate what is to be done in the present. The archivist becomes the guardian of truth, and the philologist becomes the prophet. It is thus that exegesis replaces the more arduous task of taking stock of and theorising from contemporary ecosocial processes and struggles and the outcomes of past practices and politics in order to develop new forms of struggle for socialism. Moreover, a historical conjuncture (coeval material conditions and ideological processes) shapes how Marx is re-read and reinterpreted. Saito's ahistorical understanding of philological practice, lacking in any critical self-reflection and social contextualisation, effectively eviscerates Marxism of historical materialism altogether and becomes bourgeois idealism.

More prosaically, but just as importantly, it is unclear how helpful a strategy it is to make of Marx into what would amount to a rather marginal communist figure, as he would be if located within a still marginal and mainly academic movement called degrowth, which teems with anti-communists. Nor does it seem politically sagacious to engage in Manichaean divisions into good (dualist, degrowth) and bad (monist, geo-constructivist) Marxism (see also White et al. 2017). Saito, for instance, should explain what there is to be gained by what to outsiders will likely look like an exercise in splitting hairs. Nevertheless, Saito's treatment of Marx's notebooks is important in tracing the evolution of Marx's thought, as one part of the building blocks from which to develop an Eco-Marxist framework. Beyond this, however, there lurks a realm of strained, if not fanciful speculation.

Venturing beyond Re-readings of Marx (and Sometimes Engels)

From its beginnings, Eco-Marxism has largely been confined to debating the ecological content in the theorisations of Marx and, to a still exiguous extent, Engels. The task undertaken by Eco-Marxists has consisted in building on Marx's ideas, revealing allegedly suppressed aspects of Marx's framework, or correcting perceived inadequacies in the works of Marx and Engels or in Marxism generically. This concern is understandable to some degree, most obviously given the roots of Marxism in the ideas explicated by Marx and Engels.[21] Marxism, though, is much more than Marx and Engels and has evolved into several contrasting paradigms, with at times mutually antagonistic variants having developed within each.

In some respects, the different viewpoints expressed in the Eco-Marxist debates draw from such historical currents or their subsets. Only a few of them are directly represented or central to Eco-Marxism as it has evolved so far. The Eco-Marxism of Tanuro and Löwy, for instance, reflects a neo-Trotskyist perspective (Löwy and Tanuro 2022), while Kovel's tends towards combining aspects of Frankfurt School, Eco-Feminist, and Liberation Theology frameworks, and that of O'Connor, Foster, and Burkett is influenced by the Monopoly Capital school associated with Baran and Sweezy (Foster 2021). Hence, with such varied influences encapsulated, at times implicitly, within different Eco-Marxist currents, the narrow, almost pedantic attention to Marx's works is surprising.

A few authors have recently drawn attention to the writings of Rosa Luxemburg (1871–1919), who played a major role in the German Social Democratic Party. She was a radical anti-war Marxist theoretician and organiser, co-founder of the Communist Party of Germany in 1919. She paid dearly with her life, summarily executed by the paramilitary Freikorps (later serving as Nazi Party thugs) along with hundreds of other comrades, for her involvement in the 1918–1919 German Revolution. Among the ideas she developed were the dialectical relationship between spontaneity and organisation as inseparable moments in class

[21] Arguably a major factor is the location of most Eco-Marxist intellectuals in anti-communist educational institutions, where Marx is usually targeted for disparagement when it comes to environmental issues. This may explain, in part, Eco-Marxists' efforts being overwhelmingly dedicated to proving that Marx's work has, at the very least, something important to convey about environmental degradation.

struggle and the imperialist expansionism of capitalist states induced by chronic underconsumption (too many commodities relative to working-class purchasing power). With the latter theory (developed in her 1913 book *The Accumulation of Capital*), she articulated a basis for anti-colonial and anti-militarist politics and the notion of an eventual demise of capitalism with the disappearance of all non-capitalist economies. These ideas and those expressed in other writings, including by way of correspondence, point to a sensibility, rare at the time, to the suffering of other species and to the resistance of "nature" to capitalist prerogatives (Aldridge 2023; Buhle 2013). The importance Luxemburg gave to colonial resource exploitation and the socially destructive effects of colonialism also prefigured aspects of what would now be recognised as ecologically unequal exchange.

Other influential Marxists, such as Lenin, have yet to receive much scrutiny. A limited exception is in Wallis' appraisal of Lenin's evolving politics and the unintended consequences of Bolshevik strategies and post-revolutionary policies, especially with respect to necessary compromises to enable eventual socialist advances. For instance, the partial allowances for profit-making in the USSR's New Economic Policy in the 1920s is argued to have enabled the ideological acceptability of privileged social strata and the diminution of worker control (Wallis 2020, 129–130).[22] The focus, however, is not on Leninist or Bolshevik forms of environmental practices and ideas on ecology. It is for Wallis an occasion to discuss one among several examples that speak to signposts useful for developing ecosocialist politics.

Otherwise, the work of movements-oriented Andreas Malm (2021) provides another rare example of moving beyond Marx and Engels. Malm sees a more worthwhile way of combining Marxism and ecology in a recuperation of a form of Leninism, which he calls Ecological Leninism. To him, Leninism is a politics for catastrophic times (climate change) and

[22] Ironically, this argument aligns with Stalin's evolving position on the matter, leading him to strive for ending the New Economic Policy (NEP). Although Wallis finds the Stalin administration disagreeable, perhaps a travesty, he may wish to explain how the NEP's negative effects should have been countered under the specific conditions of the USSR in the 1920s, following the ravages of the World War I and of the Civil War in a country that had been invaded multiple times by armies from 14 parliamentary democracies or constitutional monarchies, that was under siege thereafter, and that had already suffered from poor infrastructure and inadequate productive capacity even prior to the Russian Revolution.

for urgency of intervention due to reforms not being viable any longer (thresholds are being crossed). A transition also must feature the state as a central aspect of intervention to have the possibility of undermining capitalist relations and thereby transition to ecological sustainability. This is because there is little chance for the development dual power at this historical juncture, given working-class fragmentation and a dearth of political organising and militancy that enables the creation of parallel anti-capitalist institutions of the likes that emerged in Russia, Germany, Hungary, or Italy in the 1917–1919 period. Regrettably, Malm conflates theory with conjunctural strategy, that is, Leninism is flattened to the political necessities of a specific historical period. The contributions to theories on political party structure, political practice and organising, theories of imperialism are entirely ignored, along with much else of Lenin's works (cf. Redclift 1984, 8–10, 13–19).

This impoverished and superficial view does little to address Gare's notion that Lenin's environmentalism, which also led to establishing and expanding many ecological preserves (*zapovedniki*), amounted to a conventional conservationism (Gare 1993b). To date, Lenin's copious writing has not been thoroughly re-examined for any ecological content. Eco-Marxists seem generally unaware that "Lenin had … made incredible contributions to the field of ecology in several key areas: with legislation, with political support of the sciences, and with his theoretical contributions to both scientific and ecological discourse" (Stahnke 2020, 64). In fact, such ecological sensibility and indirect, yet decisive contributions to scientific ecology are evident in different ways across much of the early Bolshevik leadership, including Bogdanov, Bukharin, Lunacharsky, and (most inconveniently for Eco-Marxists) Stalin, and spanning the history of the USSR (Brain 2010; Foster 2015a).

On the other hand, another influential Bolshevik leader, Gramsci, greatly admired, especially in Global North leftist circles, for his novel theories (especially on hegemony), has found little echo within Eco-Marxism and more by way of allusions than substantive engagement. The concept of historical bloc, for instance, as discussed above, is used to invoke the possibility of forming a self-conscious environmental proletariat but no analytical application is carried out using the concept. There is no attempt yet to comb through all of Gramsci's writings. It is more within the academic field of political ecology (maligned by metabolic rift theorists as "monists" or "post-modernists") that there has been an interest in applying Gramscian concepts (e.g., civil society, hegemony) to the study of environmental politics or struggles and to notions of human nature (Jakobsen 2022; Mann 2009). What has mainly ended up being demonstrated is just how little Gramsci had to say about environmental issues as well as nonhuman nature generally (Ekers et al. 2012).

Regrettably, the range of Gramsci's texts considered are confined to those put to paper while languishing in Fascist prison. The Bolshevik core of Gramsci's politics is typically eschewed. Political ecology, in any case, as a generically critical framework, offers too vague or ambiguous a politics and is too wedded to abstract academic questions to be relevant to the development of Eco-Marxism or to Ecosocialist strategy more broadly. This is also why political ecology features tangentially or not at all in Eco-Marxist debates or Ecosocialist writings, unless it is the sort of political ecology based on some sort of Marxism (Moore 2017; Napoletano et al. 2019; White et al. 2016).

In contrast, there are examples of more direct uses of ideas from one of Gramsci's contemporaries, György Lukács, and from Lukács' student, István Mészáros, discussed above, as well as other Marxists theorists. But their works are not systematically analysed to develop Eco-Marxist theory in general but to lend support to existing ones, especially metabolic rift theory. The social background of the Marxists considered should be itself a subject of examination. It is disconcertingly rare for non-European Marxists to feature in Eco-Marxism. The missed opportunities resulting from this will be explored in a later chapter, in a discussion on socialist Pan-Africanism, including Amílcar Cabral's work. Particularly surprising is the virtual absence of Mao Zedong, whose copious writings are barely touched except for portions of canonical texts, selected out of context. The result is a caricature of a leader who is obsessed with the development of the "forces of production," mischaracterised as technology (Gare 2012). Or, in certain Marxist Geography renditions, Mao (with Lenin) becomes a generic state-centred strategy, likened to that of Robespierre, to be deployed in the interest of avoiding further climate catastrophe (Mann and Wainwright 2018). Concepts and political practices like the mass line and primary contradiction are beyond the intellectual capacities of such forms of Marxism.

All the above-mentioned Marxist theoreticians and communist leaders had something to say about the environment, but they based their understanding on the latest scientific theories ad findings in the biophysical sciences. Bukharin, for instance, understood from Marx the ontological primacy of nature, within which humanity effects changes but also adapts to the rest of nature, in a dialectical materialist framework. However, he regarded such modifications in terms of shifting conditions of equilibrium, including by means of what Foster identifies as "universal metabolism" and in concert with prevailing views in contemporary ecology and affine scientific fields (Bukharin 1925). Today, in contrast, one might think of the relationship in terms of multiple factors, threshold conditions, complexity, non-equilibrium conditions, and the like (see, e.g., Dickens 1996).

Hence, Bolshevik ecological sensibility stemmed from none of the above-discussed Eco-Marxist theories on what Marx meant or omitted, but from the further development and application of Marxist method (dialectical and historical materialism). As communist mathematician and crystallographer John D. Bernal articulated the relationship between Marxism and the technical and biophysical sciences,

> Marxism is not scientific method, nor is it in any sense an alternative method; it is at the same time more comprehensive and more advanced. Both the method of science as hitherto understood and the content of scientific discovery can be incorporated in the Marxist scheme. They need, however, to be criticized and extended. Marxism is no substitute for science, but because of its wider scope it can see the limitations of existing methods and indicate where in the past these have been used in fields in which they have no competence. (Bernal 1937, 58–59)

One might quibble about what constitutes science, but the relationship as explicated by Bernal is clear, as it was clear to Marx and Engels, who never confused the study of scientific work with actually conducting scientific work, as Eco-Marxists tend to do. Marxism provides the general framework for scientific work. It cannot serve as the basis for the practise of any specific scientific field.

Eco-Marxist Tempests in a Capitalist Teacup

Differences between Eco-Marxist schools of thought have tended to be exaggerated (Gellert 2019, 187). Some have rightfully pointed this out early on relative to the debate over Foster's theory about Marx's metabolic rift (for example, Wallis 2001; cf. Wallis 2018). Zhang (2022) demonstrates that, especially in light of persistent environmentalist anti-Marxism, all such theoretical positions can be usefully folded into a

"Marxist ecology," as they all draw from the same insights from Marx and Engels. Saito (2022, 15) also recognises that both the second contradiction and metabolic rift currents draw from Marx's framework to explain the socio-ecological crisis. One should extend this argument to Moore's "capitalist world-ecology." Though I stand corrected in earlier overestimating theoretical differences (as a contributor to the 2001 symposium on Foster's *Marx's Ecology* in *Capitalism Nature Socialism*), I would now go further to state that metabolic rift theory amounts to specifying where the second contradiction thesis could be argued to exist in Marx's work. To some extent, the interpretational divergence between O'Connor and Foster is due to the former's faulty presentation of his own work (as Polányian) and to others' misrepresentations of the second contradiction thesis (Rudy 2019). It seems more appropriate to view the second contradiction thesis as virtually the same as that subsequently proposed by Foster in the guise of metabolic rift theory. This can be concluded by returning to Foster's earlier writings, while he was part of *Capitalism Nature Socialism*. Before parting ways with O'Connor by the end of the 1990s, Foster (1992) had ably demonstrated the basis of O'Connor's theory in Marx's original text. According to Foster, the first contradiction equates to Marx's "absolute general law of capitalist accumulation," while the second contradiction is representable as "the absolute general law of environmental degradation under capitalism." The latter was at the time made evident by drawing more from Engels' than Marx's writings (Foster 1992, 79–80). It should also be noted that O'Connor insisted on using the concept of conditions of production precisely because it was used by Marx (O'Connor 1998, 160, 174, fn 12).

Later, Foster (1999a) would find that Marx himself had expressed this second general law in the form of metabolic rift, which Foster would claim to be foundational of Marx's work (Foster 2000). He would go further in trying to distance himself from O'Connor's ideas by claiming that "there is no natural feedback mechanism that automatically turns environmental destruction into increasing costs for capital itself" (something O'Connor never claimed)[23] and that the notion of two contradictions is "too undialectical" and "forces a dualistic economic perspective on us that once

[23] The late Marxist economist Paul Burkett (2006) adds that O'Connor's thesis fails to distinguish between the ecological conditions needed for profitable capitalist production and those needed for alternative ecologically sustainable production and thereby O'Connor's thesis reduces environmental to economic crises. The reductionism, however, is of

entered into is difficult to escape from." Furthermore, to Foster, the second contradiction thesis leads to demote class struggle to secondary status as the ecological crisis deepens (Foster 2002). This latter view, however, only holds if one ignores O'Connor's insistence that the two contradictions are always intertwined and equally important (O'Connor 1991).

The misrepresentation of O'Connor's work has been effective in creating the impression that second contradiction theory differs substantively from its metabolic rift counterpart (a recent example is Saito 2022, 125–129). The same can be said of "monism" in contrast to "dualism" or "capitalist world ecology" and alleged "Cartesian dualism." Further, imputing any of such concepts or theories as foundational to Marx's thought implies that divergence is un-Marxist, if not anti-Marxist. This can in part explain the acrimonious nature of Eco-Marxist debates and the factionalism that tends to characterise Eco-Marxism. This is only compounded by declaring one's interpretation as representing the most advanced stage of ecosocialist thought (Burkett 2006; Foster and Burkett 2016; Saito 2017a, 11–12).

Ultimately, the much-touted breakthroughs claimed in all Eco-Marxist currents are redundant with other and more notable Marxian concepts and analytical categories, such as the processes of value forms differentiation, capital circulation, fetishism, and alienation. This has been repeatedly and amply demonstrated in explaining the socio-ecological crisis generally, by Eco- and plain Marxists alike (Castree 2000; Foster 1992; Harvey 1996; Huber 2013, 9–19; Kovel 2007; Levins and Lewontin 2007; O'Connor 1988; Suing 2018; White et al. 2016). More incisively, there have been effective treatments using anything but metabolic theory in the study of finer-grained contradictory and ecologically destructive

Burkett's own making. Capitalist production, like all forms of production, require ecological conditions that enable human life, something that mainly bourgeois thinkers find conceptually challenging. But there is more. Perhaps unwittingly, Burkett discovers in the second contradiction thesis exactly the same failure characterising the metabolic rift theory he champions, where it is impossible to distinguish analytically between soil phosphate enrichment and soil lead pollution or between monocultural tree plantations and clear cutting, as they are all forms of metabolic rift or second contradiction (in fact, both Eco-Marxist positions are undialectical and socially reductionistic; see Engel-Di Mauro (2014). In this, lamentably, even bourgeois environmental scientists have an analytical edge over Eco-Marxists.

social processes, like biodiversity conservation (Büscher et al. 2012) and "offsetting" policies (Apostolopoulou et al. 2018).

Finally, it has yet to be shown that the field of ecology gains anything by applying metabolic rift theory at all, as indicated by the latter's absence in scientific ecology research and publications. A recent assertion that metabolic rift is, compared to capitalist world-ecology theory, closer to the dialectical biology of Levins and Lewontin (Bergamo 2023) does not make for an any more biology-relevant Eco-Marxism. This is a particularly odd situation that begs explanation, since Marx's work is insistently considered foundational to that scientific endeavour (Foster and Burkett 2016; Foster 2000, 2020b; Saito 2017a).

Eco-Marxism's Current Challenges

The more recent kerfuffle over "monism" and "dualism" constitutes another instance of exaggerating differences. On one side are the dualists, largely the metabolic rift school. On the other are the rest, the monists, or those who accuse dualists of what amounts to holding a (bourgeois) society-nature dichotomy discussed above. This, anyway, is how adherents to the metabolic rift schools characterise the debate. In contrast, I prefer to describe the groupings in terms of metabolic rift and multi-tendency Eco-Marxism. Much more fundamental departures exist with the sort of academically popular "immanentism" capable of little more than capitalism-compatible critique by offering no socialist alternative and by worrying more about recognising non-capitalist causes than about developing anti-capitalist politics (Henderson 2016, 284–289). Thus, relative to theoretical content and prevailing politics in liberal democratic institutions, the philosophical contention over monism and dualism seems misdirected and to obfuscate a more important matter of meanings or levels of abstraction ascribed to the same terms.

For instance, Marx, all Eco-Marxists would surely agree, understood nature as inclusive of humanity and humanity as a force of nature, among other forces. This is one level of abstraction, with nature as wider, not juxtaposed to or on the same plane as humanity. On the other hand, Marx would also refer to nature in juxtaposition to or as contrast to humanity or to social relations:

> nature does not produce on the one hand owners of money or commodities, and on the other hand men possessing nothing but their own

> labour-power. This relation has no basis in natural history, nor does it have a social basis common to all periods of human history. It is clearly the result of a past historical development, the product of many economic revolutions, of the extinction of a whole series of older formations of social production. (Marx 1990 [1867], 273)

Nature here is intended as the part of nature outside of society. By natural history, obviously, it is meant history not determined by human action, even though elsewhere Marx is clear about human action being part of nature. In other words, there are multiple ontological levels of abstraction or meaning about nature in Marx's writings. One is "monist" regarding humans as internal to nature and the other is "dualist" relative to recognising distinctions between humanity and the rest of nature. To simplify, Marx and Engels understood that we are part of nature yet distinct from the rest of nature. Both levels of abstraction are necessary to grasp people-environment relations by means of dialectical and historical materialism.

In this sense, the metabolic rift thesis is a reading of Marx's work focused on the broad effects of capitalist impacts on the part of the rest of nature on which we depend for sheer survival, and which is being undermined (the socio-ecological crisis). Instead, concepts like "socio-nature" (hybridity of the social and natural) or "production of nature" in other (Eco-)Marxist discourse imply an investigation of the parts of the rest of nature that are independent of human action ("first nature") and now exist as remote exceptions in terms of ecosystems relatively untouched by human action. As Marx had himself observed while critiquing Feuerbach's inconsistent materialism,

> nature, the nature that preceded human history ... is nature which today no longer exists anywhere (except perhaps on a few Australian coral-islands of recent origin). (Marx 1932 [1845–1846])[24]

This is the sort of nature, otherwise known as pristine nature, that environmentalists often invoke and that is sharply critiqued by multi-tendency Eco-Marxists and kindred Marxists. They critique conventional understandings of nature, within capitalist countries, because such notions

[24] https://www.marxists.org/archive/marx/works/1845/german-ideology/ch01b.htm, accessed 20 December 2023.

conceal a conservative or reactionary politics, one that pretends away social inequalities and relations of power and directs attention away from the struggle for socialism (see, e.g., White et al. 2016). This should be agreeable as well to proponents of metabolic rift theory, who reject any capitalism-accommodating view of nature or environmentalism (see, e.g., the sharp critiques of environmentalist thought in Foster 2000). Regardless, it will be more beneficial to the development of Eco-Marxism to evaluate analytical frameworks according to the degree to which they help towards or express commitment to building ecosocialism, rather than according to one or another interpretation of Marx's texts.

Another, politically more substantive difference is more about what aspect of the social struggle for ecosocialism is being emphasised and how to convey the motivation for the necessity of ecosocialism. Benton (2023, 683) may have spotted a major distinction over "strategic political thinking," which is more developed in the coalition-building orientation of multi-tendency Eco-Marxism compared to the paucity of such thinking in the metabolic rift school. Relative to emphasis, the more abstract take on capitalist causes by way of metabolic rift seems more useful in tracking and explaining general or ideological processes and tendencies. This level of applicability pertains to other framings like O'Connor's second contradiction, Kovel's egoic relations, Salleh's meta-industrial work, and Moore's capitalist world-ecology. Too often, though, such broad frameworks are applied at inappropriate scales of analysis, as rightfully pointed out by White et al. (2017). When one confronts much finer details or wants to explain shifts within capitalist relations relative to biophysical changes, involving, for example, combinations of non-capitalist and capitalist processes and wider and unpredictable ecological relations, it should be more apt to adopt a more "hybrid" approach that can account for relations that are not reducible to capitalism or human impacts (see also Henderson 2016). This is also crucial to understand when engaging in everyday political struggles, especially outside of the imperialist countries and in First Peoples communities.

With respect to arguments for ecosocialism, multi-tendency Eco-Marxists opt for analyses and discussions that consider different kinds of situations and contradictions and bring out complexities that can help identify ways of promoting ecosocialism that make sense for specific contexts (then the problem is scaling up and not falling fall moralistic judgements at international or national scales, as when the entirety of a country like the People's Republic of China is evaluated). The metabolic rift Eco-Marxists instead elect for instilling a sense of urgency in general. It is in some respects a difference in scale of analysis (global) and of exhortative emphasis. An example is the following admonitory statement:

> When we speak today of the world ecological crisis, however, we are referring to something that could turn out to be *final*, that is, there is a high probability, if we do not quickly change course, of a *terminal crisis*—a death of the whole period of human dominance of the planet. Human actions are generating environmental changes that threaten the extermination of most species on Earth, along with civilization, and conceivably our own species as well. (Foster et al. 2011, 424; italics in original)

A similar level of alarm is expressed in Wallis (2020, 14) by equating the current "environmental crisis" with "the absolute limits of resource-extraction ... beyond which human survival is impossible." He goes further: "Our species is at a moment of reckoning. Collective death—human extinction—is a real prospect within the next few generations" (ibid., 227). Vettese and Pendergrass (2022), opponents of what they call a once-dominant "Promethean Marxism," propose a form of ecosocialism that revitalises Utopian antecedents to establish world-level socialist planning within planetary boundaries (with half of the planet to be devoted to leaving ecosystems devoid of humans). They deem this a stark choice to make in the present to avoid doom.

There is an uncanny discrepancy in such writings, in contrast to such non-Marxist groups like Extinction Rebellion, between invoking a capitalism-caused massive and present danger to human survival and falling short of calling for actions aiming for the immediate destruction of capitalist institutions. White et al. (2017), among others, would anyway regard this sort of argumentation as catastrophist, which understandably does not sit well with metabolic rift Eco-Marxists, who might consider such criticism a form of denialism (Saito 2022; Wallis 2018) perhaps alongside Kovel (2007), who advocates for the development of parallel ecosocialist institutions from the bottom up. A less polarising perspective is to be found among metabolic rift Eco-Marxists when they demonstrate how their framework furthers, rather than disproves other Marxist ideas about nature or environmental degradation (Napoletano et al. 2019, 1815–1816). Schwartzman (2009) similarly takes issue with the notion of catastrophe only insofar as the gravity of the situation is not met with commensurate urgency in political organising. This could even imply forging strategic alliances with fractions of the ruling classes, such as with the Green New Deal (Schwartzman 2011), so as to avoid the worst effects and buy some time to have the chance to overcome capitalism for good. Perhaps eventually what now appears to be an insurmountable difference will be swept away by the realisation of substantive commonality and objective necessity.

Nonetheless, even if all Eco-Marxist schools of thought were to converge over a common interpretation of Marxism or canonical texts, they share the challenge of making their ideas politically and scientifically relevant. It is unclear what difference it makes, for the purpose of political action and ecological science, to know that Marx conceived of the metabolic rift long ago or that Marx was an ecosocialist or degrowth communist. It muddles politics even more to reduce Marxism to "the systematic search for an alternative to capitalism" (Wallis 2020, 64) or to render "the goal of socialism/communism" to be "the free development of all" (ibid. 89), rather than heed Marx and Engels' advocacy for scientific socialism in the struggle for communism, which in itself implies "the full flowering of everyone's individual capacities" (ibid.).

In the case of the metabolic rift school, there are several inconsistencies and paradoxes that require attention. Perhaps a minor one is the insistence on Marx's concept of "irreparable metabolic rift" and then contradicting Marx by claiming that the rift is reparable or healable. The issue could be easily resolved by applying a dialectical materialist understanding to ecological change in terms of whether or not it is a reversible transformation. That is, ecosystems are historical processes as well. They change with or without humans and change into different sets of relations among organisms and physical processes. Hence, human impacts, not just under capitalism (as Engels also recalled for other modes of production), induce ecological changes that are permanent, that cannot be undone, even if often they may be imperceptible or negligible (at least over the short term). They are irreparable modifications if one views the permanent change as negative. A forest may return after being logged or burned, that is, the process may be reversible, but the forest never returns in the same way. One can determine whether it is liveable or not for people and other organisms pre-existing the logging or burn, but it would be folly to expect that it be repaired in the sense that the same sort of metabolic relations is re-established. Harvey (1996) had a more useful conceptualisation on this matter when he argued for understanding environmental impacts as forms of "ecological transformation." Some transformations are harmful to people or to other life forms (e.g., mountain top removal), or they may make a negligible difference (e.g., chopping down a couple of dying trees), or they may be life enhancing (e.g., the formation of dark earths). But they are all transformations and there are always transformations going on (change is constant), involving mutually affecting and directional changes for all species and physical processes involved in the interactions (see Engel-Di Mauro 2014). Capitalist impacts tend to be quite harmful, though not always. Under capitalism, the issue is transformation of a qualitatively different sort, identified by environmental scientists as unprecedented in terms of frequency, magnitude, scale, and cumulative effects. It is irreparable transformation in ways unparalleled compared to all other modes of production in human history. This is anyway a suggested way to resolve a contradictory Eco-Marxist view where the irreparable is claimed to be reparable.

There is a more serious, political problem than what to make of the concept of "irreparable metabolic rift." It is one of accusing both "Western" and "Soviet" Marxisms of suppressing Marx's ecological thought (e.g., Foster 2023), while, on the other hand, recognising advances in ecological thought by Marxists in countries like the USSR (e.g., Foster 2015a) and building on Western Marxists like Lukács and Mészáros (e.g., Saito 2022). How both suppression and further development of ecological thought have been simultaneously possible under "Soviet" Marxism or by using canonical Western Marxist sources remains to be explained. More perplexing is the implication that correct textual interpretation or exhumations of archived notes[25] would change the course of socialist politics or history. In relation to this border-line idealism is the paradox of thousands of dedicated Marxist thinkers over more than a century failing to recognise the centrality of metabolic rift or degrowth in Marx's framework. Thus far, resort has instead been made to an evidence-challenged conspiratorial narrative, ranging from invoking an ever-present (presumably capillary) institutional suppression to conjuring up wilful editorial interventions to distort Marx's ideas and thereby deceive all subsequent readers, as if Marxism were composed of adepts to a religion with a sacred original text. Such a low regard for other Marxists' intellectual abilities is especially surprising given that many Marxists, especially influential communists, have had little difficulty in developing explicitly ecological

[25] This is aside from equating notes with published manuscripts and note taking with theorisation.

theories on the destructive tendencies of capitalism. Even more importantly, many communists with institutional power in post-revolutionary societies have developed and implemented ecologically informed policies, in spite of great difficulties. They have arrived at such theories and policies not by performing philological acrobatics or by scouring unpublished notes, but by applying Marxist method, that is, dialectical and historical materialism. This is evident not only among many of the communist party leaders' writings and policies throughout the history of the USSR, but also in the work of such disparate figures as Amilcar Cabral (1951; see also César 2023), Italian Communist Party member Laura Conti (1977, 46–47; see also Barca 2019), and the above-discussed Marxist philosopher Howard Parsons (1978).

Multi-tendency Eco-Marxists may have spared themselves the insistence on textual analysis as a main, if not sole guide for a Marxist politics, but a greater explanatory effectiveness through "hybrid" theorisation does not seamlessly translate into any systematic or clear political platform, the sort necessary for political struggle in the world at large. Neither does it facilitate an understanding of how constructive environmental practices in socialist countries have come about without any concern for theoretical hybridity, but by sticking to basic Marxist principles reworked through and developed for different contexts. Neither does it contribute to developing or modifying scientific ecology, a main basis to detect deleterious changes in ecosystems and to figure out what sort of impacts or activities are ecologically constructive. Thus, there are yet unacknowledged chasms dividing Eco-Marxism from everyday socialist organising and struggles in capitalist countries, from environmental policy formation and implementation in socialist countries, and from scientific ecology in any country. A major challenge will be to refocus Eco-Marxist attention to the task of overcoming those chasms.

References

Albritton, Robert. 2019. *Eco-Socialism for Now and the Future.* Geneva: Springer International Publishing.

Aldridge, Casey. 2023. "Heroic Illusions of Nature in Revolt: Luxemburgian Ecosocialism Between William Morris and the Book of Exodus." *Capitalism Nature Socialism* 34 (2): 88–106

Altvater, Elmar. 1989. "Ecological and Economic Modalities of Time and Space." *Capitalism Nature Socialism* 1 (3): 59–71.

Amberger, Alexander. 2023. *Dissident Marxism and Utopian Eco-Socialism in the German Democratic Republic*. Leiden: Brill.

Anderson, Perry. 1976. *Considerations on Western Marxism*. Bristol: New Left Books.

Apostolopoulou, Evangelia, Elisa Greco, William M. Adams. 2018. "Biodiversity Offsetting and the Production of 'Equivalent Natures': A Marxist Critique." *ACME: An International E-Journal for Critical Geographies* 17 (3): 861–892.

Baer, Hans A. 2018. *Democratic Eco-Socialism as a Real Utopia: Transitioning to an Alternative World System*. New York: Berghahn Books.

Bahro, Rudolf. 1978. *The Alternative in Eastern Europe*. Translated by David Fernbach. Thetford: NLB.

Barca, Stefania. 2019. "Labour and the Ecological Crisis: The Eco-Modernist Dilemma In Western Marxism(s) (1970s–2000s)." *Geoforum* 98: 226–235.

Barrillon, Michel. 2013. «Les marxistes, Marx et la question naturelle. Notes sur l'improbable écomarxisme». *Écologie & politique* 47 (2): 115–143.

Barry, John. 1999. "Marxism and Ecology." In *Marxism and Social Science*, edited by A. Gamble, D. Marsh, and T. Tant, 259–279. London: Palgrave.

Benton, Ted. 1988. "Humanism = Speciesism. Marx on Humans and Animals." *Radical Philosophy* 50: 4–18.

Benton, Ted. 1989. "Marx and Natural Limits: An Ecological Critique and Reconstruction." *New Left Review* I (178): 178–186.

Benton, Ted. 1992. "Ecology, Socialism, and the Mastery of Nature: A Reply to Reiner Grundmann." *New Left Review* 194: 55–74.

Benton, Ted. 1993. *Natural Relations. Ecology, Animal Rights & Social Justice*. New York: Verso.

Benton, Ted, ed. 1996. *The Greening of Marxism*. New York: The Guilford Press.

Benton, Ted. 2023. "Socialism and Ecology." In *The Cambridge History of Socialism, Volume II*, edited by Marcel van der Linden, 664–686. Cambridge: Cambridge University Press.

Bergamo, Jacopo N. 2023. "Pandemic Capitalism: Metabolic Rift, World-Ecology Crossing Dialectical Biology." *Historical Materialism* 31 (1): 93–121.

Bernal, John D. 1937. Dialectical Materialism and Modern Science. *Science & Society* 2 (1): 58–66.

Betto, Frei, and Michael Löwy. 2010. "Ecosocialism and Spirituality." *Capitalism Nature Socialism* 21 (2): 87–99.

Blackledge, Paul. 2020. "Engels vs. Marx? Two Hundred Years of Frederick Engels." *Monthly Review* 72 (1): 21–39.

Brain, Stephen. 2010. "Stalin's Environmentalism." *The Russian Review* 69 (1): 93–118.

Boggs, Carl. 2020. "Was Karl Marx an Ecosocialist?" *Fast Capitalism* 17 (2): 67–94.

Boxley, Simon. 2022. "Green Marxism." In *Encyclopaedia of Marxism and Education*, edited by Alpesh Maisuria, 302–320. Leiden: Brill.
Breen, Sheryl D. 2014. "Green Views of Marx: Reinterpreting, Revising, Rejecting, Transcending." *SAGE Open* 4 (1). https://doi.org/10.1177/2158244013520609.
Brentlinger, John. 2010. "Revolutionizing Spirituality: Reflections on Marxism and Religion." In *Marx Today: Selected Works and Recent Debates*, edited by John F. Sitton, 241–256. New York: Palgrave Macmillan US.
Brien, Kevin M., and Arianna Bove. 2020. "Rethinking Marx and the Spiritual." In *Marx, Marxism and the Spiritual*, edited by Anjan Chakrabarti, Anup Dhar, and Serap A. Kayatekin, 15–28. New York: Routledge.
Brownhill, Leigh, and Terisa E. Turner. 2020. "Ecofeminist Ways, Ecosocialist Means: Life in the Post-capitalist Future." *Capitalism Nature Socialism* 31 (1): 1–14.
Buhle, Paul. 2013. "Rosa Luxemburg and the Birds." *Capitalism Nature Socialism* 24 (2): 55–57.
Bukharin, Nikolai. 1925. *Historical Materialism: A System of Sociology*. Moscow: International Publishers. https://www.marxists.org/archive/bukharin/works/1921/histmat/index.htm. Accessed 19 December 2023.
Burkett, Paul. 1999. *Marx and Nature. A Red and Green Perspective*. New York: St. Martin's Press.
Burkett, Paul. 2006. "Two Stages of Ecosocialism?" *International Journal of Political Economy* 35 (3): 23–45.
Büscher, Bram, Sian Sullivan, Katja Neves, Jim Igoe, and Dan Brockington. 2012. "Towards a Synthesized Critique of Neoliberal Biodiversity Conservation." *Capitalism Nature Socialism* 23 (2): 4–30.
Cabral, Amílcar. 1951 [1988]. "O Problema da Erosão do Solo. Uma contribuição para o seu estudo na região de Cuba (Alentejo)." In *Estudos Agrários de Amílcar Cabral*. Lisboa and Bissau: Instituto de Investigacao Cientifica Tropical.
Castree, Noel. 2000. "Marxism and the Production of Nature." *Capital & Class* 24 (3): 5–36.
César, Filipa. 2023. "Meteorizações: Agropoesia de Libertação de Amílcar Cabral." *digitAR-Revista Digital de Arqueologia, Arquitectura e Artes* 9: 200–214.
Chappell, John E. 1975. "The Ecological Dimension: Russian and American Views." *Annals of the Association of American Geographers* 65 (2): 144–162.
Chen, Hong-Rui. 2015. "The Influence of David Pepper's Ecological Theory of Socialism on China's Era." In *Energy, Environment and Green Building Materials: Proceedings of the 2014 International Conference on Energy, Environment and Green Building Materials (EEGBM 2014), November 28–30, 2014, Guilin, Guangxi, China*, edited by Ai Sheng, 447–450. Boca Raton: CRC Press.

Clark, Brett, and John B. Foster. 2021. "Marx's Ecology and Metabolic Analysis." In *The Routledge Handbook on Ecosocialism*, edited by Leigh Brownhill, Salvatore Engel-Di Mauro, Terran Giacomini, Ana Isla, Michal Löwy, and Terissa Turner, 89–98. New York: Routledge.

Clausen, Rebecca. 2007. "Healing the Rift: Metabolic Restoration in Cuban Agriculture." *Monthly Review* 59 (1): 40–52.

Clausen, Rebecca, Brett Clark, and Stefano B. Longo. 2015. "Metabolic Rifts and Restoration: Agricultural Crises and the Potential of Cuba's Organic, Socialist Approach to Food Production." *World Review of Political Economy* 6 (1): 4–32.

Commoner, Barry. 1971. *The Closing Circle: Nature, Man, and Technology*. New York: Knopf.

Conti, Laura. 1977. *Che Cos'è l'Ecologia. Capitale, Lavoro e Ambiente*. Milano: Gabriele Mazzotta.

Deléage, Jean-Paul. 1989. "Eco-Marxist Critique of Political Economy." *Capitalism Nature Socialism* 1 (3): 15–31.

Dickens, Peter. 1992. *Society and Nature. Towards a Green Social Theory*. Philadelphia: Temple University Press.

Dickens, Peter. 1996. *Reconstructing Nature. Alienation, Emancipation, and the Division of Labour*. London: Routledge.

Dussel, Enrique. 2020. "Liberation Theology and Marxism." In *Marx, Marxism and the Spiritual*, edited by Anjan Chakrabarti, Anup Dhar, and Serap A. Kayatekin, 138–162. Routledge.

Eckersley, Robyn. 1992. *Environmentalism and Political Theory. Toward an Ecocentric Approach*. London: UCL Press.

Ehara, Kei. 2023. "Reconstructing Marxian Theory of Ground Rent: Based on Japanese Development of Marxian Political Economy." *Capitalism Nature Socialism* 34 (4): 119–138.

Ekers, Michael, Gillian Hart, Stefan Kipfer, and Alex Loftus, eds. 2012. *Gramsci: Space, nature, politics*. Oxford: Wiley.

Engel-Di Mauro, Salvatore. 2014. *Ecology, Soils, and the Left: An Ecosocial Approach*. New York: Palgrave Macmillan.

Engel-Di Mauro, Salvatore. 2024. "Capitalism Nature Socialism." In *Encyclopedia of Critical Political Science*, edited by Clyde W. Barrow. Edward Elgar Publishing, Ltd.

Engels, Friedrich. 1876 [2010]. "The Part Played by Labour in the Transition from Ape to Man." In *Marx and Engels Collected Works, Volume 25*, 452–464. London: Lawrence & Wishart.

Enzensberger, Hans Magnus. 1974. "A Critique of Political Ecology." *New Left Review* NLR I (84): 3–31.

Faber, Daniel, and James O'Connor. 1988. "The Struggle for Nature: Environmental Crises and the Crisis of Environmentalism in the United States." *Capitalism Nature Socialism* 1 (2): 12–39

Feldman, Jonathan Michael. 2016. "Technology, Power and Social Change: Comparing Three Marx-Inspired Views." *Socialism and Democracy* 30 (2): 28–72.

Fitzsimmons, Margaret. 1989. "The Matter of Nature." *Antipode* 21: 106–120.

Foster, John B. 1992. "The Absolute General Law of Environmental Degradation Under Capitalism." *Capitalism Nature Socialism* 3 (3): 77–81.

Foster, John B. 1999a. "Marx's Theory of Metabolic Rift: Classical Foundations for Environmental Sociology." *American Journal of Sociology* 105 (2): 366–405.

Foster, John B. 1999b. *The Vulnerable Planet: A Short Economic History of the Environment*. New York: Monthly Review Press.

Foster, John B. 2000. *Marx's Ecology*. New York: Monthly Review Press.

Foster, John B. 2002. "II. Capitalism and Ecology: The Nature of the Contradiction." *Monthly Review* 54 (8). https://monthlyreview.org/2002/09/01/capitalism-and-ecology/. Accessed 17 November 2023.

Foster, John B. 2015a. "Late Soviet Ecology and the Planetary Crisis." *Monthly Review* 67 (2): 1–20.

Foster, John B. 2015b. "The Great Capitalist Climacteric." *Monthly Review* 67 (6): 1–18.

Foster, John B. 2015c. "Marxism and Ecology: Common Fonts of a Great Transition." *Monthly Review* 67 (7): 1–13.

Foster, John B. 2016. "Marxism in the Anthropocene: Dialectical Rifts on the Left." *International Critical Thought* 6 (3): 393–421.

Foster, John B. 2017. "The Return of Engels." *Monthly Review* 68 (10): 46–50.

Foster, John B. 2020a. "Misrepresenting Marx's Ecology. A Response to Daniel Tanuro's 'Was Marx an Ecosocialist?'" *Monthly Review*. https://mronline.org/2020/01/14/misrepresenting-marxs-ecology-a-response-to-daniel-tanuros-was-marx-an-ecosocialist/#_ednref4. Accessed 24 November 2023.

Foster, John B. 2020b. *The Return of Nature: Socialism and Ecology*. New York: Monthly Review Press.

Foster, John B. 2021. "What Is Monopoly Capital?" In *Global Political Economy: A Critique of Contemporary Capitalism*, 1st ed., edited by V. Upadhyay and P. Pramajit Singh, 242–252. New York: Routledge.

Foster, John B. 2023. "Engels and the Second Foundation of Marxism." *Monthly Review* 72 (2): 2–12.

Foster, John B., Brett Clark, and Richard York. 2011. *The Ecological Rift: Capitalism's War on the Earth*. New York: Monthly Review Press.

Foster, John B., and Paul Burkett. 2000. "The Dialectic of Organic/Inorganic Relations: Marx and the Hegelian Philosophy of Nature." *Organization & Environment* 13: 403–425.

Foster, John B., and Paul Burkett. 2016. *Marx and the Earth: An Anti-Critique*. Chicago: Haymarket.

Foster-Carter, Aidan. 1974. "Neo-Marxist Approaches to Development and Underdevelopment." In *Sociology and Development*, edited by Emmanuel de Kadt and Gavin Williams, 67–108. London: Tavistock.

Franco, Marco P. Vianna. 2021. "Ecological Utopianism in *Narodnik* Thought: Nikolay Chernyshevsky and the Redemption of Land." *Capitalism Nature Socialism* 32 (4): 24–42.

Frolov, Ivan T. 1986. "Genes or Culture? A Marxist Perspective on Humankind." *Biology and Philosophy* 1: 89–107.

Gare, Arran. 1993a. *Beyond European Civilization. Marxism, Process Philosophy, and the Environment*. Bungendore: Eco-Logical Press.

Gare, Arran. 1993b. "Soviet Environmentalism: The Path Not Taken." *Capitalism Nature Socialism* 4 (4): 69–88.

Gare, Arran. 1995. *Postmodernism and the Environmental Crisis*. London: Routledge.

Gare, Arran. 2012. "China and the Struggle for Ecological Civilization." *Capitalism Nature Socialism* 23 (4): 10–26.

Gare, Arran. 2021. "The Eco-socialist Roots of Ecological Civilisation." *Capitalism Nature Socialism* 32 (1): 37–55.

Gellert, Paul K. 2019. "Bunker's Ecologically Unequal Exchange, Foster's Metabolic Rift, and Moore's World-Ecology: Distinctions With or Without a Difference?" In *Ecologically Unequal Exchange*, edited by R. Scott Frey, Pauk K. Gellert, Harry F. Dahms, 107–140. Cham: Palgrave Macmillan.

Goldman, Marshall I. 1972. *The Spoils of Progress: Environmental Pollution in the Soviet Union*. Boston: MIT Press.

Gorz, André. 1974. Leur écologie et la nôtre. *Le Sauvage*, reproduced in http://www.ieb.be/IMG/pdf/leur-ecologie-et-la-notre_andre-gorz-1974-2.pdf. Accessed 27 December 2023.

Gorz, André. 1975. *Écologie Politique*. Paris: Galilée.

Grundmann, Reiner. 1991. "The Ecological Challenge to Marxism." *New Left Review* I (187): 103–120.

Guha, Ramachandra. 1988. "Ideological Trends in Indian Environmentalism." *Economic and Political Weekly* 23 (49): 2578–2581.

Gunderson, Ryan. 2011. "The Metabolic Rifts of Livestock Agribusiness." *Organization & Environment* 24 (4): 404–422.

Haila, Yrjö, and Richard Levins. 1992. *Humanity and Nature: Ecology, Science and Society*. London: Pluto Press.

Hammond, John L. 2014. "Mística, Meaning and Popular Education in the Brazilian Landless Workers Movement." *Interface* 6 (1): 372–391.

Haraway, Donna. 1988. "Situated Knowledges: The Science Question in Feminism and the Privilege of Partial Perspective." *Feminist Studies* 14 (3): 575–599.

Harribey, Jean-Marie. 2008. "Ecological Marxism or Marxian Political Ecology?" In *Critical Companion to Contemporary Marxism*, edited by Jacques Bidet and Stathis Kouvelakis, 187–207. Leiden: Brill.

Harvey, David. 1974. "Population, Resources, and the Ideology of Science." *Economic Geography* 50 (3): 256–277.

Harvey, David. 1996 *Justice, Nature and the Geography of Difference*. Oxford: Blackwell.

Harvey, David. 1998. "Marxism, Metaphors, and Ecological Politics." *Monthly Review* 49 (Fasc. 11): 17–31.

Haug, Timothée. 2022. *La rupture écologique dans l'oeuvre de Marx : analyse d'une métamorphose inachevée du paradigme de la production*. [Ecological Rupture in Marx's Work: Analysis of an Unfinished Metamorphosis in the Paradigm of Production] Philosophie. Université de Strasbourg, 2022. In French, https://theses.hal.science/tel-03774950/document. Accessed 24 November 2023.

Henderson, George. 2016. "Marxist Political Economy and the Environment." In *A Companion to Environmental Geography*, edited by Noel Castree, David Demeritt, Diane Liverman, and Bruce Rhoads, 266–293. Oxford: Wiley.

Holleman, Hannah. 2015. "Method in Ecological Marxism: Science and the Struggle for Change." *Monthly Review* 67 (Fasc. 5): 1–10.

Huan, Qingzhi. 2010. "Eco-socialism in an Era of Capitalist Globalisation: Bridging the West and the East." In *Eco-Socialism as Politics: Rebuilding the Basis of Our Modern Civilisation*, edited by Qingzhi Huan, 3–12. Dordrecht: Springer.

Huan, Qingzhi. 2014. "Development of the Red–Green Environmental Movement in China: A Preliminary Analysis." *Capitalism Nature Socialism* 25 (3): 45–60.

Huan, Qingzhi. 2020. "Socialist Eco-Civilization as a Transformative Politics." *Capitalism Nature Socialism*. https://doi.org/10.1080/10455752.2020.1802498

Huan, Qingzhi. 2022. "Ecosocialism and Socialism with Chinese Characteristics." In *The Routledge Handbook on Ecosocialism*, edited by Leigh Brownhill, Salvatore Engel-Di Mauro, and Terran Giacomini, Ana Isla, Michael Löwy, and Turner Terisa, 339–345. Oxon and New York: Routledge.

Huber, Matthew T. 2013. *Lifeblood: Oil, freedom, and the Forces of Capital*. Minneapolis: University of Minnesota Press.

Hughes, Jonathan. 2000. *Ecology and Historical Materialism*. Cambridge: Cambridge University Press.

Jakobsen, Jostein. 2022. "Beyond Subject-Making: Conflicting Humanisms, Class Analysis, and the "Dark Side" of Gramscian Political Ecology." *Progress in Human Geography* 46 (2): 575–589.

Katz, Cindi, and Andrew Kirby. 1991. "In the Nature of Things: The Environment and Everyday Life." *Transactions of the Institute of British Geographers* 16 (3): 259–271.

Kovel, Joel. 1991. *History and Spirit: An Inquiry into the Philosophy of Liberation*. Boston: Beacon Press.

Kovel, Joel. 1995. "Ecological Marxism and Dialectic." *Capitalism Nature Socialism* 6 (4): 31–50.

Kovel, Joel. 2001. "A Materialism Worthy of Nature." *Capitalism Nature Socialism* 12 (2): 73–84.

Kovel, Joel. 2005. "The Ecofeminist Ground of Ecosocialism." *Capitalism Nature Socialism* 16 (2): 1–8.

Kovel, Joel. 2007. *The Enemy of Nature. The End of Capitalism or the End of the World*. London: Zed Books.

Kovel, Joel. 2014. "Ecosocialism as a Human Phenomenon." *Capitalism Nature Socialism* 25 (1): 10–23.

Lamy, Jérôme. 2016. «Les palimpsestes de Marx. L'émergence de la sociologie marxiste de l'environnement aux États-Unis» *Écologie & politique* 53 (2): 149–164.

Layfield, David. 2008. *Marxism and Environmental Crises*. Bury: Arena Books.

Lee, Keekok. 1989. *Social Philosophy and Ecological Scarcity*. London: Routledge.

Leff, Enrique. 1993. "Marxism and the Environmental Question: From the Critical Theory of Production to an Environmental Rationality for Sustainable Development." *Capitalism Nature Socialism* 4 (1): 44–66.

Levins, Richard, and Richard Lewontin. 1985. *The Dialectical Biologist*. Cambridge: Harvard University Press.

Levins, Richard, and Richard Lewontin. 2007. *Biology Under the Influence. Dialectical Essays on Ecology, Agriculture, and Health*. New York: Monthly Review.

Li, Zhiyan. 2021. "Some Thoughts on Ecological Marxism." *Open Journal of Social Sciences* 9: 212–219.

Liu, Yunshan. 2018. "Ecological Significance in Marx's Theory of Metabolic Rift and Targeted Poverty Alleviation in Rural Areas." *Asian Agricultural Research* 10 (8): 21–23, 27.

Losurdo, Domenico. 2017. *Il marxismo occidentale: Come nacque, come morì, come può rinascere*. Roma: Laterza.

Löwy, Michael. 2015. *Ecosocialism. A Radical Alternative to Capitalist Catastrophe*. Chicago: Haymarket Books.

Löwy, Michael. 2017. "Marx, Engels, and Ecology." *Capitalism Nature Socialism* 28 (2): 10–21.

Löwy, Michael, and Daniel Tanuro. 2021. "Introduction. Le XXIe Siècle Sera le Siècle de la Question Écosociale." In *Luttes Écologiques et Sociales dans le Monde. Allier le Vert et le Rouge*, edited by Michael Löwy and Daniel Tanuro, 7–35. Paris: Les Éditions Textuel.

Löwy, Michael, and Daniel Tanuro. 2022. "The Fourth International's Contribution to Ecosocialism." In *The Routledge Handbook on Ecosocialism*, edited by Leigh Brownhill, Salvatore Engel-Di Mauro, Terran Giacomini, Ana Isla, Michal Löwy, and Terissa Turner, 346–356. New York: Routledge.

Magdoff, Fred, and Chris Williams. 2017. *Creating an Ecological Society: Toward a Revolutionary Transformation*. New York: Monthly Review Press.

Malm, Andreas. 2016. *Fossil Capital: The Rise of Steam Power and the Roots of Global Warming*. New York: Verso.

Malm, Andreas. 2018. *The Progress of This Storm: Nature and Society in a Warming World*. New York: Verso Books.

Malm, Andreas. 2021. *How to Blow Up a Pipeline*. New York: Verso Books.

Mann, Geoff. 2009. "Should Political Ecology Be Marxist? A Case for Gramsci's Historical Materialism." *Geoforum* 40 (3): 335–344.

Mann, Geoff, and Joel Wainwright. 2018. *Climate Leviathan: A Political Theory of Our Planetary Future*. New York: Verso Books.

Marx, Karl. 1978 [1844]. "Economic and Political Manuscripts of 1844." In *The Marx-Engels Reader*, 2nd ed., edited by Robert C. Tucker, 66–126. New York: W.W. Norton and Company.

Marx, Karl. 1932 [1845–1846]. "The German Ideology. Critique of Modern German Philosophy According to Its Representatives Feuerbach, B. Bauer and Stirner, and of German Socialism According to Its Various Prophets." https://www.marxists.org/archive/marx/works/1845/german-ideology/index.htm.

Marx, Karl. 1990 [1867]. *Capital. A Critique of Political Economy. Volume One*. Translated by Ben Fowkes. London: Penguin Books.

Marx, Karl. 1998 [1883]. *Capital. A Critique of Political Economy. Volume 3. Marx & Engels Collected Works Vol 37*. London: Lawrence & Wishart.

Mattley, Ian M. 1966. "The Marxist Approach to the Geographical Environment." *Annals of the Association of American Geographers* 56 (1): 97–111.

Merchant, Carolyn. 1980. *The Death of Nature. Women, Ecology, and the Scientific Revolution*. New York: HarperCollins.

Moore, Jason W. 2000. "Environmental Crises and the Metabolic Rift in World-Historical Perspective." *Organization & Environment* 13 (2): 123–157.

Moore, Jason W. 2011. "Transcending the Metabolic Rift: A Theory of Crises in the Capitalist World-Ecology." *The Journal of Peasant Studies* 38 (1): 1–46.

Moore, Jason W. 2015. *Capitalism in the Web of Life: Ecology and the Accumulation of Capital*. New York: Verso Books.

Moore, Jason W. 2017. "Metabolic Rift or Metabolic Shift? Dialectics, Nature, and the World-Historical Method." *Theory and Society* 46: 285–318.

Napoletano, Brian M., John B. Foster, Brett Clark, Pedro S. Urquijo, Michael K. McCall and Jaime Paneque-Gálvez. 2019. "Making Space in Critical Environmental Geography for the Metabolic Rift." *Annals of the American Association of Geographers* 109 (6): 1811–1828.

Napoletano, Brian M., Pedro S. Urquijo, Brett Clark, and John B. Foster. 2023. "Identifying the Revolutionary Agent in the Radical Project of Autogestion." *Dialogues in Human Geography* 13 (3): 468–472.

O'Connor, James. 1988. "Capitalism, Nature, Socialism. A Theoretical Introduction." *Capitalism Nature Socialism* 1 (1): 11–38.

O'Connor, James. 1991. "On the Two Contradictions of Capitalism." *Capitalism Nature Socialism* 2 (3): 107–109.

O'Connor, James. 1998. *Natural Causes: Essays Ecological Marxism*. New York: Guilford.

Ofreneo, Rosalinda Pineda. 2022. "Marxism and Ecofeminism in the Era of Climate Change: Convergence and Divergence." *Jurnal IndoPROGRESS* II (1): 31–47.

Parsons, Howard. 1978. *Marx and Engels on Ecology*. Westport: Greenwood Press.

Peet, Richard. 1979. "Contradiction and Marxist Geography." *Annals of the Association of American Geographers* 69 (1): 164–169.

Pepper, David 1984. *The Roots of Modern Environmentalism*. London: Croom Helm.

Pepper, David. 1993. *Eco-Socialism. From Deep Ecology to Social Justice*. London: Routledge.

Plumwood, Val. 1993. *Feminism and the Mastery of Nature*. London: Routledge.

Pryde, Philip R. 1972. *Conservation in the Soviet Union*. Cambridge: Cambridge University Press.

Redclift, Michael. 1984. *Development and the Environmental Crisis: Red or Green Alternatives*. London: Routledge.

Ren, Qiaohua, and Caisheng Lin. 2019. "John Bellamy Foster's Ecological Marxism and the Enlightenment to China's Ecological Civilization." *Advances in Social Science, Education and Humanities Research* 376: 702–705.

Rockhill, Gabriel. 2023. "The Myth of 1968 Thought and the French Intelligentsia: Historical Commodity Fetishism and Ideological Rollback." *Monthly Review* 75 (2). https://monthlyreview.org/2023/06/01/the-myth-of-1968-thought-and-the-french-intelligentsia-historical-commodity-fetishism-and-ideological-rollback/. Accessed 17 December 2023.

Roos Andreas, and Alf Hornborg. 2024. "Technology as Capital: Challenging the Illusion of the Green Machine." *Capitalism Nature Socialism*. https://doi.org/10.1080/10455752.2024.2332218

Royle, Camilla. 2021. "Ecological Marxism." In *Routledge Handbook of Marxism and Post-Marxism*, edited by Alex Callinicos, Stathis Kouvelakis, and Lucia Pradella, 443–450. New York: Routledge.

Rudy, Alan P. 2019. "On Misunderstanding the Second Contradiction Thesis." *Capitalism Nature Socialism* 30 (4): 17–35.

saed. 2019. "Nature Is Beyond Value Because We Are Part of Nature." *Capitalism Nature Socialism* 30 (2): 143–156.

Saito, Kohei. 2014. "The Emergence of Marx's Critique of Modern Agriculture: Ecological Insights from His Excerpt Notebooks." *Monthly Review* 66 (5): 25.

Saito, Kohei. 2017a. *Karl Marx's Ecosocialism: Capital, Nature, and the Unfinished Critique of Political Economy*. New York: Monthly Review Press.

Saito, Kohei. 2017b. "Marx in the Anthropocene: Value, Metabolic Rift, and the Non-cartesian Dualism." *Zeitschrift für kritische Sozialtheorie und Philosophie* 4 (1–2): 276–295.

Saito, Kohei. 2021. "The Legacy of Karl Marx's Ecosocialism in the Twenty-First Century." In *The Routledge Handbook on Ecosocialism*, edited by Leigh Brownhill, Salvatore Engel-Di Mauro, Terran Giacomini, Ana Isla, Michal Löwy, and Terissa Turner, 346–356. New York: Routledge.

Saito, Kohei. 2022. *Marx in the Anthropocene: Towards the Idea of Degrowth Communism*. Cambridge: Cambridge University Press.

Salleh, Ariel. 1995. "Nature, Woman, Labor, Capital: Living the Deepest Contradiction." *Capitalism Nature Socialism* 6 (1): 21–39.

Salleh, Ariel. 2005. "Moving to an Embodied Materialism." *Capitalism Nature Socialism* 16 (2): 9–14.

Salleh, Ariel. 2009. "Ecological Debt: Embodied Debt." In *Eco-sufficiency and Global Justice: Women Write Political Ecology*, edited by Ariel Salleh, 1–41. Melbourne: Spinifex.

Salleh, Ariel. 2010. "From Metabolic Rift to 'Metabolic Value': Reflections on Environmental Sociology and the Alternative Globalization Movement." *Organization & Environment* 23 (2): 205–219.

Salleh, Ariel. 2017. *Ecofeminism as Politics: Nature, Marx and the Postmodern*, 2nd ed. London: Zed Books.

Satgar, Vishwas. 2018. "The Climate Crisis and Systemic Alternatives." In *The Climate Crisis: South African and Global Democratic Eco-Socialist Alternatives*, edited by Vishwas Satgar, 10–16. Johannesburg: Wits University Press.

Schmidt, Alfred. 1971. *The Concept of Nature in Marx*. Translated by Ben Fowkes. London: NLB.

Schneider, Mindi and Philip McMichael. 2010. “Deepening, and Repairing, the Metabolic Rift.” *The Journal of Peasant Studies* 37 (3): 461–484.

Schwartzman, David. 1996. “Solar Communism.” *Science & Society* 60 (3): 307–331.

Schwartzman, David. 2009. “Ecosocialism or Ecocatastrophe?” *Capitalism Nature Socialism* 20 (1): 6–33.

Schwartzman, David. 2011. “Green New Deal: An Ecosocialist Perspective.” *Capitalism Nature Socialism* 22 (3): 49–56.

Schwartzman, David, and Salvatore Engel-Di Mauro. 2022. “Prefiguration and the Emergence of the Global Subject.” *Science & Society* 86 (4): 564–583.

Schwartzman, David, and Quincy Saul. 2015. “An Ecosocialist Horizon for Venezuela: A Solar Communist Horizon for the World.” *Capitalism Nature Socialism* 26 (3): 14–30.

Sheehan, Helen. 1985. *Marxism and the Philosophy of Science: A Critical History.* Atlantic Highlands, NJ: Humanities Press.

Skirbekk, Gunnar. 1973. “Ecology and Marxism.” *Praxis* 2–3: 283–290. https://www.marxists.org/subject/praxis/praxis-international-edition/1973-2-3/skirbekk.htm. Accessed 23 November 2023.

Skirbekk, Gunnar. 1994. “Marxism and Ecology.” *Capitalism Nature Socialism* 5 (4): 95–104.

Smith, David N. 2002. “Accumulation and the Clash of Cultures: Marx’s Ethnology in Context.” *Rethinking Marxism* 14 (4): 73–83.

Smith, Neil. 1984. *Uneven Development: Nature, Capital, and the Production of Space.* Oxford: Blackwell.

Smith, Neil, and Phil O’Keefe. 1980. “Geography, Marx and the Concept of Nature.” *Antipode* 12: 30–39.

Stahnke, Ben. 2020. “Lenin, Ecology, and Revolutionary Russia.” *Peace, Land, and Bread* 1: 63–80.

Stoner, Alexander M. 2021. “Critical Reflections on America’s Green New Deal: Capital, Labor, and the Dynamics of Contemporary Social Change.” *Capitalism Nature Socialism* 32 (2): 77–94.

Suing, Guillaume. 2018. *L’Écologie Réelle. Une Histoire Soviétique et Cubaine.* Paris: Éditions Delga.

Tanuro, Daniel. 2010. “Marxism, Energy, and Ecology: The Moment of Truth.” *Capitalism Nature Socialism* 21 (4): 89–101.

Tanuro, Daniel. 2020a. *Trop Tard our Être Pessimiste! Écosocialisme ou Efondrement*. Paris: Textuel.

Tanuro, Daniel. 2020b. Was Marx an Ecosocialist? A Reply to Kohei Saito. *International Viewpoint*. https://internationalviewpoint.org/spip.php?article6358.

Tello, Enric. 2016. “Manuel Sacristán at the Onset of Ecological Marxism After Stalinism.” *Capitalism Nature Socialism* 27 (2): 32–50.

Vettese, Troy, and Drew Pendergrass. 2022. *Half-earth Socialism: A Plan to Save the Future from Extinction, Climate Change and Pandemics*. New York: Verso Books.

Vlachou, Andriana. 1994. "Reflections on the Ecological Critiques and Reconstructions of Marxism." *Rethinking Marxism* 7 (3): 112–128.

Vogel, Steven. 1988. "Marx and Alienation from Nature." *Social Theory and Practice* 14 (3): 367–387.

Wall, Derek. 2010. *The Rise of the Green Left. Inside the Worldwide Ecosocialist Movement*. London: Pluto Press.

Wallis, Victor. 2001. "Letters." *Capitalism, Nature and Socialism* 12 (4): 133

Wallis, Victor. 2016. "Socialism and Technology: A Sectoral Overview." In *Eco-Socialism as Politics*, edited by Qingzhi Huan, 45–62. Dordrecht: Springer.

Wallis, Victor. 2018. *Red-Green revolution: The Politics and Technology of Ecosocialism*. Toronto: Political Animal Press.

Wallis, Victor. 2020. *Socialist Practice: Histories and Theories*. London: Palgrave Macmillan.

Wang, Zhihe, Meijun Fan, Hui Dong, Dezhong Sun, and Lichun Li. 2013. "What Does Ecological Marxism Mean for China? Questions and Challenges for John Bellamy Foster." *Monthly Review* 64 (Fasc. 9): 47–53.

Wang, Zhihe, Huili He, and Meijun Fan. 2014. "The Ecological Civilization Debate in China: the Role of Ecological Marxism and Constructive Postmodernism-Beyond the Predicament of Legislation." *Monthly Review* 66 (6): 37–59.

Weiner, Douglas R. 1999. *A Little Corner of Freedom: Russian Nature Protection from Stalin to Gorbachëv*. Berkeley: University of California Press.

Weston, Joe, ed. 1986. "Introduction." In *Red and Green: A New Politics of the Environment*, edited by Joe Weston, 1–8. London: Pluto Press.

White, Damian, Alan Rudy, and Brian Gareau. 2016. *Environments, Natures and Social Theory: Towards a Critical Hybridity*. New York: Palgrave.

White, Damian F., Brian J. Gareau, and Alan P. Rudy. 2017. "Ecosocialisms, Past, Present and Future: From the Metabolic Rift to a Reconstructive, Dynamic and Hybrid Ecosocialism." *Capitalism Nature Socialism* 28 (2): 22–40.

Williams, Chris. 2010. *Ecology and Socialism: Solutions to Capitalist Ecological Crisis*. Chicago: Haymarket Books.

WinklerPrins, Antoinette M.G.A. 2014. "Terra Preta—The Mysterious Soils of the Amazon." In *The Soil Underfoot: Infinite Possibilities for a Finite Resource*, edited by G. Jock Churchman and Edward R. Landa, 235–247. Boca Raton: CRC Press.

Worthington, Richard. 1984. "Socialism and Ecology: An Overview." *New Political Science* 5 (1): 69–83.

Zhang, Ning. 2022. "How the Legitimacy of Marx's Ecology Is Possible: Demonstrating from the Perspective of Factuality and Effectiveness?" *TRAMES* 26 (2): 169–183.

Zhang Yunfei. 2019. "On the Historical Position of Ecological Civilization." *Capitalism Nature Socialism* 30 (1): 11–25.

Zhang, Yunfei. 1986. 生态伦理学初探 [A Preliminary Study of Ecological Ethics]. 内蒙古社会科学 [*Inner Mongolia Social Sciences*], 4: 47–51. Reproduced in *Capitalism Nature Socialism* 32 (1): 27–36.

Zimmerer, Karl S., and Kenneth R. Young, eds. 1998. *Nature's Geography: New Lessons for Conservation in Developing Countries*. Madison: University of Wisconsin Press.

CHAPTER 6

Ecofeminist Critique and Ecofeminist Ecosocialism

"an ecosocialism that is not feminist is not a socialism!" (Isla 2006, 65).

An Introduction to Ecofeminism

From the previous chapter, it may be evident that gender relations do not figure prominently in much Eco-Marxist writing, nor in wider ecosocialist ones. Yet, some decade prior to the introduction of the idea of Ecosocialism and while a red-green politics was experiencing its inchoate steps, a radical articulation of environmentalism and leftwing feminism—Ecofeminism—had already taken place. The chronological sequence alone might raise some questions regarding the lack of substantive engagement of Eco-Marxism and Ecosocialism more generally with Ecofeminist thought. In fact, one could say that the continued Eco-Marxist emphasis on re-reading mainly Marx's manuscripts and notebooks and on revisiting individual socialist luminaries' texts or concepts has overshadowed investigations over the relationship of the ecosocial crisis to the multiple forms of oppression constitutive of capitalist society, a theme dear to, among others, Marxist and Ecosocialist Ecofeminists (Brownhill and Turner 2020; Kaara 2010; Mallory 2006; Salleh 1997).

One form of oppression in particular—patriarchy—still finds too little resonance in the Eco-Marxist world and arguably ecosocialism more broadly. Patriarchy is a system where people are forced into different

S. Engel-Di Mauro, *Ecosocialism*, Marx, Engels, and Marxisms,
https://doi.org/10.1007/978-3-031-74288-0_6

social roles based on culturally highlighted differences linked to biological reproduction and associated biological traits. It is usually a binary female-male gender system, where anyone or anything associated with femaleness is viewed socially as inferior to anyone or anything associated with maleness. Patriarchy is a set of social institutions privileging maleness and largely, though not exclusively, benefiting men economically (Buckingham 2020; Salleh 1995).

In Ecosocialist works, it is rare to find patriarchy featured as central to any analysis. As remarked above, Salleh's ideas on meta-industrial labour's potential to reduce the metabolic rift (Salleh 2010) have met negligible to no interest among Eco-Marxists. Instead, Foster and associates, for example, opt for the concept of environmental proletariat, whose commonality is defined by the experience of environmental degradation and consequent potential for organising against capitalism. In any case, with few exceptions (e.g., Kovel 2005; Norgaard and York 2005), it is Ecofeminists who have primarily taken up the task of investigating the gendered character and causes of the ecosocial crisis (e.g., Diamond and Orenstein 1990; Griffin 1978; King 1983; Mellor 1997; Merchant 1980; Mies and Shiva 1993; Plumwood 1993; Salleh 1997). Given the still prevailing gendered division of scholarship, perhaps it is to be expected that it would be a feminist approach that would focus on the connections between patriarchy and the ecological crisis.

There is in any case a historical and material basis to pay attention to gender relations and gender-based modes of oppression. Environmental degradation affects women disproportionately in terms of personal health and workload (such as care work), especially working-class and racially minoritised women. In contrast, it is wealthier men who tend to have more decision-making power over environmental policies and over activities eventuating in substantive and often negative environmental impacts. Gendered economic inequalities combine with other modes of oppression to concentrate women in lower social ranks in their respective communities. Gender-based inequalities, buttressed by violence, have been repeatedly found to explain a tendency for women's greater ecological sensitivity and propensity for leading or initiating environmental struggles (Bell 2016; Buckingham 2020; Mohai 1997; Perkins 2013; Tindall et al. 2003). The linkage between patriarchal relations and environmental degradation forms a main reason for the development of Ecofeminism and, more recently of Ecosocialist Ecofeminism.

Ecofeminist Ideas, Debates, and Contributions

There are multiple meanings and politics to Ecofeminism, with a level of contentiousness that at times resembles that of Eco-Marxism, but which will not be fully explored here (see, instead, Buckingham 2020; Carlassare 1994; Foster 2021; Gaard 2017; Oksala 2018). There are several kinds of Ecofeminism, ranging from social materialist (including Marxist) to spiritualistic and even liberal strands. Closest to Ecosocialism are forms of Ecofeminism that call attention to gendered forms of social inequalities or patriarchy in relation to class differentiation and other forms of oppression in capitalism (Warren 2000, 21).

A brief historical overview could be helpful in highlighting the main ideas espoused by Ecofeminists so as to make more obvious overlaps and differences with Ecosocialism. By at least the early 1970s, feminists were connecting with environmental concerns, mainly by way of Socialist Feminism. Some have shown the linkages to have occurred already by the 1890s (Adams and Gruen 2014). The convergence of specifically Socialist or working-class Feminism with environmentalism, including sensitivity to other animals' welfare, eventually inspired what has come to known as Ecofeminism (Mellor 2022).

The term Ecofeminism came about from the milieu of French communist feminists, as expounded by Françoise d'Eaubonne (1920–2005). A member of the Communist Party of France between 1945 and 1957, she became from the 1970s onwards an influential activist in the fight against heteronormativity, making what would now be called LGBTQ struggles a centrepiece of her politics. She coined the term Ecofeminism in the early 1970s, a concept merging ecologism (environmentalism) and feminism. It was included in a collection of essays (*Le Féminisme ou la Mort*) where she also introduced the notion of *phallocracy* to describe a male-domination system (D'Eaubonne 1974).

Her main thesis was that there are parallels between the patriarchal suppression of women and the suppression of nature, and that such suppression leads to environmental destruction. As Oksala suggests, this destructive tendency can be more accurately understood relative to more recent developments in terms of both exploitation (formal subsumption) and capitalisation through physical manipulation (real subsumption), two processes that may contradict or magnify each other (Oksala 2018, 226–227). To overcome patriarchy, d'Eaubonne advocated for the abolition of the nuclear family, the end of heteronormativity, autonomous councils as basic political decision-making units, and ecologically friendly land management. She surmised that women's and men's perspectives on nature are not necessarily different because women, too, are socialised into patriarchy. The importance of how D'Eaubonne formulated her thesis cannot be sufficiently emphasised, given the oft-repeated charge of essentialism levelled at Ecofeminist thinkers.[1] From its inception, Ecofeminism was defined by an understanding of patriarchy as a social phenomenon that is not reducible to any one social group.

Out of this insight on the relationship between patriarchy and Nature have emerged diverse Ecofeminist currents. One might expect that Ecofeminist schools of thought would be at odds with each other. In fact, for the most part, the diversity of Ecofeminism is encouraged because differences in subjectivity are valued, or at least that is the intention. Different kinds of Ecofeminism have also developed that tend to be parallel to the various strands of feminism, including those liberal strands that are not necessarily antagonistic to capitalism. What unifies all Ecofeminists is that a single perspective is impossible and undesirable because people always have different subject positions. There is also an emphasis on the existence of multiple forms of oppression (racism, patriarchy, homophobia, ablism, agism, speciesism, etc.) and about striving to overcome them. Whether all Ecofeminist currents are mutually reconcilable is an open question (an example of this not being so is shown in Gaard 2017), but the boundaries of what counts as Ecofeminism are relatively clear in terms of the tenet that patriarchy leads inevitably to harming women and the rest of nature (Buckingham 2020; Mellor 1994; Plumwood 2001; Turner and Brownhill 2006).

Nevertheless, it is peculiar that Ecofeminism was shorn of its socialist roots by the 1980s, especially in North America (e.g., Adams and Gruen 2014; Valera 2017; Warren 2000; Warren and Erkal 1997),[2] where some Ecofeminist currents developed and promoted outwardly essentialist notions (Gaard 2017; Oksala 2018) that plainly contradict D'Eaubonne's

[1] Essentialism, as applied to society, is a belief that categories of people (say, females and males) have fixed or unchangeable characteristics.

[2] The editors of the volume in question have achieved a most remarkable feat in erasing d'Eaubonne entirely from the history of Ecofeminism. Within the same edited work, in classical liberal fashion, class is reduced to a "socioeconomic" category and capitalism is treated only in terms of consumption patterns.

original framing. Cultural Ecofeminism, a main example, has at its basis the argument that women are intrinsically closer to nature because of their biological traits and roles. This strand of Ecofeminism, heavily critiqued by and often dismissed by other feminists (Carlassare 1994), came to enjoying a perhaps exaggerated prominence in core capitalist countries and among white feminist communities by the 1980s (even at the United Nations, to some extent).

Regardless, Materialist Ecofeminists could also be accused of essentialism when they assert "that across history and cultures women's work has formed a central element of the 'underlaboring' work that makes human society possible" (Dordoy and Mellor 2000, 46).[3] Similarly, it could be taken as an essentialising argument to state that because "the great majority of women around the world are organically and discursively implicated in life-affirming activities, and ... develop gender-specific knowledges grounded in this material base," "women across cultures" articulate perspectives on the environment very different from those of men (Salleh 2003, 67). These kinds of sweeping claims, not uncommon in Ecofeminist literature, are contradicted by instances of gender egalitarian cultures in both present and past. Moreover, just as in the case of gendered oppression, the gender binary (female-male) is, thankfully, not a universal phenomenon (Goettner-Abendroth 2018; Isife 2023; Leacock 1992). Ecofeminists doubtless recognise this historical variability and geographical variety as well, so the problem of essentialism seems largely a displacement of the question of how Ecofeminism relates to socialism, if at all, and what kinds of politics are to be favoured to achieve what aims (see, e.g., Dordoy and Melor 2001, 96).

In part, the concern over the essentialist ideas promulgated through Cultural Ecofeminism is traceable to the contextual differences between the mainly academic critics and the often activism-based articulations of slogans or programmes. Often, "essentialism" is strategic, a first attempt to counter predominant masculinist ideas. It has been an appropriation

[3] Compare to a subsequent remark by the same authors, responding to John Clark's notion of humanity-ecology dialectics: "The expression of the dialectic between humanity and its physicality can and does, however, vary substantially across history and cultures" (Dordoy and Mellor 2001, 93). They further state that women's "mediating between nature and culture" is socially constructed (ibid., 96). It should follow that the gendered division of labour and the relative weight of women's work in sustaining society varies "across history and cultures." This would also be consistent with those scholars' commitment to a materialist dialectics.

and subversion of ideas about women being deemed closer to nature and therefore inferior. A common example in political practice is the use of the concept of "motherhood" to get women to be listened to and to include people not considered mothers, which goes against traditional roles and definitions of womanhood (Buckingham 2020, 47–49; Godfrey 2005; Mallory 2006).

Put differently, essentialism, deployed in direct political struggle, can be a form of resistance, a context-specific strategy related to material circumstances. This is a way in which Ecofeminists have argued the matter (e.g., Godfrey 2005; Mies and Shiva 1993; Salleh 2003). For example, if women are deemed closer to nature (understood as external), then they ought to be regarded as more sensitive to nature than men. Therefore, it can be argued, on the basis of prevailing patriarchal logic, that women must be in leadership regarding environmental issues, not men. Many women activists use this strategy because of its greater political effectiveness in the patriarchal societies in which they live. Such strategy contributes more towards turning patriarchal ideology on its head, rather than reproducing it. At the same time, other social givens and environmental practices can be called into question (see also the "progressive naturalism" as articulated in Plumwood 2001).

The No MUOS movement in Sicily is one of numerous examples. Since 2009, women in eastern Sicily have led protests explicitly as mothers, with the aim of bringing about the closure of the US military's Mobile User Objective System station in Niscemi. The station is cited as a source of carcinogenic risks and environmental damage in the nearby town and natural preserve. All at once, a strategy that is on the surface essentialising about gender roles not only undermines patriarchal beliefs in women's political passivity, but also even brings to light the connection between environmental degradation and militarism, and, not far behind, US imperialism (Benadusi et al. 2021; della Porta and Piazza 2016).

Pointing to similarities in gendered outcomes in different places in terms of women taking key roles in environmental struggles, however, does not bring to light the mechanisms whereby this happens except in such an abstract sense as to create a theory that covers everything and thereby nothing. Nor can taking stock of such gender-differentiated strategies explain the many cases where women do not take up such strategies or why gender would be a main constitutive process instead of or in a less explanatory way than racialised class relations. This is where

Marxist or Ecosocialist Ecofeminists (see below) have been able to articulate a more encompassing and theoretically more explanatory framework. That is, the gendered nature of environmental impacts and repercussions does not emerge from some transhistorical patriarchal mode of domination, but out of the specificity of capitalist relations in connection with reproductive labour, for which mainly women are exploited (Buckingham 2020; Oksala 2018, 220).

This critique notwithstanding, Ecofeminists have made several salient contributions, aside from developing practicable strategies. One is exposing masculinism in philosophy and science, explaining the intellectual origins of patriarchal dualism. It is a dichotomous view of the world, where one side is inferior to the other (Plumwood 1993). The dualism can also be expressed as the ideological syllogism that since nonhuman nature is inferior to the human, women, being closer to nonhuman nature, are therefore an inferior human relative to men. It is an ideology that is wider than gender relations and used to justify domination, exploitation, and oppression of women, the colonised, the racialised, workers, the rest of nature, etc. Dualistic ideology is an outcome of patriarchal and other oppressive relations in the production of knowledge. Women, as well as racially minoritised people more broadly, have only recently been included in science and philosophy in most European societies, with the problem persisting in the biophysical and technical sciences, with the notable exception of socialist countries. This is the material basis for masculinism in the production of scientific and philosophical knowledge, which is that male experiences serve as default viewpoint, as normality. The same kind of process pertains to the exclusion of people deemed not white. The result is biased science or philosophy, as feminist and other critical scholars have long demonstrated (Buckingham 2020).

Another Ecofeminist contribution is showing the limits of liberal feminism as ultimately exclusive of most women. As well known to Marxists, among the foundational principles of liberalism are formal (not substantive) equality, equal opportunity, individualism (individual rights above community needs; focusing on individual blame for problems instead of examining social causes), and individual private property, including over the means to live. Liberalism, though, is a denial of basic reality. We are always part of and cannot survive without a community that is also always within an ecosystem. We are never in the same situation or set of conditions, especially in societies founded on multiple forms of oppression.

Treating people as if they are equal or calling for equal rights or opportunities ignores the material basis (including environmental) and systemic causes of social inequalities. Ecofeminism, like Marxism, is premised on analysing people's material conditions, rather than formal disembodied rights, freedoms, and the rest (see also Clark 2000).

These flaws extend to Liberal Feminism. Marxist and Materialist Ecofeminists point out that calling for formal gender equality is like asking women to behave more like hegemonic men, reaching the same destructive levels of environmental impact. As d'Eaubonne and other Ecofeminists put it, in a patriarchal system, women in powerful positions will likely behave like powerful men (see also a case study in Sweden described in Buckingham 2020, 7, 50–51). Other forms of oppression are ignored in liberal thought (and action), like racism and compulsory heteronormativity; hence, it is mainly white upper-class women benefiting from liberal policies. There is no accounting for community or collective needs and for society-wide or international forms of oppression, like racism and colonialism, because the focus is on individuals, unless it is politically inconvenient, since groups' rights are also recognised under liberalism, under the guise of "business," "country," "national interest," etc.

Ana Isla, for example, has demonstrated such deceptiveness in liberal thought by showing how privatising commons in Costa Rica for the ostensible purpose of biodiversity preservation has not only been a cloak for biopiracy, but also a process, at times violent, of pauperisation especially for local peasant and Indigenous women (Isla 2005). Kovel has pointed, more abstractly, the fact that in the struggle for women's liberation

> bourgeois feminism languishes, ultimately, I should think, because it has never been able to imagine an alternative that encompasses a better world than that of capital, and therefore succumbs to the system's extremely potent ways of absorbing its partial negations. (Kovel 2005, 6)

Perhaps another way of describing the problem is that liberalism is fundamentally self-contradictory, if not delusional, as a result of being an effort to deny the destructive effects inherent to capitalism.

These are reasons for Ecofeminists to stand for replacing the social structure in which gender inequalities are a constitutive part. The aim is to establish a different kind of society where everyone's work is valued and more equally distributed, especially care work. Just as any socialist formation, Ecofeminism, save for its liberalism-leaning variants, is a striving for a society free of oppression and in harmony with the rest of nature. The two are intrinsic to each other, in Ecofeminist thought. However, one could also argue that harmony with the rest of nature, even if one could define its meaning and concrete manifestations (an open question), is a redundant statement for any political philosophy that takes humanity as part of Nature (and, to be clear, does not call for humanity's annihilation). In fact, there are Materialist Ecofeminists who reject the notion that "there is a harmonious balance with nature that humans can 'discover' or 'recover'" (Dordoy and Mellor 2001, 98; cf. Plumwood 2001). The challenge, as with any socialist formation, is clarifying what characterises such a new kind of society, what constitutes progress towards its attainment, and identifying the most effective ways to reach such a desired ultimate objective beyond making wish lists.

Ecofeminism to Overcome Perceived Deficiencies in Socialism

Like Ecosocialists, Ecofeminists of a socialist persuasion call for another form of socialism, but one that places the issue of and the struggle against patriarchy in front and centre. To Mellor, "A socialism that does not challenge the economic and sexual domination of women by men will never achieve an egalitarian society, nor one which is ecologically sustainable" (Mellor 1992, 62). Eventually, Ecosocialism itself would come under critical fire. Ecofeminists such as Salleh would re-delineate the problem in terms of "reaching for an eco-socialism that is at once feminist and decolonial" (Salleh 2017, 34). To be sure, the sort of feminism from which to redefine Ecosocialism would be any but its liberal variant, which "remains phallic in orientation" (ibid., 105). Similarly, it is not any post-structural or postmodern form of decoloniality to which Ecosocialism should turn, but one that accounts for gender ethnicised class exploitation (Turner and Brownhill 2006) or meta-industrial labour (Salleh 2010) and builds on "a multitude of prefigurative efforts" (Brownhill et al. 2022, 2; see also Barca 2020).

At bottom, in an Ecofeminist lens, all forms of socialism have succumbed to patriarchal practices and ideologies. To overcome the "imposed altruism of nurturing and caring work" (Mellor 1992, 59) of patriarchal relations, where men generally benefit, socialist struggles must prioritise women's experience to bring about such a desired society.

> By reintegrating the divided worlds of men and women, such a society will pull men back to the pace of biological time in which women live. It will create a society in which political and economic decisions will be local and accountable. It will equalize the resource of time, so that we can begin to slow down the pace of human development to sustainable levels. (Mellor 1992, 59)

To achieve this, Mellor and others at first described the effort in terms of Ecosocialism embracing feminism and Ecofeminism embracing socialism. Doing so entails not only addressing modes of production, but also material conditions beyond the economy, that is, biological differences and ecological limits (see also Oksala 2018, 231). That is, Ecosocialism must be underlain by the "deep materialism" that Ecofeminism provides (Salleh 1997; Dordoy and Mellor 2000, 44).

Addressing the extra-economic material conditions in relation to the capitalist mode of production, Salleh argues, in continuity with much of socialist feminism, that women have a distinctive structural situation (see, for example, Kollontai 1977 [1920]). Women face two forms of exploitation, understood as the imposition of and extraction of surplus labour for others' economic benefit. In addition to workplace exploitation through capitalist surplus labour extraction, females' surplus labour is extracted mainly in the household or through the family. Thus, females' surplus labour benefits men collectively. Capitalists benefit in particular through females' unpaid work in reproducing and maintaining workers (Salleh 1995).

This is something that, for example, Flora Tristan (1843) had indicated long ago but, unlike many Ecofeminists (or many feminists generally), understood as a jumping point to unify the working classes for the purpose of attaining socialism, the sort of socialism that Ecofeminists would find inadequate. The call for working-class unity, incidentally, anticipated a political project encapsulated in a slogan for which the *Communist Manifesto* eventually became famous (Blomberg 1998). Relative to ultimate political aims, Tristan's and many other women's ideas of socialism down to this day depart substantially from at least some aspects of Ecosocialist Ecofeminism. This substantive political divergence should be confronted directly by Ecofeminist thinkers, at least more than so far has been done. Be that as it may, even to unify working classes, the specificity of workers' material conditions must be understood. Ecosocialist or Material Ecofeminists' framework is premised on the same understanding as Marxists regarding humanity being part of nature (Dordoy and Mellor 2000, 45).

In contrast to much of socialist or Marxist feminism, however, Ecofeminists link gendered workers' material conditions not just to capitalist processes but also to bodily differences and the rest of nature. Salleh (1995) identifies four processes affecting women's relation to the rest of nature under capitalism: (1) biological reproduction (such as birthing, suckling); (2) caring and maintenance work; (3) manual work in the gender division of labour; and (4) symbolic dualistic representations of women as closer to or part of nature, understood as passive, inert, separate from society or culture (Salleh 1995; 2003, 67; see also previous chapter regarding bourgeois ideologies of Nature). Noteworthy is the life-affirming work that is mainly accomplished by women, or so it is asserted.

This is also the basis for Salleh's above-described meta-industrial labour concept, which includes some men (Salleh 2003, 69–71; 2010), and should be reconciled with the notion that women mediate nature on behalf of men (Salleh 2003, 68). This is unless one wants to embark on a mediation ranking that starts from women to non-capitalist Indigenous and/or peasant men, all the way to wealthy and mainly white heterosexual men. Obviously, not all women would be among the meta-industrial labouring "class," since that would include the minority high-ranking female purveyors of global harm. Conversely, the work of patriarchal non-capitalist Indigenous and peasant men may be more ecologically reproductive or nature-mediating than, as intimated by Salleh (2003, 72), that of most women in the Global North. This would need to be made to cohere with the claim that patriarchy is the deepest contradiction of capitalism and, more concretely, with the prospects for any unification of the meta-industrial workforce to overcome capitalist patriarchy. It is also unclear whether it is admissible to recognise that ecologically sustainable food provisioning done by men mediates nature for women. If such is acknowledged, perhaps a balance sheet of gender-differentiated nature mediation from case studies across the world could be tabulated to demonstrate the validity of the thesis. For the moment, empirical evidence can be found both supportive and contradictory of the thesis.

Ecofeminist Critique of Eco-Marxism

The above Ecofeminist critiques of socialism and Ecosocialism carry over to Marxism and Eco-Marxism, a main theoretical backbone of Ecosocialism. As discussed above, Ecofeminists regard what they deem as "traditional" socialism, if they consider it at all, as being stuck in a "social" form of materialism that, in Materialist Ecofeminist terms, does not account for "the special role that women and the natural world play in creating the basic structures of human social life" (Dordoy and Mellor 2000, 45). As Salleh has expressed the issue,

> Ecofeminists view the humanity versus nature dualism, and the split between productive versus reproductive labors, as reflecting a profound alienation embodied in the social construction of masculine gender identity and the social construction of its thought products. (Salleh 2003, 95)

One "would agree with Marx that humans are continually subject to forces and relations of (re)production but that these are much deeper and wider than capitalist relations" (Dordoy and Mellor 2001, 98). Though useful to start from Marxian concepts and general framework, Marx's own critical analysis of capitalism is found to be deeply wanting by Ecofeminists. Not only does it offer little guidance over embodied and ecologically embedded processes but—so Ecofeminists insist—promotes anti-ecological notions such as the development of the forces of production and implicitly gendered ideas about the proletarian nature of the revolutionary subject, assumed to be white and male (Dordoy and Mellor 2000; Salleh 1997; Turner and Brownhill 2004).

Anticipating Salleh's concept of meta-industrial labour, Dordoy and Mellor assert that Ecosocialists must look into provisioning and care activities that subtend the formal economy and that remains the only model of non-commodified labour, even as it is expropriated. These would be the main sources from which to draw out the potential to achieve Ecosocialism (Dordoy and Mellor 2000, 58–59). To demonstrate how Ecosocialism needs to go beyond "traditional socialism," by which they mostly mean Marxism, they point to Marx's sole focus on the capitalist economy at the expense of examining "the whole of embedded and embodied human existence" (Dordoy and Mellor 2000, 45).

These ideas follow from understanding capitalism as a recent form of patriarchy, in line with the earliest works in Ecofeminism, and related forms of socialist and radical feminism. Ecofeminist theories are in this and other ways direct critiques of Eco-Marxism in drawing attention to and claiming that the primary contradiction in capitalism is not captured by the classical Marxist understandings, whether in the guise of second contradiction, metabolic rift, or possibly world-ecology or web of life. Those kinds of contradictions are certainly acknowledged by Ecofeminists taking historical materialist positions, but to Ecofeminists the deepest contradiction revolves about reproduction, both social, biological, and ecological, and the consequent alienation of mainly but not exclusively men from the social, biological, and ecological basis of their existence. Hence the call for a "deeper materialism," an "embodied materialism," attentiveness to "reproductive" or "meta-industrial labour" (as revolutionary subject, even), and like conceptual apparatuses to address the fundamental contradiction (Salleh 1995, 2005, 2009).

Consequently, Ecofeminist politics involves struggles for representation (liberal, socialist feminisms), the critique of representation (radical, post-structural feminisms), and the reimagining and prefiguring of ecologically sustainable ways relating to and living with the rest of nature (Ecofeminism).[4] Salleh (2003, 66) deems this a "dialectical openness or indeterminacy" characterising Ecofeminism, one that accounts for "strategic essentialism" and for the heterogeneity of Ecofeminist interventions. It is a paradigm where many perspectives live because there are many different situations. In this way, one can also see how Ecofeminism modifies the Marxist concept of differentiated unity. Nevertheless, two major strands of Socialist Ecofeminism have developed specifically in relation to Ecosocialism over the course of the 1990s and early 2000s. They are perspectives and theories that dovetail yet are in tension with Ecosocialism

[4] The third movement is a bit of a leap, as it overlooks the processs of constructing alternative knowledge systems to supplant conventional ones and of developing practicable and liveable alternatives in a context of highly and permanently modified ecosystems and at times poisonous environments. The weight placed on social change (the first two forms of politics) is, to make an understatment, disproportionate compared to the importance of processes that are much greater than humanity, such as millions of interactions among species, ecosystems, the biosphere, the planet, etc., all lumped under a generic "nature." It is also not just politically that one can resolve environmental degradation (climate change is hardly just a function of social change), since the outcome of human intervention is contingent on what other organisms do and on how physical processes shift. But this is a chronic, unacknoweldged problem in most all leftist politics especially in liberal democracies, where under-equipped theories relative to technical knowledge repeatedly fail to address biophysical phenomena even while claiming the opposite.

in all its tendencies. One is the Embodied Materialism framework advocated for by Ariel Salleh. The other is Terisa Turner and Leigh Brownhill's Gendered Ethnicised Class Analaysis.

Embodied Materialism

According to Salleh, there are identifiable concrete processes making women's relation to Nature (biological differences and wider ecological relations) distinctive under capitalism. Life-affirming activities make most women more sensitised to environmental problems. Thus, women under capitalism become the main mediators of nature for men (Salleh 1995, 33). It is because of what Salleh has called the nature-women-labour nexus in capitalism that women tend to have insights and sensitivity to processes that are harmful to life (Salleh 1997).

The nexus results from a gendered value system predicated on material conditions. One is given by most property being owned by men globally and women mainly in Global South countries usually producing and processing food for household subsistence and no wages. In any event, women tend to be paid much less for the same kind of work and often hired for work associated with unpaid work (teaching children, social assistance, etc.). Capitalists tend to hire women part-time or as casual workers, with little to no benefits. Moreover, women's housework and care work (creating use-value) are not paid and are omitted from official economic statistics.

One way to describe gender relations in capitalist patriarchy is that "women's work makes accumulation possible for all kinds of men, and the 'surplus' women generate is quite crucial to the operation of capitalist patriarchy" (Salleh 1995, 26). This, in an embodied materialism perspective, is the primary contradiction of capitalism, the exploitative, gender-based division of labour and the parallel destructive impact on Nature (human bodies and environments). Women's livelihoods and work (the use-values produced by women) are externalised just as the destructive impacts on Nature are externalised. As the environment is destroyed for profits mainly going to men, women are expected to care for and heal the people harmed by the environmental damage. This is how women become mediators of Nature for men.

Hence, the first contradiction of capitalism, contrary to Marxist theory, is to be found in the Nature-woman-labour nexus. The patriarchal exploitation of women makes possible class exploitation (*the first contradiction of capitalism*) possible (Salleh 1995, 31). The second contradiction of capitalism, in contrast to O'Connor's thesis, is the treatment of women, like Nature, as a "natural resource." They are objectified, as if devoid of subjectivity, as part of the capitalist conditions of production.

The way out of capitalist patriarchy is defined by four principles of Ecofeminist politics (Salleh 1995, 37). First, it is to understand Nature and history as a material unity. Second, Nature, women, and men are at once active subjects and passive objects. Third, the woman-nature metabolism holds the key to historical progress. Fourth, reproductive labours guided by care are models to be put into practice to arrive at ecological sustainability. These principles emerge from lived conditions. The material unity of Nature and history is given by women's special historical role. Patriarchal relations force women into situations that make them more attuned to environmental harm and to damage to life. This is why women are temporarily the more important historical agents to overcome capitalism for an ecologically sustainable egalitarian society.

With respect to strategy, Salleh describes three main political tasks (Salleh 1995, 38). One is to establish the right of women to a political voice. This is a strategy that makes common cause with liberal feminism. Another is to dismantle the capitalist-patriarchal relation of "man to nature," an effort in tune with radical and socialist feminist positions. Finally, struggles must be focused on showing how women (and, in future, men as well) can live differently and constructively with Nature. This third political task is identified with Ecofeminism and is aimed at transcending both capitalism and socialism. Importantly, Salleh, as already mentioned, has subsequently included men who are part of "meta-industrial" labour. This may be of consequence to how the final task is to be defined, which would not be focused on women alone. In any event, Salleh describes the ultimate objective as the creation of self-sufficient, decentralised relations of production, where women and men work together in joy and reciprocity with external nature. There will be no alienation or distortion by a gendered division of labour and by capital accumulation, but how this entails the supersession of socialism has not yet been explicated (see also Salleh 2022).

Gendered Ethnicised Class Analysis

A parallel and coeval theorisation came to be offered by Brownhill, Kaara, and Turner by way of a "gendered class analysis" framework, which draws from Autonomist Marxist Feminism (e.g., Mariarosa Dalla Costa, Silvia Federici, Selma James) and subsistence perspectives (Brownhill et al. 1997; Turner and Brownhill 2004; Isla 2018). The subsistence perspective means producing for living (not profit making), finding happiness, and realising self-determination "within the limits of necessity" (Bennholdt-Thomsen and Mies 2000, 19–20) and to reinstate or create commons (Giacomini et al. 2018). The perspective is premised on redefining capitalism as dependent on non-capitalist relations "directly concerned with producing and maintaining food and life." That is, capitalism depends on subsistence work, without which neither workers nor capitalists can exist. Capitalism entails the exploitation of much more than wage workers.

This has been alleged to be a blind spot in Marxism and that expounds on and goes beyond the critical insights of Rosa Luxemburg (Bennholdt-Thomsen and Mies 2000, 11; see also Mies 1986, and the previous chapter on Luxemburg's misinterpretation of Marx's work). Von Werlhof additionally characterises it as "the self-evidence of existence without domination" (von Werlhof 2007, 22). The perspective is virtually the same as the concept of "meta-industrial labour" developed by Salleh a bit later and with the same level of generalisation (and problems) regarding who is included under that category and the assumption of an automatic environmental friendliness to meta-industrial labour.

In the "gendered class analysis" framing, the same assumptions are made, but it is asserted that most unpaid workers are peasant women. Unpaid workers, when they undermine the mainly male exploitation of their work, effectively challenge global capital and help mitigate or preempt ecologically destructive activities. As the authors explain,

> Central to the problem addressed by gendered class analysis is the global resistance to capitalist privatization and state capitalism. This resistance specifically affirms subsistence "commoning," by which we mean the affirmation of social relations that realize all the requirements for life on earth and the full development of human capacities. (Turner and Brownhill 2004, 22–23)

Such resistance and its spread, otherwise put, is the fount of the Ecosocialist Ecofeminist (or Ecofeminist Ecosocialist) future, where commons are restored or globally expanded, production is for subsistence, everyone achieves self-realisation, and ecological sustainability reigns.

Unlike that of most other Ecofeminists, however, this framework is an elaboration on and expands the horizons of O'Connor's second contradiction thesis by analytically privileging "unwaged peoples," who

> (a) are the primary producers of unwaged labor, nature, social services and built space; and (b) depend most immediately on these values for their survival. These gendered class facts explain much of why unwaged people including women, peasants, indigenous peoples, the urban poor and others are at the forefront of contemporary social struggle against capitalist war and imperial rule. Waged workers are also engaged in this struggle. Gendered class analysis recognizes however, and seeks to theorize, the realities of the frequent refusals of waged workers to act in solidarity with social movement initiatives and the structural collaboration of selected waged workers in male deals to enforce the exploitation of the unwaged, especially women. (Turner and Brownhill 2004, 23)[5]

This kind of analysis was initially elaborated out of a case study in Kenya, where the impacts of neoliberal farming policies undermined ethnic ties, drew peasant women together economically, and exacerbated antagonisms among women and their husbands over cropping systems. Tensions eventually resulted in the dissolution of peasant men's attachment to state and capitalist entities, thereby sundering the "male deal," and the formation of more subsistence-oriented cooperatives by peasant women. (Brownhill 2009; Brownhill et al. 1997; Brownhill and Turner 2002).

Such successful peasant women's resistance to capitalism illustrates the basis for the sort of Ecofeminism (revolutionary Ecofeminism) predicated on gender ethnicised class analysis. The ethnicity axis would be further expounded theoretically in subsequent writings (Brownhill and Turner

[5] Paradoxically, Turner and Brownhill would, by definition, exclude themselves from being among those at the forefront of the struggle, along with many, if not most of the activists to whom they refer in their case studies, if they receive wages. Yet many waged women and men are often involved as the main protagonists, so the notion of unwaged workers being more likely to be at the forefront of such struggles should be tempered and extensive empirical studies should be brought forth in support of such an argument.

2019; Isla 2022). The general aim associated with this framework is one of unifying all exploited people (including those who receive wages), "every segment of the global working class," to achieve Ecosocialism (Brownhill and Turner 2020, 8; italics in original). It is thus not too dissimilar from broad Marxist principles of privileging the sort of politics that facilitate working-class unity.

Further refinement to this political project has been developed more recently by connecting it to rape culture (Brownhill 2022; Brownhill and Turner 2020). This is the process of justifying men's violence against women that defines how most women and men relate to environments and to each other. It is the basis for world capitalism, where a hierarchy is imposed through the above-described male deal to have most men reinforcing the violent basis of capitalist exploitation. The connection is explained thus:

> Rape culture … is a … mechanism of capital accumulation and the preservation of private property. … Without the pervasive violence of misogyny and its imminent threat, women would not be producing labor power, cannon fodder, and cheap wage goods or allowing the enclosure of our commons. … Male violence against women … serves on a cultural level to limit the mobility, freedom, health, and safety of victims and survivors, but also of all other women and girls, who, to varying degrees, must calculate and adjust their lives to reduce the daily risk of violence. Rape culture further fundamentally serves the purposes of capitalist profit extraction, because it maintains and entrenches division, isolation, and the alienation of gendered and ethnicized members of the dispossessed class. (Brownhill and Turner 2020, 5)

This reinforces the need for an Ecosocialist Ecofeminist framework based on "ethnicised gendered class analysis" (ibid. 4), as sexism and racism combine to enforce the social divisions useful for capitalist exploitation of both people and the rest of nature (ibid. 9). From this, it follows, as already stated, that political strategies must prioritise the unification of oppressed peoples in a common struggle.

This analytical framework has featured prominently in Isla's work in Costa Rica and the Amazonian areas within Peruvian borders (Isla 2005, 2018), where resistance has met with failure. The framework has also informed the study of alliances involved in developing international farmers alliances (Giacomini 2022), fighting for community garden space in Passaic, New Jersey (Blumberg et al. 2018), as well as the analysis of interrelated modes of oppression in the violent displacement of Adivasi communities along the Narmada Valley for exclusive hydropower development projects (Chattopadhyay (2022).

One could quibble about whether some of the arguments proposed in the quoted text above are tenable, such as that built space is mostly made by unwaged workers. Certainly, the notion that "nature" is produced by unpaid people suggests a very narrow and reductionistic notion of nature to whatever humans do. Otherwise, it is a preposterous notion glaringly contradicting any perspective positing that we are part of nature, including Marxism. The authors likely know all this and doubtless would agree that at the very least the wording should be improved. This would be a pedantic matter were it not for the fact that Ecofeminists, like Eco-Marxists, subject the works of Marx, Engels, and Marxist to exactly this kind of critical textual scrutiny, sometimes predicating arguments and even politics on the outcomes of such analysis (see below on Ecofeminist critiques of Marxism).

To return to the analytical framework generally, there are substantive aspects to gendered ethnicised class analysis that elicit a few questions. As in Salleh's case of meta-industrial labour and Foster's environmental proletariat, it is unclear whether temporary alliances or the convergence in material conditions described or claimed suffice to create a subject capable of overcoming capitalist relations. To date, this has not been demonstrated, as there have hardly been any revolutions arising from such allegedly interrelated multitudinous social positions in relation to environmental issues.

Moreover, the applicability of the gender ethnicised class framework and associated political strategies is questionable, based as they are on particular case studies. For instance, similar gendered struggles featuring successful resistance to neoliberal policies by Mandinka peasant women in The Gambia have not undermined male deals, nor have such successes resulted in the formation of peasant women subsistence-privileging cooperatives (Carney 1993; Schroeder 1999).

It is anyway difficult to tell apart temporary alliances (say, the recent "Cowboy and Indian" anti-pipeline alliance in parts of the US) from a lasting convergence of interests (multiple communities affected by the same type of environmental degradation). Equally problematic is the assumption that radicalisation is of the same sort for all affected and leads necessarily to revolutionary potential, especially in any Ecofeminist or Ecosocialist direction. Instead, it appears that the protest movements considered, touted as signalling some promising upsurge, are largely temporary and politically disorganised, if not incoherent, or at the very least inadequate to the task of toppling capitalism or patriarchy over the long term.

This tendency to see revolutionary tendency or potential where very little exists is in part due to the imprecise or inconsistent social categories proposed. In the case of gender ethnicised class analysis, the category of "unwaged" is so elastic as to include lumpen proletarians and, in the case of peasants (never actually defined), part of the petit bourgeoisie (land-owning peasants). This is made evident in a list of disparate and disconnected actions, mainly led by women from different class positions in different countries, ranging from protests against discrimination and misogynistic violence to land reappropriation and electoral politics (Giacomini et al. 2018). More recently, the category of children has been included without specifying age range, which varies culturally, and without stating whether this includes adolescents (Brownhill and Turner 2019, 8). Expanding yet again the breadth of the "unwaged" and "global working class" brings even more ambiguities relative to gendered, ethnicised class positions and interests (and one should add sexuality- and age-linked ones as well to this).

In fact, the strategy developed is tantamount to asking for inter-class collaborationism, where, for instance, what is good for the settler male worker and female smallholder is good for the female settler and Indigenous male. They are all victims of capitalist patriarchy. Similarly, the notion of a "male deal" and "deepest contradiction" (Salleh) needs to be clarified further to understand, for example, what to make of female (mother) transnational corporate officers' impositions on male and female state agents in a country under neocolonial rule. Are corporate officer mothers to be categorised as enforcers or resisting victims of the male deal or as part of meta-industrial labour? Are male and female (non-mother) state officers in a colonised country to be deemed class enemies? It would be more effective, both theoretically and practically, to develop terms and analyses more consonant with actually existing social relations of power.

To collapse the category of unwaged with "dispossessed" only adds to the confusion, since plenty of dispossessed people are wage-dependent or seek waged employment, sometimes precisely because of being dispossessed. Yet, the category of dispossessed could include anyone ranging from pastoralists being denied access to pastures to capitalists losing property by eminent domain or revolutionary action. Similarly, it is difficult to argue about children being at the forefront of struggle on account of their dispossessed or unwaged status (one might also wish to conduct a study about children who receive allowances and those who do not). Imprecision regarding subject positions, material conditions, and interests impedes political clarity and working-class unification. A presumption of Ecofeminism-compatibility in any eruption of mass protest involving women, children, peasants, Indigenous communities, etc. says nothing about anti-capitalist content and potentials or organising possibilities relative to reaching Ecosocialist Ecofeminist goals.

Furthermore, the gendered class analysis is premised on the disputable notion of "autonomously organized multitudes of indigenous peoples, women, the urban poor and peasants in battles against imperial rule" (Turner and Brownhill 2004, 24; cf. Isla 2005).[6] To show these multitudes' effectiveness, two case studies from Kenya and Nigeria are discussed, where revolts against "imperial rule" featured a prominence of female activists using the tactic of shaming by exposing themselves naked to repressive authorities. The authors interpret those actions against land privatisation in Kenya and oil extraction in Nigeria as instances of successful challenges to the "male deal" (see below).[7] Yet, land privatisation in Kenya has continued, even if mainly benefiting local capitalists (Bon 2021; Minoia 2020). In the Niger Delta, partly foreign-owned oil production proceeds unabated, at roughly two million barrels daily (Tsimitakopoulos 2022). In the case of Nigeria more broadly, foreign capital inflows have hardly stopped and have exacerbated gendered inequalities (Adeosun and Owolabi 2021; Ifeakachukwu 2020).

Autonomy, to be meaningful, implies having a land base or control over resources enabling independent action from dominant capitalist institutions, even more so imperial ones. The problem is that such "multitudes" are claimed to be dispossessed, so it is unclear how autonomous their actions really can be. On the other hand, political organisation

[6] This penchant for exuberantly sanguine thinking is reflected in claiming that there is a "global shift in power away from corporate rule toward alternative, life-centered social forces" (Turner and Brownhill 2004, 22).

[7] A comparison with the liberal feminist Femen organisation, who are bankrolled in part by several capitalists, would be useful towards identifying how similar tactics are used for different political aims.

denotes the existence of coherence and the capacity for sustained collective action. Given what has transpired in Kenya and Nigeria, the most that could be is that political coherence and/or organisational capacity have been inadequate. The wherewithal to combat "imperial rule" (also problematically undefined) is too often absent with respect to the material basis of autonomy and too often fleeting in terms of political engagement. The evidence for those case studies anyway points to a largely ineffective "multitude" relative to combating imperial rule.

Assumptions of autonomy and organisational capacity are ultimately superfluous to a gendered class analysis framework. What a gendered class analysis brings to the fore, unlike other Ecofeminist perspectives so far considered, is the concept of the "male deal" (see also Federici 2004). According to this view, the fragmentation of workers ("the dispossessed"), unwaged and waged, results from male and some female workers cooperating with capitalists in the exploitation of other workers, especially unwaged female ones. This is the "male deal," which, it must be emphasised, includes a minority of women (male or masculinised women, one might call them). Such complicit males and females gain in material ways from participating in capital accumulation and engaging in relations of domination over other workers. Notably, capital accumulation is impossible without such cooperation, so that those rebelling against the "male deal" essentially undermine the "capitalist exploitation at the root of ecocide" (Turner and Brownhill 2004, 25; see also Isla 2018). At bottom is a social struggle over "fertility," which is "defined broadly as the capacities to produce and sustain human, other animal and plant life [the authors surely mean all life forms]" (ibid.). The struggle unfolds among "(1) capitalists, (2) male dealers, and (3) women united with men in a gendered class alliance" (ibid.). This cross-gender class alliance is a noteworthy departure from the above-discussed Ecofeminist viewpoints, where there is a lack of clarity about men's role or how men are supposed to be transformed (but the transformation timing and processes are unclear).

Consequently, the task is to identify the subjects involved in struggles against the "male deal" enabling the exploitation of "fertility" and who thereby develop potentials for the unity of the exploited, or a "gendered class alliance" (ibid.). This is how a gendered class analysis not only specifies the primary relations of oppression, but also sheds light on the political strategies to follow. In some respects, this diverges from O'Connor's greater accent on drawing together environmentalists and unionised workers and on democratising the state as part of such a process (O'Connor 1998). What unites all these theories is the reductionistic treatment of biophysical processes to social struggles and change. For example, plant or mycorrhizal reproduction is not necessarily contingent on any social struggles over "fertility." In this respect, not much progress has been made compared to Marx and Engels' understanding of "the wealth of nature" as much bigger than humanity.

Relating, Linking, Debating Ecofeminism, and Ecosocialism

Because of its greater affinities with socialism, the materialist and Marxism-inspired variants of Ecofeminism, socialist Ecofeminism, would find more overlap with Ecosocialism, though not without difficulty or perceived impediments. The two frameworks were initially introduced to each other as separate (non-capitalised, hyphenated) schools of thought, Ecofeminism and the chronologically younger eco-socialism, as if Ecofeminism had no socialist roots (by the early 2000s, the hyphenation would become optional for both terms). The journal *Capitalism Nature Socialism*, whose very founding involved Ecofeminists,[8] rapidly emerged as a prominent outlet to discuss the relationship (Carlassare 1994; Faber and J. O'Connor 1991; Isla 2006; Mellor 1992; Noble et al. 1989; Salleh and M. O'Connor 1991).

What the late 1980s through the 1990s spawned, though, was some unnecessary tension and miscommunication, including among some of the journal founders (e.g., Isla 2006; Salleh and M. O'Connor 1991). An impediment to constructive dialogue or to finding complementarities was an early inadequate treatment of Ecofeminism,[9] for which apologies were

[8] An Ecofeminist Collective was eventually formed to partake of managing said periodical.

[9] Regrettably, unwarranted generalisations about men and socialist histories would also tinge the reply by remarking that "in a future post-patriarchal scenario men may engage in caring labor themselves" and accusing "brother socialists" of using "the gender blind categories of Marxism" (Salleh and O'Connor 1991, 134), allegations decidedly disrespectful of women Marxist luminaries like Alexandra Kollontai, Clara Zetkin, or 向警予 (Xiàng Jǐngyǔ, 1895–1928) and 康克清 (Kāng Kèqīng, 1911–1992). There would be similar kinds of preconceptions, coupled this time with an unhelpful self-isolating ultimatum, exhibited more than a decade later: "There's little point in running thoughtful essays by women if men among the readership don't engage with these fresh [Ecofeminist]

made (Faber and J. O'Connor 1991). Although there exists, even now, a masculinist tendency to underestimate or ignore the foundational importance of, for instance, sexed bodies and care work, the Ecosocialist critique in those early days was not about Ecofeminism generally, but about the "radical" (or Cultural) form of Ecofeminism prevailing at that time in the US, with its politically counterproductive essentialism and spiritual neo-Romanticism abound. Nevertheless, the problem was (is) the lack of specification regarding the sort of Ecofeminism being addressed and the disinterest shown about Ecofeminist scholarship and activism (Noble et al. 1989). This adds to an already complicated discussion resulting from the variety of contrasting currents represented in Ecofeminism and Ecosocialism respectively. As Mellor put it: "Eco-feminists range from New Age thinkers to socialists and eco-socialists range from Marxists to anarchists" (Mellor 1992, 45).

By the early 2000s, there would be a more mindful treatment of Ecofeminism. It appears this arose first by way of a defence of Ecofeminism from the criticisms of the founders of Social Ecology and an exposition of an alternative Ecofeminism-aligned dialectical perspective to reform Social Ecology (Clark 2000). Some Materialist Ecofeminists responded positively, but also critiqued the idealistic reliance on ethics as the foundation of dialectics and the idealisation (or lack of concrete specificity) of women's work in that articulation (Dordoy and Mellor 2001). Clark (2001) would point out that, among other problems and unlike other Ecofeminist currents, such Materialist Ecofeminism itself does not venture far enough. It does little to call into question the gendered care work underlying the prerogative of meeting basic needs in Marxist frameworks, especially in what are deemed as productivist variants of Marxism. It seems the differences between two perspectives were too subtle to warrant a continuation of the debate.

The middle of the early 2000s saw the return of direct dialogue and debate within the journal *Capitalism Nature Socialism*, much of it

insights and the intellectual challenge of recognizing how gender often enters into the very formulation of a political concept" (Salleh 2005, 9). One could equally accuse Ecofeminists of refusing to engage with the insights and challenges on these issues from Marxist Communist women intellectuals like Claudia Jones and Jodi Dean. Salleh's accusatory and condescending tone is certainly understandable when contextualised in historical experiences of masculinism in leftist movements in liberal democracies (Salleh 2005, 10), but from that specificity a generalisable politics does not follow. There may also be reasons unrelated to masculinism for any lack of engagement with such "fresh insights."

overseen by Ariel Salleh (2006a). The series of exchanges brought into greater relief several questions that in some ways define an ongoing, often fruitful tension between Socialist Ecofeminism and Ecosocialism. These include the relationship between reproduction and production, the contradictions within or beyond capitalism, the concrete repercussions and foundations of patriarchal ideological constructs, the matter of essentialism, and defining revolutionary subject positions (Salleh 2006a, 36–37). The process has regrettably replicated gendered divisions in academic work, with Ecosocialist male critical commentary to Ecofeminist female scholarship and theorisation followed by Ecofeminist female replies. The discussion would nevertheless shed further light on key points of contention but also substantive agreement or overlap. At times, the exercise would expose some mutual misunderstandings that would induce additional elaboration and sharpening of ideas.

For instance, the dialogue between Godfrey (2006) and Goldstein (2006) about the issue of essentialism would clarify the important role of historical materialism in Ecofeminism. Rosewarne (2006) would argue that Ecofeminism, as evinced in Turner and Brownhill's analysis of struggles in Kenya and Nigeria (see above), is a manifestation of the social movements arising from capitalist undermining of the conditions of production. Understandably, Turner and Brownhill (2006) would reiterate a critique of O'Connor's theory, which is ill equipped to deal with movements forming to challenge capitalist attacks on social and ecological reproduction. The current author exchanged ideas with Sivia Federici regarding the role of historical specificity and regional variation in the relationship between the onset of capitalism and witch hunts (Federici 2006). Chapman and Isla would debate the issue of ecocentrism and the feminine principle, with Isla refocusing the issues towards the patriarchal contradiction of capitalism and its implications for the rest of nature (Chapman 2006; Isla 2006).

Feisty exchanges also emerged, some of them pointing to unresolved (perhaps unresolvable) ideological divides and debates traceable to the early 1990s. Rudy's depiction of Salleh's embodied materialism as ahistorical ("transhistorical biologism") and a reification of patriarchal constructs would be interpreted by Salleh as proof of Rudy succumbing to capitalist patriarchy (Rudy 2006; Salleh 2006b). A more delineative positioning relative to political objectives came about through another debate, one centred about technology. Wallis would express broad agreement with the subsistence perspective (Bennholdt-Thomsen and Mies 2000). He would raise doubts, though, about that perspective's efficacy in addressing most people's lived circumstances and the structural processes shaping people's current and variegated needs (Wallis 2006, 41). The subsistence perspective, according to Wallis, offers a view of technological development solely according to its negative repercussions (rather than also the positive potentials) and leaves out the development of strategy to raise ecological sensibility and political awareness. He finds it unhelpful to take "first-epoch" socialism as the basis to reject socialist strategies and Marxist perspectives, which he recognises need to be redefined.

Mies would reply that there is no attempt to build any generalisable strategy by means of the subsistence perspective (Mies 2006), something Wallis had not actually argued. However, her critique of capitalist technological innovation as stifling or negating prior technologies, especially less polluting or destructive ones, serves as a corrective to Wallis' view. Furthermore, the ecosocial crisis increasingly compels the adoption of a subsistence perspective, regardless of how needs are defined. Drawing from the work of Max Neef (where needs are distinguished from satisfiers), she understands the concept of needs in terms of basic human necessities, those that are truly universal. This would approximate Marx's notion of species-being if it were not for a lack of historical materialist or evolutionary understanding about needs. For Wallis (2008), Mies' discussion of the concept of needs was instead an unnecessary detour from the issue of the process needed to reach the objectives Ecofeminists and Ecosocialists share. Finally, she finds Wallis' view of socialism to preclude future creations of social relations that will overcome older socialist paradigms and build what may or may not be called socialism. To her, "simply adding 'ecofeminism' or 'ecology' to the old socialist paradigm will not do" (Mies 2006, 47). The entire patriarchal worldview, including "socialist patriarchy," must be disputed.

To Wallis (2008), Mies as well as, subsequently, von Werlhof (2007) were unjustifiably lumping together different Marxist and socialist histories and movements into a single entity hopelessly tainted by patriarchy and "technophilia." Regrettably, as Farraclas (2008) would point out, Wallis (2008, 111) mischaracterised Mies' remarks as biologically deterministic and showed to be ill-informed on Ecofeminist theory. Just as regrettable is Farraclas' denunciation of the "intrinsically patriarchal nature of Marxism and socialism" (Farraclas 2008, 116), deemed power-hungry worldviews incapable of understanding anything other than in terms of "domination and hegemony" as "natural facts of life" (ibid.,

117). Somewhat less mutual caricaturing and contrived differences would be evident in the exchange between von Werlhof (2007) and Wolf (2007).

Drawing Ecofeminism and Ecosocialism Closer

Recall that Ecofeminism emerged out of a socialist milieux, at least in France, but was virtually eviscerated of that background by the 1980s in mainly white Anglophone intellectual circles in settler colonial regimes (Australia, North America). It seems that to some degree the encounter with Eco-Marxism reignited the socialist origins of Ecofeminism. At first, this was expressed in terms of Socialist or Marxist Ecofeminism from the early 1990s through the early 2000s (Isla 2005; Salleh 1995; Turner and Brownhill 2004), giving way more and more to Ecosocialist Ecofeminism or Ecofeminist Ecosocialism.

Out of the above debates or displays of discord, aside from clarifications and demarcations, some common ground has been brought to light with the acknowledgement that both Ecofeminism and Ecosocialism are concerned with questioning and refuting ideologies justifying oppressive relations, or "reified naturalism" (Salleh and M. O'Connor 1991, 130). One could further posit for Ecosocialism, which presents a heterogeneous politics, what Carlassare eloquently described for Ecofeminism:

> Ecofeminism derives its cohesion not from a unified epistemological standpoint, but more from the shared desire of its proponents to foster resistance to formations of domination for the sake of human liberation and planetary survival. (Carlassare 1994, 52)

In this regard, Ecofeminism is scarcely in divergence from Ecosocialism. Both are premised on the same overarching goals of social equality. However, as discussed below, and just like in the histories of socialism, sameness in what is desired does not necessarily imply or translate into paradigmatic sameness or political strategy convergence. Identifying the principal locus of struggle in the male deal instead of the working class (even when broadly conceived) or the state is one example of substantive divergence.

The above-described debates also stimulated a broader concern over the relationship between leftist scholarship and political action, manifested in terms of the question of regarding environmental activism by women as necessarily Ecofeminist (Faber and J. O'Connor 1991; Mallory

2006, 47). This implies the superimposition of categories at odds with how direct participants identify or understand as their own politics (see, e.g., Giacomini et al. 2018). This remains arguably an open question or a potential mischaracterisation problem with potentially negative implications (one being of overestimating the popularity of Ecofeminist ideas or political strategies). The above-described No MUOS movement is an instance where the women leaders involved may be putting into practice principles that could be characterised as Ecofeminist, yet they are not identifying as Ecofeminist at all. A social movement's practices and theories are distinct from academic or external theorisation about that social movement. The former are responding to, thinking through, and acting to change social (and perhaps also ecological) circumstances, thereby reshaping them and not necessarily to their advantage or according to expectations. The latter draws from the former to understand wider social implications and processes and what can be applicable to other situations.

The two aspects are certainly not, and arguably should never be viewed as mutually exclusive (Godfrey 2005). For greater and anti-systemic political effectiveness and to avoid reinforcing the bourgeois divide between intellectuals and masses or thought and execution, one ought to do both. It is a hallmark of Ecofeminism, as well as Marxism and socialism generally, that people formulating theoretical generalisation tend to be involved in organising and/or participating in concrete actions. This dedication to combine practice in particular situations with larger-scale theorisation and to make theory work relevant to concrete social struggles could also be recognised as a potentially shared characteristic between Ecofeminism and Ecosocialism.

Prospects for an Ecofeminism-Ecosocialism Complementarity or Unity

There may be much overlap between socialist versions of Ecofeminism and Ecosocialism, but ultimately it might be more difficult to claim an Ecosocialist Ecofeminism or an Ecofeminist Ecosocialism without jettisoning some features of Ecosocialism or of Ecofeminism as currently exist. Theories like embodied materialism and gendered ethnicised class analysis recentre Ecosocialist objectives squarely on Ecofeminist terms, for example. It is part of what Ecofeminists have repeatedly pointed out as the limits of Marxism regarding the inability of accounting for most forms of reproductive labour and perhaps also for the rest of nature or ecological relations. On the latter, Eco-Marxists claim that the framework developed by Marx and Engels did in fact address Nature (with humans as part of it) or that it is ecological. The debate is about the extent to which the framework suffices to explain or analyse mainly the ecosocial crisis. However, there is not much room for social reproduction or patriarchy as central preoccupations of Ecosocialism, except in additive terms.

An expanded view of Ecosocialism outlined by Kovel offers a different reading that could bridge the two political philosophies. According to Kovel, Ecofeminism is "The third foundational line of ecosocialism," alongside socialism and the development of an ecological mode of production (Kovel 2005, 3). Building on classical Ecofeminist theses, Kovel sees patriarchal dualism as a main stumbling block to recognising intrinsic value (see above chapter on this topic) and the ecological crisis that inevitably results from such "a male parthenogenetic fantasy" (ibid., 5). The capitalist exacerbation of patriarchal oppression through labour-power commodification led to the development of both socialism and feminism. These must be intertwined as Ecosocialism and Ecofeminism by means of "alliances and associations" to overcome capitalism and replace it with Ecosocialism. This supersession entails restoring the "female principle ... immemorially regarded in terms of caring for, or holding, or gathering, or providing, or of staying with, or of weaving together" (ibid., 7).

Two kinds of problems arise from this fusion of Ecofeminism within Ecosocialism. First, Kovel regrettably takes sexed differentiation as a transhistorical phenomenon, in accord with some Ecofeminist positions, some of which do not align with socialism (see above regarding the history of Ecofeminism). The female-male binary typical of patriarchal societies is thereby and paradoxically taken for granted, collapsing gender to sex and thereby social construct to bodily difference.[10] This self-defeating slippage from the social to the biological in attempts to undermine patriarchy has been frequently pointed out by Materialist, Socialist, or Marxist Ecofeminists, including in the very journal that Kovel once edited. Second, and arguably more importantly, at least some Ecofeminists might not view favourably the subsumption of Ecofeminism under Ecosocialism (as one of its foundations). This subsumption has been indirectly critiqued

[10] "Gender remains the form of difference authentically grounded in nature, insofar as sex is a basic natural category... The primary division of labor is therefore organized by gender" (Kovel 2005, 4).

by reframing the linkage between Ecosocialism and Ecofeminism in terms of Ecosocialist Ecofeminism or Ecofeminist Ecosocialism, about which more below.

There is another way to facilitate the joining of the two broadly socialist and philosophically ecological schools of thought. Salleh identifies a three-phase process to arriving at an Ecosocialism worthy of Ecofeminism, with each moment characterised by a concomitant politics. The first is to realise that the patriarchal division of humanity shapes and limits people's bodily experiences and perceptions and with those all sorts of social practices and relations. The second is to prioritise efforts to overcome "gendered habits." The third moment is sustained dialogue to enable the modification of Ecosocialists' conceptual apparatus or viewpoints (Salleh 2005, 9–10). The dialogue, as described above, has been sustained since the early 1990s and, arguably, the second phase might be in the making.

In some respects, Salleh's call for embodied materialism in Ecosocialism elucidates on how the intertwining of Ecofeminism and Ecosocialism may be facilitated. The process entails adopting the view of privileging the nature-woman-labour nexus (Salleh 1997). By doing so, it is possible to build a framework useful towards uniting, but on an equal footing, movements typically perceived as disparate, namely those "engaged in workers', women's, indigenous' and ecological struggles" (Salleh 2005, 12). This is to be accomplished by relying on the insights of meta-industrial workers ("mothers, small farmers, hunter-gatherers on the fringes of capital") because they have the most knowledge about the socially and ecologically destructive aspects of capitalism. It is out of those worldviews that the subsistence perspective has been developed (Bennholdt-Thomsen and Mies 2000). In this manner, Marxism can be updated and thereby make for an inclusive and more effective Ecosocialism (Salleh 2005, 12–13). The same process could be claimed for the adoption of a gender ethnicised analysis that aims to unify all "unwaged" and "dispossessed" people (Turner and Brownhill 2004).

Gaps nevertheless remain between the two philosophically ecological camps. Complicating the forging of a linkage, if not union, is how to understand and account for biological differences and ecological limits, a problem continuous with the historical tensions between aspects of (Socialist) Feminism and Marxism. In other words, the debate is about recognising the material specificities and constraints of "women and nature" without any recourse to essentialism (Mellor 1992, 44; see also

Carlassare 1994). Yet a historical materialist analysis can explain the attraction of essentialism as well, were that even to be the question. The tendency for male over-representation in the most impacting institutional decision-making processes, in sexual violence, in warfare, and in environmental destruction is an empirically well-documented or concrete condition. It should be small wonder that many women may wish to distance themselves from masculinity, when predominantly defined and enacted that way (but see Hultman 2017), and consequently identify more with the allegedly natural as they become critically aware of patriarchy. The patriarchally imposed material basis underpinning most women's lives is what is often still missed by Ecosocialists and much Eco-Marxist theorisation.

The issue, then, is not about essentialism (if it ever really was), but about a masculinist view of the world. Ideological outcomes of material conditions are in any case but one possible analytical avenue to join Ecofeminism and Ecosocialism. The issue is ultimately about how production, reproduction, and the rest of nature relate to each other. In both capitalism and "male-dominated productivist" socialism, Mellor and other Ecofeminists have argued, production has been rendered male by severing it from biological reproduction and from ecology, a practice that eventuates in gender-differentiated social harm and environmental destruction, as Ecofeminists have long contended (Mellor 1992, 51; Dordoy and Mellor 2000, 42–43).[11]

[11] At this level of abstraction, however, it is not possible to distinguish between forms of patriarchy and, in terms of socialist politics, there is here a failure to recognise the tangible improvements to women's lives made in socialist countries precisely as a result of such allegedly "male-dominated productivist" socialism. Regrettably, like many scholars from imperialist countries, Mellor repeats standard anti-communist liberal environmentalist rants when stating that "In fact, the *theory* of historical materialism based on the primacy of (formal) economic relationships as defined by capitalism has produced the catastrophe of the command economies which, by mimicking the productive power of capital, continued to deny the reality of much of women's lives and produced ecological devastation" (Mellor 1992, 51). Such unsupported claims presented as self-evident truth flatter not the academic living in an imperialist country notorious for (eco)feminist and socialist movements' consistent historical failures to overcome capitalist patriarchy. In any case, the author plunges into self-contradiction and betrays a tinge of liberal politics by pointing to theory, not praxis as bringing about change in the world. The author then merely confuses bourgeois lucubration for socialism when remarking that "A socialism conceived in the industrial center of a colonial power, based on a dominant race and sex, is not equipped to transport us to a feminist or a green future" (Mellor 1992, 61).

Expressed another way, capitalist patriarchy is based on treating both women and nature as exploitable. The ensuing contradiction is not just generated by and for capitalism, but by and for males in capitalism and capitalists at the same time. This is a main meaning of capitalist patriarchy in relation to the ecosocial crisis. In contrast to socialist politics, Ecofeminism is about overcoming any form of patriarchy, including capitalist forms (Chapman 2006; Isla 2006). This primary concentration on patriarchy is also at odds with all the most influential Eco-Marxist theories discussed in the preceding chapter. It is also in tension with, among other political currents, self-determination struggles that posit racism as the foremost problem.

Nevertheless, Ecofeminists of Gender Ethnicised Class Analysis persuasion remain steadfast with the view that the two broad political philosophies belong together. They understand the process in terms of specifically revolutionary Ecofeminism being "at the heart of ecosocialism" (Brownhill and Turner 2019, 8; see also Mellor 2022), rather than one of its pillars, as Kovel would have had it. They name it interchangeably Ecofeminist Ecosocialism and Ecosocialist Ecofeminism. This linkage process manifests itself through "the convergence of global popular struggles that both resist capitalist extractivism and strengthen commoning alternatives that will carry humanity into a post-fossil age" (ibid.).

It remains to be more fully explicated how such a formula sits with other forms of Ecofeminism (including those presumably not regarded as revolutionary), where the objective is ultimately the end of patriarchy, not just its capitalist form (e.g., Barca 2020; Mies 2006). This is not a pedantic point, but the seeking of a clarification of objectives. It is consistent with thinking of capitalism as the latest form of patriarchy. As Kovel has phrased it, "Capitalism, far from being a rational system of market relations, is a phase in the evolution of un-free, gendered labor" (Kovel 2005, 5). It is not an understanding of the social world that necessarily resonates with Ecosocialists, for whom the terminus would be an Ecosocialist society that assumes the end of capitalism, which is assumed to include patriarchy along with other forms of oppression. The differences in emphasis are not epiphenomenal, as illustrated in the debates described above.

Another tension is within Ecofeminism more broadly, where the majority of Ecofeminists do not subscribe to any form of socialism. The end point for most Ecofeminists, especially as represented by Adams and

Gruen (2014), is certainly not Ecosocialism, as argued in Gender Ethnicised Class Analysis (Brownhill and Turner 2020). Salleh's theory of three political tasks for Ecofeminism (see above) could be brought in to resolve this, perhaps. However, according to that perspective, both capitalism and socialism are to be overcome (Salleh 1995). This adds to the difficulty in finding mutually compatible political aims. More light should be shed about the status of Ecosocialism in Salleh's three-task political formulation. Likewise, the notion of an Ecosocialism with an Ecofeminist heart (Brownhill and Turner 2019) posits the end goal as the attainment of an Ecosocialist society with Ecofeminist characteristics, yet it remains to be demonstrated how this coheres with most Ecofeminism, where socialism is shunned and class is regarded as just another axis of difference. Even more importantly, it is unclear how such a conceptualisation relates to other forms of Ecosocialism without demanding of other Ecosocialist currents to take a specifically Ecofeminst path to achieving Ecosocialism (as recommended, for instance, in Brownhill and Turner 2020; Mellor 2022).

References

Adams, Carol, and Lori Gruen. 2014. "Groundwork." In *Ecofeminism. Feminist Intersections with Other Animals and the Earth*, edited by Carol Adams and Lori Gruen New York: Bloomsbury.

Adeosun, Oluyemi T., and Kayode E. Owolabi. "Gender Inequality: Determinants and Outcomes in Nigeria." *Journal of Business and Socio-economic Development* 1 (2): 165–181.

Barca, Stefania. 2020. *Forces of Reproduction: Notes for a Counter-Hegemonic Anthropocene*. Cambridge: Cambridge University Press.

Bell, Karen. 2016. "Bread and Roses: A Gender Perspective on Environmental Justice and Public Health." *International Journal of Environmental Research and Public Health* 13 (10): 1005.

Benadusi, Mara, Alessandro Lutri, and Laura Saija. 2021. *Si Putìssi: Riappropriazione, Gestione e Recupero dei Territori Siciliani*. Firenze: Editpress.

Bennholdt-Thomsen, Veronika, and Maria Mies. 2000. *The Subsistence Perspective. Beyond the Globalised Economy*. London: Zed Books.

Blomberg, Gisela. 1998. "Flora Tristan: A Predecessor of Marx and Engels." *Nature, Society, and Thought* 11 (1): 5–16.

Blumberg, Renata, Rosa Huitzitzilin, Claudia Urdanivia and Brian C. Lorio. 2018. "Raíces Del Sur: Cultivating Ecofeminist Visions in Urban New Jersey." *Capitalism Nature Socialism* 29 (1): 58–68.

Bon, Bérénice. 2021. "Making Railway Land Productive: The Commodification of Public Land in Kenyan and Indian Cities." *Geoforum* 122 (2021): 118–128.

Brownhill, Leigh. 2009. *Land, Food, Freedom: Struggles for the Gendered Commons in Kenya, 1870 to 2007*. Trenton: Africa World Press.

Brownhill, Leigh. 2022. "The Ecology of Misogyny." In *The Routledge Handbook on Ecosocialism*, edited by Leigh Brownhill, Salvatore Engel-Di Mauro, Terran Giacomini, Ana Isla, Michael Löwy, and Terisa E. Turner, 109–116. New York: Routledge.

Brownhill, Leigh, and Daniel Faber. 2023. "Common Strategies for Eco-Feminists and Eco-Socialists: An Introduction to the New Co-Editors in Chief of *CNS*." *Capitalism Nature Socialism* 34 (1): 1–10.

Brownhill, Leigh S., Wahu M. Kaara, and Terisa E. Turner. 1997. "Gender Relations and Sustainable Agriculture: Rural Women's Resistance to Structural Adjustment in Kenya." *Canadian Woman Studies/les cahiers de la femme* 17 (2): 40–44.

Brownhill, Leigh S., and Terisa Turner. 2002. "Subsistence Trade and World Trade: Gendered Class Struggle in Kenya, 1999–2002." *Canadian Woman Studies* 21/22 (4/1): 169–177.

Brownhill, Leigh, and Terisa E. Turner. 2019. "Ecofeminism at the Heart of Ecosocialism." *Capitalism Nature Socialism* 30 (1): 1–10.

Brownhill, Leigh, and Terisa E. Turner. 2020. "Ecofeminist Ways, Ecosocialist Means: Life in the Post-capitalist Future." *Capitalism Nature Socialism* 31 (1): 1–14.

Brownhill, Leigh, Salvatore Engel-Di Mauro, Terran Giacomini, Ana Isla, Michael Löwy, and Terisa E. Turner. 2022. "Preface: An Introduction to Ecosocialism." In *The Routledge Handbook on Ecosocialism*, edited by Leigh Brownhill, Salvatore Engel-Di Mauro, Terran Giacomini, Ana Isla, Michael Löwy, and Terisa E. Turner, 1–12. New York: Routledge.

Buckingham, Susan. 2020. *Gender and Environment*. London: Routledge.

Carlassare, Elizabeth. 1994. "Destabilizing the Criticism of Essentialism in Ecofeminist Discourse." *Capitalism Nature Socialism* 5 (3): 50–66.

Carney, Judith. 1993. "Converting the Wetlands, Engendering the Environment: The Intersection of Gender with Agrarian Change in the Gambia." *Economic Geography* 69 (4): 329–348.

Chapman, Robert. 2006. "The Eco-Pre-Fix: Reading 'Conservation as Enclosure'." *Capitalism Nature Socialism* 17 (4): 48–57.

Chattopadhyay, Sutapa. 2022. *Politics of Development and Forced Mobility: Gender, Indigeneity, Ecology*. New York: Springer Nature.

Clark John. 2000. "The Matter of Freedom: Ecofeminist Lesson for Social Ecology." *Capitalism Nature Socialism* 11 (3): 62–80.

Clark John P. 2001. “Alan Dordoy and Mary Mellor’s ‘Ecosocialism and Feminism’.” *Capitalism Nature Socialism* 12 (1): 87–91.

Clark, John P. 2017. Foreward, in *Ecofeminism as Politics: Nature, Marx and the Postmodern*. Ariel Salleh, London: Zed Books.

D’Eaubonne, Françoise. 1974. *Le Féminisme ou la Mort* [Feminism or Death]. Paris: P.Horay.

della Porta, Donatella, and Gianni Piazza. 2016. “Il cambiamento di scala del Movimento No MUOS: oltre la protesta contro l’inquinamento elettromagnetico.” *StrumentiRes. Rivista online della Fondazione RES* 2 (7): 1–28.

Diamond, Irene and Gloria Feman Orenstein. 1990. *Reweaving the World: the Emergence of Ecofeminism*. San Francisco, CA: Sierra Book Club.

Dordoy Alan & Mary Mellor. 2000. Ecosocialism and Feminism: Deep Materialism and the Contradictions of Capitalism.” *Capitalism Nature Socialism* 11 (3): 41–61.

Dordoy, Alan, and Mary Mellor. 2001. “John P. Clark’s Ecofeminist Lessons for Social Ecology.” *Capitalism Nature Socialism* 12 (1): 92–98.

Faber, Daniel, and James O’Connor. 1991. “Reply.” *Capitalism Nature Socialism* 2 (1): 137–140.

Farraclas Nicholas. 2008. “On the Birthmarks of the Old Society: A Reflection on the Exchange Between Maria Mies and Victor Wallis.” *Capitalism Nature Socialism* 19 (1): 112–118.

Federici, Silvia. 2004. *Caliban and the Witch*. New York: Autonomedia.

Federici Silvia. 2006. “Capital and the Body: A Rejoinder to Salvatore Engel-Di Mauro.” *Capitalism Nature Socialism* 17 (4): 74–77.

Foster, Emma. 2021. “Ecofeminism Revisited: Critical Insights on Contemporary Environmental Governance.” *Feminist Theory* 22 (2): 190–205.

Foster, John Bellamy, and Paul Burkett. 2001. “Marx and the Dialectic of Organic/Inorganic Relations: A Rejoinder to Salleh and Clark.” *Organization & Environment* 14 (4): 451-462.

Gaard, G. 2015. “Ecofeminism and Climate Change.” *Women’s Studies International Forum* 49: 20–33.

Gaard, Greta. 2017. *Critical Ecofeminism*. Lanham: Lexington Books.

Giacomini, Terran. 2022. “Agroecology as Ecofeminist Activism.” In *The Routledge Handbook on Ecosocialism*, edited by Leigh Brownhill, Salvatore Engel-Di Mauro, Terran Giacomini, Ana Isla, Michael Löwy, and Terisa E. Turner, 189–200. New York: Routledge.

Giacomini, Terran, Terisa Turner, Ana Isla, and Leigh Brownhill. 2018. “Ecofeminism Against Capitalism and for the Commons.” *Capitalism Nature Socialism* 29 (1): 1–6.

Godfrey, Phoebe. 2005. “Diane Wilson vs. Union Carbide: Ecofeminism and the Elitist Charge of ‘Essentialism’.” *Capitalism Nature Socialism* 16 (4): 37–56.

Godfrey, Phoebe. 2006. "Essentialism and the Semantics of Resistance: A Rejoinder to Jesse Goldstein." *Capitalism Nature Socialism* 17 (4): 103–104.

Goettner-Abendroth, Heide. 2018. "Re-thinking 'Matriarchy' in Modern Matriarchal Studies Using Two Examples: The Khasi and the Mosuo." *Asian Journal of Women's Studies* 24 (1): 3–27.

Goldstein Jesse. 2006. "Ecofeminism in Theory and Praxis." *Capitalism Nature Socialism* 17 (4): 96–102.

Griffin, Susan. 1978. *Women and Nature: The Roaring Inside Her*. New York: Harper and Row.

Hultman, M. 2017. "Green Men? Exploring Industrial, Ecological, and Ecomodern Masculinity." In *Routledge Handbook of Gender and Environment*, ed. S. MacGregor, 243–257. New York: Routledge.

Ifeakachukwu, Nwosa P. 2020. "Globalisation, Economic Growth and Income Inequality in Nigeria." *Indian Journal of Human Development* 14 (2): 202–212.

Isife, Evaristus Emeka. 2023. "Matrirachy in Traditional Africa and Its Relevance to Contemporary Feminism." *GWEBUIKE: An African Journal of Arts and Humanities* 9 (4): 1–21.

Isla, Ana. 2005. "Conservation as Enclosure: An Ecofeminist Perspective on Sustainable Development and Biopiracy in Costa Rica." *Capitalism Nature Socialism* 16 (3): 49–61.

Isla, Ana. 2006. "Women, Enclosure, and Accumulation: A Rejoinder to Robert Chapman." *Capitalism Nature Socialism* 17 (4): 58–65.

Isla, Ana. 2018. "The 'Greening' of Costa Rica. The War against Subsistence." In *Climate Chaos. Ecofeminism and the Land Question*, edited by A. Isla, Chapter 15, 418–443. Toronto: INANNA Publications and Education.

Isla, Ana. 2022. "'Greening,' the highest stage of extractivism in Latin America." In *The Routledge Handbook on Ecosocialism*, edited by Leigh Brownhill, Salvatore Engel-Di Mauro, Terran Giacomini, Ana Isla, Michael Löwy, and Terisa E. Turner, 57–66, 67–80. New York: Routledge.

Kaara Wahu. 2010. "Reclaiming Peoples' Power in Copenhagen 2009: A Victory for Ecosocialist Ecofeminism." *Capitalism Nature Socialism* 21 (2): 107–111.

King, Ynestra. 1983. "The Eco-feminist Imperative." In *Reclaim the Earth: Women Speak Out for Life on Earth*, edited by Léonie Caldecott and Stephanie Leland, 9–14. London : Women's Press.

Kollontai, Alexandra. 1977 [1920]. "Communism and the Family." In *Selected Writings of Alexandra Kollontai*, translated by A. Hilt, 250–260. Westport: Lawrence Hill.

Kovel Joel. 2005. "The Ecofeminist Ground of Ecosocialism." *Capitalism Nature Socialism* 16 (2): 1–8.

Leacock, Eleanor. 1992. "Women's Status in Egalitarian Society: Implications for Social Evolution." *Current Anthropology* 33 (S1): 225–259.

Mallory, Chaone. 2006. "Ecofeminism and Forest Defense in Cascadia: Gender, Theory and Radical Activism." *Capitalism Nature Socialism* 17 (1): 32–49.

Mellor, Mary. 1992. "Eco-Feminism and Eco-Socialism: Dilemmas of Essentialism and Materialism." *Capitalism Nature Socialism* 3 (2): 43–62.

Mellor, Mary. 1994. "Varieties of Ecofeminism." *Capitalism Nature Socialism* 5 (4): 117–125.

Mellor, Mary. 1997. *Feminism and Ecology*. Oxford: Polity.

Mellor, Mary. 2022. "Ecofeminist Ecosocialism." In *The Routledge Handbook on Ecosocialism*, edited by Leigh Brownhill, Salvatore Engel-Di Mauro, Terran Giacomini, Ana Isla, Michael Löwy, and Terisa E. Turner, 49–56. New York: Routledge.

Merchant, Carolyn. 1980. *The Death of Nature: Women, Ecology, and the Scientific Revolution*. London: Harper & Row.

Mies, Maria. 1986. *Patriarchy and Accumulation on a World Scale: Women in the International Division of Labour*. London: Zed Books.

Mies, Maria and Vandana Shiva. 1993. *Ecofeminism*. London: Zed Books.

Mies, Maria. 2006. "Questioning Needs: A Rejoinder to Victor Wallis." *Capitalism Nature Socialism* 17 (4): 44–47.

Minoia, Paola. 2020. "Corporate Land Grabs: Colonial Continuity and Space of Exception in Kenya." *Land Use Policy* 99: 104964.

Mohai, Paul. 1997. "Men, Women, and the Environment: An Examination of the Gender Gap in Environmental Concern and Activism." In *Women Working in the Environment: Resourceful Natures*, edited by Carolyn E. Sachs, 215–239. London: Taylor & Francis.

Noble, Charles, John Wooding, and Lori Ann Thrupp. 1989. "The Struggle for Nature." *Capitalism Nature Socialism* 1 (3): 163–174.

Norgaard, Kari, and Richard York. 2005. "Gender Equality and State Environmentalism." *Gender & Society* 19 (4): 506–522.

O'Connor, James R. 1998. *Natural Causes. Essays in Ecological Marxism*. New York: Guilford.

Oksala, Johanna. 2018. "Feminism, Capitalism, and Ecology." *Hypatia 33*(2): 216–234.

Perkins, Patricia E. Ellie. 2013. "Environmental Activism and Gender." In *Handbook of Research on Gender and Economic Life*, edited by Deborah M. Figart and Tonia L. Warnecke, 504–521. Edward Elgar Publishing.

Plumwood, Val. 1993. *Feminism and the Mastery of Nature*. New York: Routledge.

Plumwood, Val. 2001. "Nature as Agency and the Prospects for a Progressive Naturalism." *Capitalism Nature Socialism* 12 (4): 3–32.

Rosewarne, Stuart. 2006. "Socialist Ecology's Necessary Engagement with Ecofeminism." *Capitalism Nature Socialism* 17 (4): 78–86.

Rudy, Alan. 2006. "On the Ecofeminist Editorial: 'Moving to an Embodied Materialism'." *Capitalism Nature Socialism* 17 (4): 105–114.
Salleh, Ariel. 1995. "Nature, Woman, Labor, Capital: Living the Deepest Contradiction." *Capitalism Nature Socialism* 6 (1): 21–39.
Salleh, Ariel. 1997. *Ecofeminism as Politics: Nature, Marx and the Postmodern*. London: Zed Books.
Salleh, Ariel. 2001. "Sustaining Marx or Sustaining Nature? An Ecofeminist Response to Foster and Burkett." *Organization & Environment* 14: 443–450.
Salleh, Ariel. 2005. "Moving to an Embodied Materialism." *Capitalism Nature Socialism* 16 (2): 9–14.
Salleh, Ariel. 2006a. "Towards an Inclusive Solidarity on the Left: Editor's Introduction." *Capitalism Nature Socialism* 17 (4): 32–37.
Salleh, Ariel. 2006b. "Embodying the Deepest Contradiction: A Rejoinder to Alan Rudy." *Capitalism Nature Socialism* 17 (4): 115–124.
Salleh, Ariel. 2009. "Ecological Debt: Embodied Debt." In *Eco-sufficiency and Global Justice: Women Write Political Ecology*, edited by Ariel Salleh, 1–41. Melbourne: Spinifex.
Salleh, Ariel. 2010. "From Metabolic Rift to 'Metabolic Value': Reflections on Environmental Sociology and the Alternative Globalization Movement." *Organization & Environment* 23 (2): 205–219.
Salleh, Ariel. 2017. *Ecofeminism as Politics: Nature, Marx and the Postmodern*, 2nd ed. London: Zed Books.
Salleh, Ariel. 2022. "An Ecocentric Epistemology for Ecosocialism." In *The Routledge Handbook on Ecosocialism*, edited by Leigh Brownhill, Salvatore Engel-Di Mauro, Terran Giacomini, Ana Isla, Michael Löwy, and Terisa E. Turner, 57–66. New York: Routledge.
Salleh Ariel, and Martin O'Connor. 1991. "Discussion. Eco-socialism/Eco-feminism." *Capitalism Nature Socialism* 2 (1): 129–137.
Schroeder, Richard A. 1999. *Shady Practices: Agroforestry and Gender Politics in the Gambia*. Berkeley: University of California Press.
Tindall, David B., Scott Davies, and Celine Mauboules. 2003. "Activism and Conservation Behavior in an Environmental Movement: The Contradictory Effects of Gender." *Society & Natural Resources* 16 (10): 909–932.
Tristan, Flora. 1843. "To Working Men and Working Women." In *Before Marx: Socialism and Communism in France, 1830–48*, edited by P.E. Corcoran, 112–125. London: Palgrave Macmillan.
Tsimitakopoulos, Keryn. 2022. The Niger Delta Basin: Reform Has Finally Come, But Can It Revive Investment in the Oil and Gas Sector? S&P Global Commodity Insights Blog, January 19. https://www.spglobal.com/commodityinsights/en/ci/research-analysis/the-niger-delta-basin-reform-has-finally-come-but-can-it-reviv.html. Accessed 20 January 2024.

Turner, Terisa E. 2022. "Ethnicized, Gendered Class Analysis: A Theoretical-Methodological Framework for Analyzing Ecofeminist, Ecosocialist Praxis." In *The Routledge Handbook on Ecosocialism*, edited by Leigh Brownhill, Salvatore Engel-Di Mauro, Terran Giacomini, Ana Isla, Michael Löwy, and Terisa E. Turner, 39–48. New York: Routledge.

Turner, Terisa E., and Leigh S. Brownhill. 2004. "We Want Our Land Back: Gendered Class Analysis, the Second Contradiction of Capitalism and Social Movement Theory." *Capitalism Nature Socialism* 15 (4): 21–40.

Turner, Terisa E., and Leigh Brownhill. 2006. "Ecofeminism as Gendered, Ethnicized Class Struggle: A Rejoinder to Stuart Rosewarne." *Capitalism Nature Socialism* 17 (4): 87–95.

Valera, Luca. 2017. "Françoise d'Eaubonne and Ecofeminism: Rediscovering the Link Between Women and Nature." In *Women and Nature? Beyond Dualism in Gender, Body, and Environment*, edited by Douglas A. Vakoch, and Sam Mickey, 10–23. New York: Taylor & Francis.

von Werlhof, Claudia. 2007. "No Critique of Capitalism Without a Critique of Patriarchy! Why the Left Is No Alternative." *Capitalism Nature Socialism* 18 (1): 13–27.

Wallis, Victor. 2006. "Vision and Strategy: Questioning the Subsistence Perspective." *Capitalism Nature Socialism* 17 (4): 38-43.

Wallis, Victor. 2008. "On Marxism, Socialism, and Ecofeminism: Continuing the Dialogue." *Capitalism Nature Socialism* 19 (1): 107–111.

Warren, Karen. 2000. *Ecofeminist Philosophy: A Western Perspective on What It Is and Why It Matters*. Lanham: Rowman & Littlefield.

Warren, Karen, and Nisvan Erkal. 1997. *Ecofeminism: Women, Culture, Nature*. Bloomington: Indiana University Press.

Wolf, Frieder Otto. 2007. "The Missed Rendezvous of Critical Marxism and Ecological Feminism." *Capitalism Nature Socialism* 18 (2): 109–125.

CHAPTER 7

First Peoples and Ecosocialism

The Context of First Peoples

Kaianerohkó:wa, or Great Goodness (in Kanien'kéha), is a constitutional framework developed over centuries by the Rotinonhsión:ni Confederacy, handed down from the wisdom of Wendat peacemaker Dekanawida, supported by the mediator Mother of Nations Jigonhsasee and the Kanien'kehá:ka chief Ayonwentha. It is the basis of lasting peaceful relations among Rotinonhsión:ni Peoples, establishing norms of mutual respect and mutually beneficial roles, elderly women's prominent roles (with the power of instant recall of chiefs), balancing personal autonomy with multiple levels of socially integrative structures, as well as recognition and respect for all nonhuman relations. It is a paragon of combined substantive democracy with ecological sensibility, a breakthrough achieved centuries ago and unrivalled by any liberal democratic constitutions to this day.

Kaianerohkó:wa principles were among the main parameters whereby First Peoples on the north-eastern side of Great Turtle Island conducted international relations, including the first European settlers (Alfred, G.R. 1994, 77–82; see especially Kahentinetha Rotiskarewake's explanatory essay in Hall 2023, 78–100). The Kaianerohkó:wa informs the Kaswénhta

S. Engel-Di Mauro, *Ecosocialism*, Marx, Engels, and Marxisms, https://doi.org/10.1007/978-3-031-74288-0_7

treaty of cross-generational coexistence with European settlers since 1613 (Bedford and Workman 2012, 26). At the time of writing, it is the Kaswénhta's 410th anniversary year. The treaty is represented by the Teiohá:te, a belt with two parallel purple rows on a white background. It symbolises

> a river where the Indigenous people's canoe and the settler people's ship sail side by side. As an alliance belt, it suggests that both parties can only advance in the same direction if they refrain from encroaching on each other's path—thus, precluding both assimilation policies and cultural appropriation. (Hall 2023, 216)

Colonial authorities, and many settlers more broadly, most egregiously broke the Kaswénhta agreements by occupying land illegally, waging wars of conquest, expelling Rotinonhsión:ni communities, and other horrors, and imposing settler laws, languages, and belief systems, and many other actions causing enduring harms and abuses. The pellucid breach of the treaty continues in flagrant fashion.

Much of Rotinonhsión:ni Confederacy land comprises what is now New York State, from where I am writing. The settler belligerence and genocidal violence that is intrinsic to the US empire and led to the imposition of the state of New York is also a crass attempt at erasing the Kaswénhta treaty (in every sense). It is even a living breach of the US empire's own laws (Mohawk 2004). Considering the global significance of New York State, including for being the third largest economy within the US imperial metropole (and the tenth largest capitalist economy in the world), the struggle for the restoration of sovereignty by some of the Rotinonhsión:ni First Nations is of global significance. It should be considered crucial to the anti-imperialist Left and to any political formation seeking decolonisation.

The ways and experiences of the Rotinonhsión:ni Confederacy and their descendents form part of a vast universe of First Peoples.[1] Thousands of different ways of being and of belonging to life-worlds and environments persist, thanks also to hundreds of years of struggles to overcome, or just to survive the onslaught of capitalist settler colonialism, likely the most genocidal process known to human history. It is impossible within a lifetime even to grasp the most superficial aspects of so many histories, interrelations, cultural frameworks, languages, cosmologies, forms of food procurement, value systems, beliefs, legends, economic structures, and other dimensions of constantly evolving ecological relations and of social life within and between First Peoples (for an inkling of this immense variety, see, for example, Chase-Dunn and Hall 1998; Dunbar-Ortiz 2014; Warrior 2015).

Each First People situation and historical trajectory differs and is complicated by dynamic and changing interrelations with other First Peoples and settler colonial communities and with the living worlds or environments that they also shape over time. The same applies to any discussions of commensurate scale, such as in the case of African Peoples and Afro-Descendent Communities globally. Any description of such an immense universe of interactions within a work on ecosocialism can only be a rough sketch at most. The point is to appreciate the diversity and push for greater familiarity and precision, including for ecosocialists, with at least the First Peoples where they live or in the communities of struggle of which they are part. As pertinent to the scale of analysis and relative

[1] Different terms are used to refer to peoples who have inhabited a place prior to the genocidal imposition of settler colonial regimes, such as Aboriginal, Autochthonous, Native, Indigenous, Tribal, and First Nations. All such terms carry with them major problems or political disadvantages, or they may be entirely acceptable, depending on context. The phrase First People is preferred here as an attemt to move away from concepts that are redolent with colonial ideology and mainly associated with political constructs derived from European norms and histories (see Peters and Mika 2017). However, because of writing from within a Great Turtle Island context (North America), I will also use the terms First Nations and Indigenous. The first serves as a critical contrast to Tribe or Band, terms associated with US and Canadian settler colonial institutions, respectively. They denote a lesser political status compared to the term Nation. Since it is widely used internationally and in academic settings, Indigenous will also feature as a referent, but mainly when discussing global processes. Terminology is important in ideological struggles, but there is no illusion here that it does much to change the existing balance of power towards susbstantive decolonisation.

to the author's limited understanding, the name preferred by the mainstream or intellectual representatives of a First People will be used. That is, when speaking in generic terms, the term First People or Indigenous is used, otherwise specifying names.

At first, a general picture of multiple kinds of contexts of Indigenous struggles is discussed, relying on several concrete and recent examples of struggles in different places in Great Turtle Island[2] and Abya Yala.[3] This is followed by a brief overview of ecologically sustainable Mamaceqtaw (Omǣqnomenēwak or Menominee) forestry as an example of major shifts within the Mamaceqtaw Nation, including forceful adaptation to and partial adoption of capitalist relations, in their long-standing struggles to regain land appropriated by colonisers. The main reason for diving into such aspects of Mamaceqtaw history, aside from my greater familiarity with it, is to demonstrate the importance of studying specific circumstances and trajectories. It helps avoid the presumption that First Peoples are necessarily invested in anti-capitalist politics on account of their ecologically more sustainable practices and anti-colonial resistance. The same goes for the much discussed or abused concepts and perspectives of Sumak Kawsay and Buen Vivir, which are taken up after the discussion on Mamaceqtaw forestry. The final section is dedicated to an appraisal of ecosocialist views on First Peoples.

A Diversity of Struggles and Contradictions

Íŋyaŋ Woslál Háŋ, known to most as Standing Rock Reservation, gained notoriety with the 2016 protests against the Dakota Access pipeline. Police and military, even using armoured vehicles, brutally squashed the protests, enabling pipeline building to be completed in 2017. It was a

[2] This is a reference to the process whereby the Earth came into being and that features the prominent role of a Great Turtle, as understood among First Peoples such as those of the Lenape and Haudenosaunee confederacies. Since the 1970s, Great Turtle Island has been used as a metonym by many in First Peoples communities to indicate lands roughly corresponding to Central and North America.

[3] A Guna phrase that means "land in full maturity," the culmination of succeeding planetary cycles that have led to the cycle of life. Since the 1980s, it is used especially in Central and South America to indicate the Americas or sometimes the regions of Central and South America. The term is not only a geographical referent, but also a concept to signal "a differentiated indigenous locus of cultural and political expression" (del Valle Escalante 2014, 1).

show of force by the settler colonial state. Running roughshod over sacred sites and threatening local water supplies as well as the 1851 Traverse de Sioux and 1868 Fort Laramie Treaties, US government institutions succeeded in forcing piped oil flow across a landscape again defiled, as part of a continuing war against the First Nations in the Mni Sose (Missouri River) catchment (Estes 2019). This is unlike the fate of the TransCanada Keystone XL, which, to simplify, was effectively blocked with US Presidential help and by means of coalitions across the settler colonial (Canada-US) border. The intervention of many from diverse First Nations was crucial to the effort.

Yet Keystone XL was but one of four parts of a multi-tentacled project, which remains operational and, as one of countless examples, demonstrates the great odds faced by most First Peoples in just voicing opposition even with support and direct involvement from parts of the settler communities as well as people descendent from many different African and Asian societies (see, for example, Estes 2019). More broadly, the clear failure to stop all Keystone and other recently built pipelines, demonstrates the severe limits of opposition movements who remain within the framework of liberal democracies' legal norms, which are among the pillars of settler colonial institutions.

With respect to First Peoples' struggles for decolonisation or sovereignty, and often just for self-defence or sheer survival, the challenges are even more formidable because, aside from colonial oppression, political unity or concerted action within and between First Peoples is generally lacking, and there is no necessary coincidence of material interests (see, for instance, Johanson 2004). In the case of organising against the Dakota Access pipeline, for example, unity of purpose was not a foregone conclusion.

More recently, at the time of writing, there are sharp disagreements being stirred up within Diné and Pueblo communities about a buffer zone planned about Chaco Canyon National Park prohibiting any drilling for fossil fuels for 20 years on public lands. This is a major achievement from local grassroots Diné and Pueblo communities, such as The Greater Chaco Coalition and The Red Nation, and including some Diné and Pueblo private landowners, in pressurising the US Federal Government to institute greater protective measures for the sacred location and for local people's health. Much of the controversy is related to fears from some of the private landowners (allottees) from those communities who receive mineral extraction royalties and could be hurt financially by the prohibition. The matter remains divisive within the Diné and Pueblo communities as a result of misinformation from some local politicians and influential figures, magnified by right-wing media (Joe 2023). The controversy has been exacerbated by the Navajo Nation Council passing a resolution against the buffer (*Indian Country Today* 2023). In some respects, historically, the current tension is traceable to the 1887 General Allotment Act, which was an attempt to impose private property on First Peoples and thereby dissolve them socially into the capitalist settler colonial world.

Some controversy has also arisen, including among some within First Nation communities, with respect to the excessive killings of bison in a winter-time hunt in 2023 at Yellowstone National Park. Bison hunting is allowed by treaty rights to Nez Perce, Blackfeet, Confederated Tribes of the Umatilla Reservation, Northern Arapaho, Confederated Salish and Kootenai Tribes, Crow, and Shoshone-Bannock Nations. Or, one could say, treaty rights are to recognise and protect from settlers what First Nations in the area have been doing centuries before colonial invasions. The large number of bison killed, in any case, is due mainly to widespread deprivation among First Nations (a long-term structural problem of colonial dispossession) and the state of Montana refusing to have bison roam and to have members of First Nations hunting throughout the park. The latter restrictions increase the potential for the spread of disease, which is among the pretext used by settler coloniser livestock businesses to complain about bison protection measures (Planas 2023). There could be many other examples of divisions within First Nations, at times degenerating into violent confrontations within those communities, such as the US state-sponsored paramilitary terrorism inflicted on Wazí Aháŋhaŋ Oyáŋke (Pine Ridge Reservation) in the 1970s to destroy the American Indian Movement (Estes 2019).

Nick Estes recounts how, in 2014, he had felt compelled to protest his own Nation, the Oceti Sakowin Oyate (the Nation of the Seven Council Fires), who, among some other First Nation authorities, at first supported the Keystone XL pipeline (Estes 2019, 12). This is in addition to a highly selective or sparse outrage among settler colonial communities, but also from First Peoples who are not directly affected, about settler colonial institutions' ecological harms.

> Half a century ago, there were no mass protests against the dams that still wreak havoc on our river [Mni Sose], a history I have spent the more than

a decade speaking and writing about, with little interest from the outside world. (Estes 2019, 17)

A reason for this seems to be a lack of substantive interest in, if not complete disregard for First Nations' sovereignty or self-determination among those purporting to support First Nations' causes. This is aside from settler colonial institutions' suppression of any form or degree of First Nation sovereignty.[4]

There are myriad cases, both long-standing and recent, of ecosocial harm perpetrated on First Peoples that gather little to no interest beyond the Indigenous communities affected. Closer to where I am writing, on land of the Lenni Lenape Esopus (whose descendents reside among the Stockbridge-Munsee in Wisconsin and Ontario), many poorly publicised struggles dot and permeate the many landscapes to which the Rotinonhsión:ni (Haudenosonee or "Iroquois"[5]) belong. Since the 1770s, the *Onoñda'gegá'* Nation have been fighting American occupation and mass dispossession by the State of New York and major political figures, who include George Washington. The takeover was illegal under both Rotinonhsión:ni and US Congress laws, never mind the infringement of treaties. A recent decision, after nearly a decade, by the Inter-American Commission on Human Rights (Organisation of American States or OAS) formally recognised the unlawful land seizure of more than a million hectares, legitimising at the international level the *Onoñda'gegá'* claims. This was after decades of attempts at filing suits, to no avail, through the settler colonial regime's sham courts, where land title charlatanism reigns supreme by means of the "Doctrine of Christian Discovery," which is now rejected even by its original Vatican instigator (Mohawk 2004; Pietrorazio 2023). It is doubtful that any land will be recovered, regardless of the US membership to the OAS (though the US government have no qualms

[4] For those preferring to move away from what could be taken as state-centred political formulations, the issue can be reframed as upholding First People's norms based on leadership "exercised by persuading individuals to pool their self-power in the interest of the collective good" (Alfred, T. 1999, 25). This, to my understanding, pertains at least in polities like the Kahenien'kehá:ka, or in a confederation like that of the Rotinonhsión:ni (Haudanosaunee or Iroquois).

[5] This is a slanderous term referring to rattlesnakes in a pejorative sense (Alfred, T. 1999, xxv–xxvi), so it is not used here except to guide the reader unfamiliar with Indigenous Peoples' names for themselves, their political institutions, and related terms.

instrumentalising that international institution to exclude Cuba and to facilitate coups, as in Bolivia in 2019; Farthing and Becker 2021).

About 200 Km west of the *Onoñda'gegá'* reservation, another struggle—centred on tightly combined health, livelihood, and ecological concerns—is unfolding and barely covered even by the local press. The Ta:nöwö:de' Onödowá'ga: Yoindzade' (Tonawanda Seneca Nation), at the time of writing, have been leading local residents' mobilisation to defend roughly 270 hectares of wetland and grassland in the Iroquois National Wildlife Refuge as well as local waterways from being polluted by the Science & Technology Advanced Manufacturing Park (STAMP) manufacturing site. There, several businesses have been pressurising local government to allow for the building of facilities to produce hydrogen and semiconductor components, among other materials. The conservation area under threat is located in proximity to the Ta:nöwö:de' Onödowá'ga: Yoindzade' reservation. They have taken care of the nearby wetlands and grasslands for generations even if settler colonial officialdom still does not recognise those conservation areas as part of Ta:nöwö:de' Onödowá'ga: ancestral land. The business pressure is in response to the US government's showering funds to the private high-technology sector (the 2022 CHIPS and Science Act), a desperate attempt to keep those industries within the US (Owens 2023; Rivers 2023).

These are just two of innumerable Indigenous struggles that get hardly any attention generally and, it must be added, by much of the Left, including ecosocialists, and certainly not just in the case of New York State. The Dakota Access pipeline was doubtless of eminent importance. One could reason that Indigenous defence of lands they inhabit, and blockage of fossil-fuel infrastructure are paramount at this conjuncture, even as the effort against the Dakota Access pipeline was defeated. However, the possibility of overturning the basis of colonial law and restoring land to a First People, threats to the well-being of an Indigenous community and to a wildlife refuge hosting hundreds of migratory species, and the potential stream pollution leading to Lake Erie all seem worth much greater attention than relegation to sporadic coverage in a few local papers.

There is an unsettling selectivity as to which Indigenous struggles get to count and to what degree. There is no space here to undergo a search for causes but, given that it is First Peoples having initiated all the above-described actions of resistance, it seems the selectivity and attentiveness problems lie well beyond Indigenous activism. Different forms of Pan-Indigenous organising have helped spread the word in many instances and facilitate alliances otherwise not had (see examples in Estes 2019; Grossman 2017). Still, the reach is often limited. And it cannot be otherwise, as the means of communication are a ruling class monopoly, and much effort is still needed to incite most settlers to take First Peoples' self-determination struggles as crucial to everyone's future. There have been many examples of Pan-Indigenous organising, such as the more well-known Society of American Indians, the American Indian Movement, and the Women of All Red Nations (Estes 2019). One could go further back to include alliances like those spearheaded by Shawaanwa Chief Tecumseh in the early 1800s. A recent initiative is the first Congreso Indígena para la Descolonización (Indigenous Congress for Decolonisation),[6] held on Nahua (Pipil) land in what is now El Salvador, where more than a hundred leaders from First Peoples across Central and North America met to share about experiences and diverse circumstances and discuss many intersecting themes related to decolonisation. It remains to be seen whether wider, especially settler society will organise themselves to support those movements, to help topple settler colonial regimes.

It is no shortage of Indigenous activism that hampers the formation of protest movements much wider and longer lasting than what is usually the case. These problems are exacerbated by additional challenges. Within a settler colonial (and liberal democratic) system, aside from suppression or Leftist disinterest, fighting for cultural revitalisation or survival can also be fraught with conflict internal to and between First Nation communities. As Kahnawà:ke Kaniekehaka scholar Gerald Robert Alfred explained:

> Different views on the nature and meaning of tradition lead to internal factionalism and conflict; challenging the laws and structures of a colonial regime leads to reaction by, and confrontations with, the state. This is the price to be paid for engaging in a nationalist assertion—for being true to the wisdom and integrity of our tradition, for heeding the voices of our ancestors. (Alfred, G.R. 1994, 2; see also Alfred, T. 1999, 1–2)

There are contradictions among and within First Nations, mainly wrought through centuries of genocidal colonisation and capitalist encroachment

[6] See: https://www.fger.org/los-pueblos-indigenas-existimos-primer-congreso-indigena-para-la-descolonizacion/, accessed 1 August 2023.

but also long-standing ones that preceded European invasions and were exacerbated thereby.

Such processes are illustrated in the settler colonial context of Guyana, where tensions within and between Indigenous communities are in part due to diverging priorities over land use. There is “a diversity of motivations underlying indigenous lands claims, including those animated by securing and exploiting extractive wealth” (Hook 2019, 49). Many in Akawaio, Carib, Makushi, and Patamona communities engage in small-scale, artisanal gold mining and thereby contribute to local pollution problems and tensions within and between those communities, though nowhere near the level of corporate or state mining firms. In contrast, the majority of Makushi have wrought alliances with outsiders, including settlers, to resist, if not stave off plantations, logging, and mining on their lands because of their understanding of such extractive practices as having deleterious effects (Whitaker 2020).

Mamaceqtaw Forests

Similar processes can be gleaned from Mamaceqtaw[7] history to understand the current contradictory situation of ecologically sustainable forestry integrated into the global capitalist timber trade. Since being forced into a reservation in 1854, the Mamaceqtaw have taken to community-owned commercial logging over 900 km^2 of forest, applying a mix of ancestral and conventional forestry techniques and maintaining high levels of biodiversity, net primary productivity, and tree density. The forest comprises about 95% of land remaining under Mamaceqtaw administration out of an original 40,500 km^2. Ironically, what the settler colonial regime forced on the Mamaceqtaw was at the time considered of marginal economic utility. The forest of the Mamacetaw reservation comprises a large chunk of what remains of the region’s old-growth Northern Hardwoods-Hemlock-White Pine association (Dockry and Whyte 2021; Pecore 1992; Runstrom et al. 2002; Trosper 2007; Sands and Abrams 2011).

[7] Mamaceqtaw means “the people who live according to the seasons.” European colonists came to know them as Omäëqnomenëwak, a term derived from Anishinaabe that means “the wild rice people,” given their reliance on wild rice (*Zizania aquatica*) as food staple.

Thanks to selective harvesting mandated under Chief Oshkosh (1827–1858), there are currently more trees standing compared to when logging and sawmill operations started nearly 175 years ago. More than five million m^3 of wood have been taken and sold during that time, which is more than the nearby and twice-larger Nicolet National Forest. Mamaceqtaw commercial forestry has become an internationally renowned example of ecological sustainability nestled within a wider landscape appropriated and ravaged by capitalist settler colonial society (Beck 2002; Davis 2000; Dorney and Dorney 1990; Kay 1979; Mausel et al. 2017; Wikstrom 2023, 106–107; Wolfley 1998).

The outlook is not necessarily green, however (and certainly far from ecosocialist!). Deepening capitalist pressures and the consequent reshaping of Mamaceqtaw society create well-known contradictions within the Mamaceqtaw world, between ecological sustainability and a form of dependent economic growth (e.g., profits from logging and gambling) enforced by a settler colonial empire. The lumber industry, under Menominee Tribal Enterprises, remains a main source of employment but, especially with increasing mechanisation, requires ever fewer employees. Poverty and unemployment are widespread, housing is in short supply, basic services and amenities are scarce or absent, and much infrastructure is in a state of decay. This situation has led to most Mamaceqtaw having to live outside of the reservation. There is pressure to turn to clear-cutting for urban expansion and to seek larger profit margins for debt repayment, recovery of land lost through the Termination Act period (see below), and accommodation of newcomers into the wealth redistribution system (Beck 2005; Hosmer 1999; LaDuke 1994; Menominee Tribal Enterprises 1997).

There is no international outcry to relieve pressure on the Mamaceqtawak and thereby on their forest ecosystems, just as there is very little to none about most First Peoples' struggles. What should be of interest to ecosocialists is how the Mamaceqtawak have achieved a form of ecologically sustainable forestry by means of capitalist relations under settler colonial oppression. To understand this process, which contradicts any simple equation of First People with frontline ecological struggles, a brief historical overview can be helpful.

Prior to the colonial invasion and massive land theft, the Mamaceqtawak, who tended to lead a seasonal rounds (semi-nomadic) existence, had progressively enlarged forest-grassland ecotones by means of controlled fires, affected fish and ungulate migration patterns through hunting, and modified ecosystem composition through agriculture, gathering (food and medicinal plants), maple syrup production, and manoomin (wild rice) cultivation. There is a specific gendered aspect to how these ecological relations unfolded and were maintained. As described by Whyte (Potawatomi), Caldwell (Mamaceqtaw), and Schaefer (Odawa),

> women of Menominee and related groups had special leadership and expert positions in relation to activities such as those involving wild rice. Gender difference, then, was not associated with oppression but with responsibilities entrusted within a society. This is why the connection between patriarchy and settler colonialism creates distrust between men, women, and two-spirit persons in many Indigenous peoples, as trust is replaced with oppressive gender relations. This oppressive connection has played a role in what are now many morally problematic, male-dominated Tribal governments and agencies today and the heightened risks of sexual violence and murder that Indigenous women and two-spirit persons face. And even the most clearly recorded articulations of some Indigenous traditions, … are now told in a masculinist way. (Whyte et al. 2018, 161)

Nevertheless, in the face of enormous odds, the Mamaceqtaw Nation have maintained ecologically constructive relations (Flader 1983; Loop and Anderton 1998; Sands and Abrams 2011; Whitney 1996). Mamaceqtaw social relations, even as they have shifted in response to settler colonial invasions, did not support economic shifts that would lead to excessive resource extraction and environmental degradation, as in the case of European capitalist societies (Beck 2002; Bieder 1995; Holm 1992).

There are some exceptions as well that must be acknowledged. About 350 years ago, destructive resource extraction did occur from among Mamaceqtaw Nation by way of the fur trade and commodity exchange with French colonists. This was a process common to other regions of Great Turtle Island under capitalist encroachment. It was how Mamaceqtaw came to be affected by a globally expanding capitalist circuit and their oscillations. British colonial expansionism wrought direct impacts on the Mamaceqtaw with military incursions into major hunting and fishing areas and increasing colonial deforestation, along with decimating smallpox epidemics and pressures through forced western migrations of other First Peoples, fleeing wars and European colonial invasions. A gender-inflected social change followed, whereby leaders and people of mixed descent gained more political prominence, especially through dealing with colonists and colonial authorities who negotiated exclusively with males (Beck 2002; Bieder 1995; Hosmer 1997, 1999; Jaimes and Halsey 1992; Kay 1984; Whitney 1996).

Facing bleak prospects of land loss and genocide, Mamaceqtaw leadership centred foreign policy on seeking neutrality or forging alliances with more powerful forces. For instance, they avoided conflict with the British during the Seven Years War (1756–1763), despite French commercial and military ties, while, in 1730, many Mamaceqtaw warriors joined the French armies in defeating the Meskwaki. They allied with the British during the 1776 war of secession and against the Long Knives (Americans) in the War of 1812. But they did not participate in pan-Indigenous alliances against European powers, such as those led by Pontiac in 1763 and Tecumseh in 1810. Such foreign policies were ultimately unsuccessful. By 1816, Mamaceqtaw chiefs were forced to promise their loyalty to the new US empire, in continuity with a pacification-oriented foreign policy. Since then, contradictions within the US settler colonial dictatorship led to succession of policies alternating between unkept treaties and genocidal removal (1815–1887), planned extermination and forced assimilation (1887–1934), and welfare dependence and cultural annihilation (1934–present). It is a testament to the Mamaceqtaw's far-sighted, shrewd, and highly skilled diplomacy, political manoeuvres, and economic strategies that removal or annihilation were prevented. Among the major factors was the Mamaceqtaw's ultimate success in convincing the Bureau of Indian Affairs to grant commercial wood production in 1871 (Beck 2005; Deloria 1985; Hosmer 1999).

During the great social upheavals set in motion by the processes of capitalist imperialism, especially under the US empire, there developed political rivalries among Mamaceqtaw clan chiefs. Some would even occasionally use their own social and commercial contacts with European colonists to enhance their social prestige. Various spiritual currents emerged and spread as well, further splitting the Mamaceqtaw among Christian, traditionalists, and neo-traditionalists, beyond internal differences already developing along ethnic and linguistic lines among native, mixed, and European descendents. Mamaceqtaw took measures as well to adapt to the disadvantageous circumstances by renewing social structures (e.g., new clans to accommodate métis and European members) and developing new cultural norms (Beck 2002; Hosmer, 1997: 495; Jaimes and Halsey 1992). It was mainly a communal redistribution mechanism and cultural values favouring individual autonomy within the limits of community belonging that most mitigated the discord and rivalries related to the progressive reduction of resource access and territory and the hardship caused by forced permanent sedentarism (Hosmer 1997).

The mounting economic inequality within the Mamaceqtaw resulting from lucrative forestry was substantially lessened through a political breakthrough of enormous inter-generational consequence, reverberating to the present. The Council of Chiefs decided in 1873 to have incomes generated from timber sales divided among all members, reasoning that the woods constitute a resource that belongs to the entire community. In many ways, the Mamaceqtaw gravitated towards forestry as a source of both survival and spiritual renewal. It was a response to, among other factors, colonial land theft, loss of access to major subsistence resources, constant and contradictory settler colonial pressure to dispossess the Mamaceqtaw, and internal social and spiritual upheaval (Beck 2002, 2005, 116–119; Ourada 1990; Runstrom et al., 2002; Whitney 1996).

These changes, internal to the community but triggered by settler colonial pressures, resulted in a progressively stronger sense of belonging to the forest as a vital source of survival that goes well beyond the dynamics of timber markets. This feeling of belonging to the forest was reinforced by struggles to maintain cultural identity, which included community and clan loyalty (represented by local forest-associated species), an emphasis on individual autonomy, a close relationship between (forest) ecological relations and cultural processes (rituals, storytelling, etc.), and a faith in ancestral affinities with the rest of nature (Pecore 1992). Affinity with the forest is part of resisting settler colonial processes of "terraforming Indigenous territories in ways that erase Indigenous cultures, histories, and experiences" (Dockry and Whyte 2021, 99; see also Whyte et al. 2018, 158).

This can in part explain the tenacity of the Mamaceqtaw's ecologically sustainable silviculture in the face of relentless settler colonial pressures. Among them is the assimilationist aims of US Federal governments. They are the most enduring and served to justify forced absorption into settler colonial society through the 1959 "Menominee Termination Act." Through this Act, Congress ended reservation status to facilitate land privatisation. The Menominee Enterprise, Inc. (MEI) was created to convert reservation land into title shares to be distributed among the Mamaceqtaw, while retaining the same federal mandate over forest management control. In 1961, the reservation was declared a county under the administration of the state of Wisconsin, in clear violation of 1850s treaties. The administrative change entailed property tax payment and harmonisation with state laws, which led to spiralling costs that the community could not pay. A financial crisis ensued and the little infrastructure that was built in prior decades fell into disrepair, while the hospital and schools were forced to close. To cope with the crisis, the MEI began to sell land to wealthy American seasonal tourists, who had forest near the lakes and streams cut down to establish second homes on the most scenic southeastern areas. This was done without community consent, generating animosity among the Mamaceqtaw, especially as it affected the state of the forest and disrupted subsistence activities. The MEI's unpopular measure sparked the Legend Lake struggle, which united most Mamaceqtaw and rendered even more explicit the connection between decolonisation and environmental struggle (Dockry and Whyte 2021; Peroff 1982).

With the aid of American allies and activists, Mamaceqtaw opposed to the MEI plans began an eventually successful struggle and a lobbying campaign to pressure the federal government to restore reservation status. Women's leadership (including by one Ada Deer, future head of the Bureau of Indian Affairs, 1993–1997) was prominent and crucial to such success.[8] The positive outcome was partially facilitated by combining pressures by means of electoral politics (citizenship was unilaterally imposed on First Nations in 1924) and by means of extra-legal actions, such as through the American Indian Movement (AIM), a globally significant and influential formation established in 1968. When the reservation was restored in 1973, it coincided with AIM's heroic intervention at Wounded Knee village to try to liberate Wazí Aháŋhaŋ Oyáŋke (Pine Ridge Reservation) from US government-supported despots. While the Oglála Lakhóta Oyáte suffered assassinations and further repression (Leonard Peltier being among the victims and who remains imprisoned on trumped up charges), the Mamaceqtaw, having regained reservation status, returned to their subsidiary status under the Department of the Interior.

However, through settler colonial legal means, the Mamaceqtaw leadership was able to find an accord with the US government in 1975 that

[8] DRUMS (Determination of Rights and Unity for Menominee Shareholders), the organisation that played a decisive role in regaining reservation status, was founded by Mamaceqtaw leader Ada Deer (Peroff 1982; Tronnes 2002).

prohibited the commercialisation of reservation land without US government approval. This development may have been in part precipitated by the AIM-inspired Menominee Warrior Society[9] standoff with part of a nearby settler community (Gresham), aided by armed white vigilantes and the Wisconsin National Guard (Tronnes 2002). The objective was to reclaim stolen land and turn an abandoned building into a hospital. The action of the armed Mamaceqtaw caused not a few strains with the women-led Menominee Restoration Committee, what had remained of Tribal authority recognised by and negotiating with the US government (the Tribal government would be subsequently restored along with the reservation status).

Restoration marked a major historically unprecedented shift towards women leadership and the retrenchment of appeasement strategies, stymieing traditionals' influence and diluting anti-colonial efforts (Tronnes 2002, 546). Since that historical turn, the Mamaceqtaw have special dispensation for hunting and gathering within the reservation for sole subsistence purposes and for some administrative autonomy that allows for the application of traditional laws, as long as they do not contradict federal laws. Sovereignty, however, is not recognised and the FBI has, since 1985, ultimate authority over criminal cases. Under such blatantly dictatorial conditions, forestry still regained commercial strength in the early 1990s, thanks in part to aggressive marketing campaigns and niche market participation (especially through sustainable forestry labelling) and rising prices on timber worldwide. The general economic situation on the reservation improved with multiple revenue flows derived from businesses emerging around a casino opened in 1987, but this has not yet resulted in overcoming high levels of poverty, structural unemployment, and low life expectancy (Hosmer 1999; Robbins 1992; *Wisconsin State Journal* 2023).

The Mamaceqtaw mitigated colonial pressures and concomitant ecosystem alterations through pacification strategies with succeeding US governments and through the partial and controlled commodification of the forest. Mamaceqtaw self-determination and resistance is inextricable from the status of their nearby ecological relations, most importantly the trees. However, a main conduit for the intertwining of self-preservation and ecologically constructive silviculture has been community-owned yet

[9] Several of the members had earlier participated in the AIM action at Wounded Knee and AIM leadership lent direct support in the negotiations at Gresham.

profit-oriented and still extractive timber commercialisation (extractive in that most logs are not returned to the ecosystem and end up in very distant places through the global timber trade). Logging is selective and planting tree seedlings are clustered and mixed with other plants relative to ecological context and relations, but insecticides are also applied to combat such challenges as oak wilt and emerald ash borer (all made worse by climate change) and seedlings are purchased from outside vendors (Kapp 2019; Mausel et al. 2017; Olson et al. 2021; Trosper 2007). These are just some of the forms of capitalist and technically conventional practices benefiting the Mamaceqtaw as a whole, while keeping the forests thriving. It may seem contradictory to find an Indigenous communally redistributive system based in part on capitalist resource extraction (see also the above-described case of several First Peoples in so-called Guyana) until one understands the capitalist settler colonial pressures and anti-colonial resistance histories undergirding the current form of Mamaceqtaw forestry.

Sumak Kawsay, Suma Qamaña, Buen Vivir

Maybe the kind of contradictory processes identified above in the case of Mamaceqtaw resistance is relatable to the controversies over the practices and ideas associated with sumak kawsay, in Kichwa, or suma qamaña, in Aymara. They form important components to the revised constitutions of Ecuador and Bolivia since 2008 and 2009, respectively. With or without the full weight of such state institutions, the two worldviews have travelled far and wide. Intellectuals and activists, mainly from within Kichwa and Aymara societies, but also and eventually even more from some parts of non-Indigenous communities, have been promoting those frameworks mainly in its translated form of buen vivir, or the good life or good living. Acosta defines it as "the great transformation towards societies characterized by equity and justice, freedom, and equality, and living in harmony with Nature" (Acosta 2023, 305). However, "there is no clear outline of what this alternative consists of" (Acosta and Abarca 2018, 132). Multiple and at times contrasting political meanings come attached to buen vivir, meanings that have also shifted over time as political struggles have evolved (Vanhulst et al. 2020).

A rough translation has suma qamaña as buen convivir bien or living well together (Yampara 1993), or life creation (Albó 2011). The concept refers to life and coexistence in harmony and plenitude, which is not exactly encompassed by the more vague notion of good life (Altmann 2022; Ranta 2018). Additional cultural parameters related more closely to the expression sumak kawsay are explained below and make it clear that a much larger cultural framework and set of ecosocial practices are called forth when sumak kawsay is invoked. Notably, the ayllu, a relatively self-sufficient network or clan sharing land ownership, is a salient set of social relations associated with those worldviews (Cuestas-Caza 2018; Walshe and Argumedo 2016; Yampara 2001). Hence, suma qamaña refers more specifically to a harmonious time-space of community well-being in the ayllu, a term that implies political autonomy within a confederation and, notably, oneness with the rest of nature (Coral-Guerrero et al. 2021; Gudynas 2011, 444; Medina 2011, 45, 51). These contextualisations are important to bear in mind when discussing what has been adopted by some ecosocialists in the guise of buen vivir. For the sake of brevity, the focus is mainly on the Ecuadorian context, and, because of their substantive overlaps, the two frameworks are treated as similar enough as to be relatively interchangeable.

The historical context of the terminology and trajectory of its varied uses are of no small political consequence. The recent revival of interest in Indigenous thought is partly traceable to Indigenous socialist (mainly communist) struggle in the 1920s, against both capitalism and settler colonial racism. It became fully articulated in the current understandings in Ecuador by the 1980s as part of organising and mass mobilisation within Indigenous Andean communities that culminated with the 1992 march through Quito and then the uprising of 1994. Suma qamaña is a way to mobilise mainly Aymara and Kichwa-speaking Indigenous communities, a way to call attention to the better lives those First Peoples had before colonialism. It includes becoming aware of and uplifting Indigenous lifeways and knowledge, getting educated about the pre-colonial past and the sources of oppression in the present. Moreover, and crucially, it is part of a call for reappropriating land and, in some respects, reinstating or affirming Indigenous confederations. It is a deeply decolonial project, a firm rejection of assimilation into the settler colonial and thereby capitalist world (Altmann 2022).

Efforts were made shortly after the 1992 march by some Kichwa intellectuals from the Pastaza region of Ecuador to make Indigenous demands communicable to external communities, especially non-governmental organisations and development agencies within and beyond Ecuador. As reported in Altmann (2022, 106–107), this is how suma qamaña or sumak kawsay was defined in the Plan Amazanga (a study of Kichwa communities) led by Sarayaku Kichwa Alfredo Viteri Gualinga:

> It is the theory and practice that shows how one ought to live. This principle makes it evident that life is the unifying substrate of everything. It defines the equalitarian, communitarian, integrative and reciprocal sense between the communities of the Kichwa society. Present are: the true word, hospitality, reciprocity and solidarity, sharing, the ethics of communitarian work, respect for wisdom, and respect for freedom. The vigor of Sumac Causai is fed from the permanent dialogue with nature and its spiritual dimension.

It is the spiritual dimension and its contextual specificity that are often lost as sumak kawsay travels beyond the Pastaza Kichwa, and possibly even the Sarayaku Kichwa. The places where Sarayaku Kichwa live are

> defined by spirits and gods and the souls of the ancestors. They can communicate with shamans and, thus, establish the rules for the search for a *sumac allpa* [land devoid of evil] that is a necessary condition for a *sumac causai*. (Altmann 2022, 107)

It could be an insurmountable challenge outside of Sarayaku Kichwa communities to feed the vigour of sumak kawsay when the parameters for the permanent dialogue with nature are unspecified or cannot be specified in places with multiple and divergent ways of making such dialogue (think, for instance, of Queens in New York City, with one of the highest linguistic and cultural diversity in the world over a relatively small area).

Even more problematic is the spiritual dimension that makes sumak kawsay possible, since the presumption is, for sumak kawsay to be feasible beyond the Sarayaku Kichwa world, to have everyone conform to a belief system involving spirits, gods, souls of ancestors and the intercession of shamans, who will tell those who are not shamans about the rules for finding land without evil to attain life and coexistence in harmony and plenitude. Just as important to consider is the basis of sumak kawsay on principles of austerity and sufficiency, with any surplus to be shared locally or destroyed, if there is too much relative to community needs (García-García 2020). This spiritual and economic system may be workable among at least some Kichwa or other relatively self-sufficient communities, but transposing such a cosmovision and associated practices beyond that Kichwa context seems unworkable without strife and coercion.

In fact, sumak kawsay is understood differently even among the Sarayaku Kichwa, ranging from the above-described notions to the desirability of making more money or having more infrastructure (Lalander and Lembke 2018, 640). Perhaps this is why such transposition has not been attempted. Instead, the effort has been focused on translating sumak kawsay to draw multiple political forces into alliance beyond the Kichwa. This has led eventually to the worldwide spread of what ultimately is a nominal notion of sumak kawsay as alternative to conventional forms of development. For some, the idea has opened paths to post-development thought, including, in some readings, another way of justifying the rejection of state-socialist variants of development (Demaria et al. 2020) or of socialism in its entirety (Gudynas 2011, 446).

Regardless, the transfer of the terminology into non-Kichwa societies has resulted in the concept's distortion through decontextualisation. And when a Kichwa or Aymara cosmovision is promoted in the guise of buen vivir, a misleadingly vague rendition, sumak kawsay and suma qamaña become even more disembedded and disconnected from the original Indigenous communities whence the concept formed and took root. In fact, some would go as far to say that buen vivir is more related to Western philosophy (dating back to Aristotle),[10] that the concept of development is alien to Andean First Peoples, and that therefore sumak kawsay is more a way of life than an alternative to development or an alternative way of living within a capitalist world.

Nevertheless, buen vivir, even if it were ultimately a "Western" borrowing, can be useful as legal basis for decommodification and upholding the legitimacy of Indigenous views within state institutions. It has been useful, for instance in Colombia, towards mobilising people, including within state institutions, for such purposes as the protection of rivers, the defence of U'wa claims, and the prevention or cessation of oil extraction and mining (Barragán 2020; Cuestas-Caza 2018; Franco and Balaudo 2019).

There is, in any case, a less rigid way of understanding all this difficulty of translation, given that the framework emerged also from intellectual and activist milieux external to Sarayaku Kichwa society. Altmann (2022) recounts that people from other Indigenous and also non-Indigenous

[10] Contrast this with the view that "El *Buen Vivir*, en esencia, proviene del mundo indígena amazónico y andino [In essence, buen vivir comes from the Amazonian and Andean Indigenous world]" (Demaria et al. 2020, 57).

backgrounds have been involved in the recent reshaping of the concept of sumak kawsay and that the purpose is to stimulate wider society into working against presently destructive ways (see also Hidalgo-Capitán and Cubillo-Guevara 2020). That objective is eminently sensible, but the matter seems much more to be about Indigenous self-management (strengthening or bringing back the ayllu system) in adopting external knowledge on an Indigenous community's terms (Viteri 2002). It is a direct response to a settler colonial regime, and one among different kinds of possible responses that could express itself in ways that resemble those of the capitalism-immersed and profitable Menominee Tribal Enterprises, in contrast to, say, the strategies of the short-lived, suppressed Mamaceqtaw Warrior Society. This translation of frameworks like sumak kawsay into practice is evident in studies that, going beyond philosophical explications, bring out the actual applications of a Kichwa or Ixil Maya cosmovisions in everyday life within a pervasive, largely inescapable capitalist world (for instance, Rengifo Arias et al. 2022).

In this light, invoking sumak kawsay as alternative to now prevalent ways of living may in practice be mainly reformist and not as anti-colonial or anti-capitalist as claimed. Depending on circumstances, it can be accommodating or adaptational to capitalism or "Western development," rather than a perspective helpful towards overturning predominant lifeways. This is not so much a fault of the sumak kawsay framework or paradigm itself, in its original cultural context (not in its borrowed and decontextualised form), as much as the result of prevailing relations of power. These cannot be confronted solely by means of developing and promoting dissident ideas inspired by the non-capitalist lifeways of a tiny fraction of humanity that are being undone or under pressure, overwhelmed by surrounding capitalist relations. Perhaps, if it can be generalised at all from out of its specificity, suma qamaña could make sense for state-free peoples under a settler colonial regime subordinated to another capitalist state (in the case of Ecuador, it would be the settler colonial dictatorship of the United States).

Be that as it may, the plethora of published materials on the subject matter suggests something more than etymology and translatability is being contended. This is intimated by the three main currents that have been identified around those notions. These are summed up by Ataraz et al. (2021, 7) as:

> the indigenist, operating from the ground and searching for links with other similarly defined indigenous notions in other parts of the world such as *ubuntu* in Africa, the neo-Marxist discourses that, from the outside, seek in buen vivir a latent critique of modern capitalism, and the ecological post-developmentalist critique that has, they argue, provided the greatest contribution to other global debates about profound social, economic, and environmental ecological transformation.

Indigenist currents, in academia represented overwhelmingly by non-Indigenous writers, insist on specificities, valuing non-European perspectives, and aspects of decolonisation, and claim ties to Indigenous political movements. Socialist or "neo-Marxist" perspectives are more associated with leftist governments, mainly in Ecuador and Bolivia, who give more emphasis on development in terms of raising standards of living, especially for the working classes broadly and especially for First Peoples, who often comprise the majority of the poor and disenfranchised. The third current claims buen vivir as part of a move beyond development altogether, which they view as the imposition of ethnocentric, authoritarian, homogenising, masculinist, "Western" ways or worldviews (Altmann 2022, 108; Ataraz et al. 2021, 7; Avilés Irahola 2020; Bustamante et al. 2021, 126–131; Cuestas-Caza 2018; Le Quang 2020).

This third view is usually associated with the notion of post-development, which combines at least four "powerful imaginaries:" post-capitalism; post- or degrowth; post-patriarchy; and decoloniality (Demaria et al. 2020, 48). Post-developmentalism, though, seems more confined to an academic endeavour. When it comes to political struggle at wider institutional levels, the main divisions may be better captured by replacing post-developmentalism with a leftist environmentalism. This is the case of the Correa administration in Ecuador (2007–2017), through which sumak kawsay and buen vivir came to public and international attention (García-García 2022; Le Quang 2020).

Debate among these currents may have been at times acrimonious, yet there is much more commonality running through all of them compared to the capitalist systems they all want to overcome. All three ways of understanding buen vivir take for granted the community-basis and social nature of human life, stress the need to recognise the "rights of Nature" (and bring an end to domination ideology) and to turn states into plurinational institutions. They share the objective of reaching a balance of sovereignty and regional integration for a post-extractivist society (Le

Quang 2015, 9; Puente 2011). The principal underlying differences seem to amount to divergence over method and strategy.

This divergence, though, collapses into a shared process in terms of academic coverage and practices. A study of publications on buen vivir or sumac kawsay has shown that the frequency of coverage rose meteorically from 2009 to 2017 and has been falling since. Scholarly interest appears to correlate more with shifts in Ecuadorian politics than policy applications of sumak kawsay or social conditions related to practices of sumak kawsay. Disturbingly, with all the academic fanfare about Indigenous Peoples' worldviews underlying sumak kawsay or buen vivir, Indigenous scholarship on the topic is virtually absent, as are studies on Indigenous practices and notions revolving about sumak kawsay (Mantilla and Quinche 2021; García-García 2022; Le Quang 2020). The frothy debate bubbling up from such an intellectual movement suggests that what is at stake about sumak kawsay and buen vivir has little to do with any First Peoples per se, or their concerns, and more to do with the political repercussions for different leftist projects and wider settler colonial societies of the successes of Indigenous organising and movements in forcing state institutions to be more responsive to their demands, as in Ecuador and Bolivia.

Ecosocialism and the Indigenous

Ecosocialists have connected their ideas to struggles by First Peoples in differing ways. The numerous cases of frontline defence of forests, water, and other founts of life and the terminology, concepts, and rallying cries arising from Indigenous Peoples, like sumak kawsay, have captured the imagination of many ecosocialists (interestingly, world-renowned Mamaceqtaw forestry has not). Among the more salient reasons for this attraction is the association of Indigenous Peoples with an intrinsically anti-capitalist, non-dualistic worldview (oneness with the rest of nature), communal property, self-sufficiency, direct relationship with and livelihood dependence on local ecosystems, a tendency for ecologically sustainable practices, and a critical view and/or rejection of the state (Baer 2022; Brownhill and Turner 2020; Isla 2006; Kovel 2007; Le Quang 2015; Löwy 2014; Saito 2023; Salleh 2017; Wallis 2018).

Otherwise, there is virtual silence about or little more than passing reference to First Peoples (e.g., Albritton 2019; Schwartzman and Schwartzman 2019; Malm 2021; Tanuro 2015; Wallis 2015). Rarely,

Indigenous self-determination struggles are viewed as at most marginal compared to globally majoritarian working-class politics (Huber 2002). Such a viewpoint suffers from the glaring omission of more than a century of debates and political strife within socialist movements, parties, and states around the question of national self-determination. The issue is not just about the exemplary nature often claimed about First Peoples' ecologically sustainable practices, but also because of the strategic importance of decolonisation in overcoming capitalist relations globally. This is how First Peoples' struggles can be argued to be disproportionately significant relative to their sheer numbers. After all, as the Russian and many other successful socialist revolutions have amply demonstrated historically, revolutionary potential cannot be reduced to demographic matters.

To the most influential ecosocialist thinkers, though, First Peoples' struggles require much greater attention (see also the many works in Brownhill et al. 2021). For example, Ariel Salleh identifies the experience of Indigenous Peoples, who along with peasants and householders comprise the "eco-sufficient" meta-industrial labour class (Salleh 2009, 6), as the necessary grounding of radical politics (Salleh 2017). Joel Kovel, in his germinal *The Enemy of Nature*, refers to a brief but decisive mind-changing encounter with a "tribal way of life in an as-yet relatively preserved rain forest ecosystem" in Suriname (Kovel 2007, 53–54). In the same volume, as another example, there is a recognition of the substantive difference of controlled forest burning by Indigenous Peoples in the Amazon, compared to the induced mass forest fires through capitalist practices (ibid., 119–120).

Wall and Blanco (2010) see Indigenous Peoples as the vanguard of the global anti-capitalist struggle. Their self-defence would roughly translate into ecosocialism in European terms. Exposing the connivance of local government, non-governmental environmental organisations, and core-capitalist states in forcing open communal lands in Costa Rica to resource exploitation, benefiting large corporations, Ana Isla (2015) emphasises Indigenous and peasant communities' resistance to biopiracy and its gender-differentiated effects. To Brownhill and Turner (2020), it is especially Indigenous women who possess a "unique capacity to catalyze the unity of all sections of the exploited class" (Brownhill and Turner 2020, 8). Acosta and Abarca (2018, 133) make the bold generalisation that "*Buen vivir* or *sumak kawsay* is key to the philosophy of life in indigenous societies" and that "Indigenous people's struggles understand that protecting *Pacha Mama*, or Mother Earth, is a substantial part of their lives" (ibid., 135; italics in original).

Michael Löwy (2015) devotes an entire additional chapter on Indigenous struggles in the English-language edition of his original book on ecosocialism (Löwy 2011). He sees Indigenous struggles in "Latin America" as anti-capitalist almost by default. This is not only due to the harm capitalist firms and governments inflict on Indigenous communities by destroying the environments where they live, and from which draw their main sustenance. The very foundations of Indigenous Peoples' ways of living and thinking contradict capitalism.

More specifically, in Latin America, Löwy finds that "the conflict has intensified during the last few decades because of both the intensity and the extent of capital's exploitation of the environment, and also because of the rise of the alter-globalization movement—which took on this struggle—and the indigenous movements of the continent" (ibid., 15). In this understanding, Indigenous communities have an important role to play in steering Latin American leftist governments in what he sees as the right direction, presumably towards ecosocialism (Löwy 2014, 23). With exceptions, this foundational contradiction between capitalism and Indigenous lifeways facilitates the formation of wide alliances, if not affinities, between Indigenous communities and anti-capitalist movements, including socialist ones. Löwy cites the Alliance of Peoples of the Forest and the 2009 Belém World Social Forum as major examples of such alliances. Other examples are the 2007 initiative of the Ecuadorean Correa-led government to prohibit fossil-fuel extraction in Yasuní-ITT National Park (eventually retracted due to lack of international support) and the 2010 People's Conference on Climate Change in Cochabamba, Bolivia, convened by Evo Morales Ayma (Aymara), the first Indigenous president.

Problems with Ecosocialist Views of First Peoples

That voices influential within ecosocialism take First Peoples' concerns to be important, if not central, to building ecosocialism should be considered an advance over those historical and existing socialists, as well as some ecosocialists, dismissive of or at times antagonistic to First Peoples' self-determination struggles. But ecosocialists' depictions of and inspiration from Indigenous politics and objectives are far from spotless. The predominant level of abstraction eschews differences between and within First Peoples of major significance relative to political strategy. A focus on Indigenous practices and struggles aligning with (preferred) ecosocialist politics masks those that are not so easy to fit into an ecosocialist framework or set of objectives and that would require deep study, direct dialogue, and long-term commitment. Finally, the geographical and historical breadth considered is typically narrowed to parts of the Americas and over periods rarely going back further than the 1970s, missing the longer-term and wider dynamics that can help explain present-day situations and tensions and thereby debilitating analytical and political assessments for want of a comparative frame of reference. These problematic ways of considering or interpreting First Peoples' struggles elide the variability and historical evolution of politics, lifeways, worldviews, and much else that should be of great import when trying to support or seek alliances with Indigenous communities, not only those who are in open conflict with settler colonial institutions.

Given the centrality of First Peoples in the writings of major ecosocialist works, it is surprising to find a predominant level of abstraction so coarse as to make it impossible to bring into view First Peoples' scholarship, histories, social dynamics, diverse politics, and much else related. Many Indigenous intellectuals have made great contributions to social theory, political philosophy, and other fields of knowledge. There are also copious empirical studies on many Indigenous Peoples. Yet ecosocialist writings feature little to no reference to Indigenous scholarship and very little to no discussion on First Peoples' lifeways, worldviews, or politics.

The First Peoples invoked in Salleh's meta-industrial class and Kovel's Suriname voyage and controlled fires in the Amazon,[11] in Isla's,[12] Löwy's and in Brownhill and Turner's political analyses, or in Acosta and Abarca's sumak kawsay are buried under a general category that evades having to confront any concrete situation and thereby come up with context-appropriate strategies. As shown above, there are not a few instances that directly contradict such ecosocialist generalisations about First Peoples. No research is anyway even attempted to verify the claims made. Treating First Peoples' perspectives as if they were all virtually the same glosses over major differences in outlook and political strategy. As a prosaic illustration, the Indigenous politics represented by Laguna Pueblo Deb Haaland are hardly the same as those embodied by Aymara socialist Evo Morales Ayma. Or, to state the same in another way, imagine if ecosocialist perspectives, themselves diverse, were treated as little different from those of the capitalist mainstream.

The problem is not just the lack of basic empirical support for what is being claimed or for the implicit flattening of all Indigenous Peoples into a single worldview. It is a question of developing sensible and feasible political strategies and mutually enhancing alliances. Ignoring the specificities of each First People hampers a grasp of prevalent political currents within an Indigenous community and how they vary over time. It could be crucial to find out whether there are any potential conflicts within an Indigenous community or contradictions with ecosocialist objectives that impinge on organising for common projects or that affect the potential for affinities with ecosocialism.

For instance, at least two contrasting views on national liberation run through leading Kaniekehá:ka intellectuals. Taiaiake Alfred appeals to Wahsá:se (warrior dance) as "spiritual revolution and contention" (Alfred 2005, 27) as the foundation of strategies to achieve Kanien'kehá:ka

[11] The Kayapó are specified in this instance (in refernce to the work of Susanna Hecht and Alex Cockburn), but they are presented as if they could stand in for all Amazonian First Peoples, as if it made no difference whether one is referring to the Guauraní, Huaoraní, or any of the hundreds of First Peoples inhabiting the Amazon region.

[12] In Isla's study of biopiracy in Costa Rica (Isla 2015), one does not even learn of the name of the First People involved, the Tonjibe, in the struggle until page 76. There is no description of their history, relations with other nearby First Peoples, their lifeways and how they have changed (whether they are peasant or not), whether there are any internal tensions or political differences, and all such matters that would affect the Tonjibe's capacities to resist biopiracy.

People's self-determination and the end of colonialism. For this, he draws inspiration from the Satyagraha movement and Gandhian thought (ibid., 204–208),[13] but also the EZLN, and repudiates the state form, on the basis what must be pointed out as a misguided conceptualisation of the state as "an artifice of Euroamerican rationality" (ibid., 207).[14] This understanding of decolonisation strategy is quite a contrast to Louis Karoniaktajeh Hall's Pan-Indigenist call to learn from and improve upon the historical experiences of the Rotinonhsión:ni Confederacy by reviving an armed Rotisken'rakéhte (Men's Council Fire) to aid in establishing, eventually, a "Republic of Indian Nations," based on equality among all First Nation members, a cooperative economy, a confederal government, and other such institutional arrangements. And both Alfred and Hall take pains to distinguish themselves from other Onkwehón:we (Kanien'kehá:ka people) who have forsaken traditional ways and philosophies. It is a complex matter, likely requiring much effort at self-education and prolonged interactions with Kanien'kehá:ka communities, to understand which Kanien'kehá:ka movement or politics could find affinities with ecosocialism, and then which kind of ecosocialism.

The issue becomes even more problematic when the value of Indigenous struggles is made contingent on conformity with a particular strain of ecosocialism. Löwy (2014, 23) regards Indigenous communities' role as important only insofar as they remain autonomous and unsubordinated to leftist governments in Latin America "and supersede a purely local perspective and develop a broad anti-systemic, anti-capitalist, ecosocial agenda." This unflattering view of First Peoples' concerns is exacerbated by a narrow geographical imagination. For instance, Indigenous Peoples outside of the Americas, such as Adivasi Peoples, Maori, and Torres Strait Islanders, have not garnered much or any attention (but see, for instance, Baer 2022; Chattopadhyay 2014). Some Ecosocialist Ecofeminists implicitly broaden the understanding of Indigenous to include African Peoples, such as the Turkana (Brownhill and Turner 2020, 2), but this is even

[13] He separates this from what he deems as the inherently violent "Western" notion of revolution adopted by the likes of Ho Chi Minh and Che Guevara (Alfred 2005, 49), revolutions that, according to him, "have failed to produce long-term or generally successful resistances" (ibid., 51).

[14] The state is thousand of years older and geographically much more widespread set of institutions than the states that emerged in European polities. One an refer to examples from, among others, the Aztec civilisation, Babylonia, China, the Inka, Japan, the Mughal, and many more.

less frequent practice and an underexplored fact of the struggles over who counts as Indigenous, where, and under whose criteria. The narrow spatio-temporal scale of virtually all ecosocialist work so far impedes the development of a longer-term and wider view on settler colonialism, which is useful to understand the roots of current conflicts, the varying strategies required to confront different settler colonial situations, and the frictions, at times traceable to past conflicts, that may exist within and between Indigenous communities. Ecosocialist attention to First Peoples should be at least commensurate with the detailed studies often showered on settler societies or imperialist countries.

A singular example that runs counter to this general tendency in ecosocialist writings is the ethnographic study by Lalander and Lembke (2018). By taking into consideration the multiple and contrasting voices within the Kichwa communities with whom they have been in dialogue, they conclude that there exist processes of class differentiation within those communities and that a class analysis is among the tools useful to clarify the relationship between buen vivir and ecosocialism. This is how they could discern that the Kichwa struggle is not only about affirming Indigenous ways and the state of the environment they inhabit, but also about class position and wealth distribution.

There are political consequences to a chronic lack of attentiveness to the diversity in social dynamics and changes, to the variability of environmental practices, and to the evolving and differing worldviews among and within Indigenous communities. Alleged successes in uniting ecosocialist and First Peoples' objectives might consequently need to be revisited more critically, like the Alliance of Forest Peoples. Arguably, they are not as promising as would at first seem. In the first instance, there is a background of genocides committed by invading settlers as late as the 1940s that created the conditions for the very existence of caboclos (people of partially Indigenous ancestry) and the local rubber tapping economy, which forms the ecosocial backbone of the Alliance of the People of the Forest (Sedrez 2014, 149). Yet land restitution or any other form of decolonisation is not among the priorities of the Alliance, which is forest preservation in line with the needs of currently inhabiting communities. Nevertheless, the Alliance contributes to anti-capitalist struggle by impeding clear cutting for capitalist farming operations, but only in part, as it does not necessarily coincide with the self-determination of Indigenous Peoples there.

One can run through many other examples. Regarding the 2009 Belém World Social Forum, there are those pointing out unresolved tensions about tokenism or instrumentalist representation of Indigenous Peoples, Eurocentric norms of communication or deliberation, and self-serving selectivity of Indigenous voices (see, for example, Velitchkova and Choi-Fitzpatrick 2010). The Correa government's 2007 initiative to protect Yasuní-ITT National Park from fossil-fuel extraction, presented in the framework of working towards buen vivir, cannot be fully grasped without accounting for the political context of the time. The introduction of buen vivir within the new Ecuadorian constitution was not so much an effort to implement an alternative to conventional forms of development as much as to unite leftist and Indigenous movements to defeat neoliberal forces. Detractors go as far as charge the Correa government with cynically appropriating buen vivir for political gain (García-García 2022). Others omit the failure of international institutions to support the Correa government's attempt to safeguard Yasuní-ITT National Park and focus only on leftist government continuity with a "developmentalist" paradigm and distortions of the "rights of Nature" legal framework to justify opening the park to resource extraction (Gudynas 2017, 4).[15] Still others trace the debacle as part of a government decision to follow policies undermining the alliances of forces encompassing socialist, indigenist, and environmentalist concerns (Cuestas-Caza et al. 2020).

Such readings are understandable, given the political commitments of such authors, but they are also disingenuous in a continent ravaged by a couple of centuries of US imperialism, including in Ecuador (Becker 2021). Coups and orchestrated mass or trade union protests are among the means used to topple governments daring to break away from the clutches of the comprador bourgeoisie and the US capitalist interests they represent in Abya Yala. The very least to be considered when criticising

[15] For Gudynas (2017, 6), the problem is one of "an obsession with growth and developmentalism" and theoretical weaknesses "that allow for the subsequent drift toward the political practices of developmentalism." Just where do the obsession and faulty theory come from are a mystery. But he is even more ambitious than merely identifying the naughty ideas guiding "progressivist" South American governments (and, he claims, also ecosocialists who harp on the primacy of use-value): "Incorporating social and environmental justice requires changing valuation perspectives, breaking with the myth of a 'good' development, and at the same time modifying our understandings of society and Nature" (ibid., 8). In the end, for deep thinkers like Gudynas, the world's problems are to be resolved by changing our minds.

leftist governments that disappoint one's expectations are the prevailing social conditions and power relations affecting leftist government policy, including international and national ruling class pressures and threats of violent overthrow. And the very least that should be expected of such critics of leftist governments is a practicable plan addressing how to parry such internal and external pressures simultaneously and, meanwhile, how to achieve their lofty aims in practice, in technical detail.

But it is anyway in itself problematic to adopt buen vivir as part of ecosocialism, as pointed out above. This is not least because buen vivir entails divergent methods and politics. Worse, there is a tendency for non-Indigenous scholarship to represent Indigenous worldviews in the singular, claiming otherwise while demonstrating little to no interest in studying Indigenous specificities and dynamics. Even if one overlooks this essentialist tendency or views the problem as a political challenge of integrating the universal and the particular by developing unity in diversity (Cuestas-Caza et al. 2020; Lalander and Lembke 2018), aligning ecosocialism with buen vivir is no straightforward matter, given the contrasting currents within both. There are in any case schools of buen vivir that reject any form of socialism out of hand, such as post-developmentalists. Buen vivir socialists, especially those involved in government may also not be keen on ecosocialists who offer little more than critique. A more constructive prospect, because buen vivir is decontextualised and homogenising of First Peoples' perspectives, is first to explore the really existing lifeways, dynamics, and histories of Indigenous communities and then assess whether ecosocialism is even relevant to a specific Indigenous community and their struggle, and, if there is relevance, for which specific Indigenous perspectives. As in the case of the temporary or long-term tensions and splits within the Mamaceqtawak, the point is that perspectives and politics can differ among and within First Peoples as much if not more than in settler societies or societies in imperialist countries. Alternatively, ecosocialists could just prioritise decolonisation into their politics and support the sort of movement, possibly associated with buen vivir, that espouses that objective (Lalander and Lembke 2021).

There is much clarification and concreteness for ecosocialists to attain when it comes to decolonisation. This includes having more dialogue with those representing the interests of their respective First Peoples. It is a dialogue that is far from simple or straightforward, though. For instance, if one were to follow the precepts delineated by Tuck and Yang (2012, 28), ecosocialist projects would be incommensurable with decolonisation, so the issue would be of "strategic and contingent collaborations." The difficulty here is identifying who is to be collaborating with whom and over what, since there can be and there do exist political disagreements between and within First Peoples themselves, just as between and within settler colonial societies.

As Ecosocialist Ecofeminist Ana Isla (2006, 64) has reported, for example, the case of a delegation from the Talamanca People in Costa Rica, in refusing pressures to make biodiversity into a source of income, rebuffed the help offered by a delegation of Indigenous People from Canada regarding negotiations on biodiversity. Regrettably, there is no information given about who represented whom and which First Nation community from Canada was involved. It is surprising, given the exposure of such major differences, to find within the same page all Indigenous Peoples being painted as ecocentric. At the very least, there is a divergence in the substance of ecocentrism and political strategies between the communities cited that ought to be explored and explained. Recognising differences and specific commonalities would help lay the basis for building a transnational or internationalist politics of mutual aid and coordinated action in the struggle for decolonisation and ecosocialism.

Settler Colonialism, Decolonisation, and Ecosocialism

In the main, even if with qualification, the importance of First Peoples' struggles to impeding or mitigating environmental damage cannot be denied. This is even if some First Peoples are internally divided over this or actively partake of destructive forms of resource extraction or of economic activities associated with ecosocial harm, such as the mainstream of the financial sector. There are plenty of contrary examples showing the opposite tendency, based on traditional values among, for example, the Nishnaabeg (Simpson 2017). A recent study demonstrates substantive effectiveness by Indigenous movements, along with other social movements, at curbing carbon dioxide emissions and supporting efforts to implement emission reduction and energy transition policies (Thiri et al. 2022, 10–11).

But the struggles are very uneven, diverse, and not necessarily aimed at addressing planetary or even regional levels of pollution and may not consider any coordination outside of immediate environs, even as they usually contribute positively to wider political efforts. Ecosocialists are not incorrect in expressing sympathies for and writing supportively on Indigenous People generally and, even better, getting involved in aiding Indigenous Peoples' struggles materially. What is called for here is much more political caution and discernment, greater analytical precision, and clarity of criteria regarding which struggles to support, when, and how, always. Achieving this necessitates much (self-)education about the specificities of an Indigenous community and direct and sustained contact with that community.

Alliances or collaborations are obviously not a one-way process. Much like socialists or Marxists diverge on issues of Indigenous self-determination, Indigenous scholarship and activism are split on the issue of socialism (Bedford and Irving 2001; *New Socialist* 2006[16]; Simpson 2017, 76). Should one side with First Nation authorities recognised by the US or Canadian government or with traditionalists or those within First Peoples who do not recognise such authorities, or when is this not an issue? As Coulthard has demonstrated, this is not a straightforward matter. The policies of the settler colonial state of Canada, for instance, have evolved ways to "entice Indigenous peoples to come to identify, either implicitly or explicitly, with the profoundly asymmetrical and non-reciprocal forms of recognition either imposed on or granted to them by the colonial-state and society" (2007, 439). Ultimately, for there to be Indigenous self-determination, decolonisation must happen, that is, restitution of land and attaining symmetrical and substantively reciprocal relations (see also Coulthard 2014, 7).

Whyte (Potawatomi), Caldwell (Mamaceqtaw), and Schaefer (Odawa) describe settler colonialism as

> a form of oppression in which settlers permanently and ecologically inscribe homelands of their own onto Indigenous homelands. Settlers do not, as in imperial or metropolitan forms of colonialism, seek fundamentally to extract wealth and harness Indigenous labor for the sake benefiting peoples in central homelands located somewhere else. Some argue that settler colonialism is one condition for strengthening U.S. imperialism abroad because it establishes the needed land base for U.S. food security, manufacturing,

[16] This is the "Special Issue on Indigenous Resurgence," containing interventions from luminaries like Taiaiake Alfred, Jeff Corntassel, Glen Coulthard, Teiowí:sonte Thomas Deer, Lana Lowe, and Andrea Smith (who has been accused of pretending to be Cherokee), http://newsocialist.org/wp-content/uploads/2013/08/NewSocialist-Issue58%20w%20cover.pdf, accessed 3 August 2023.

> military development, and metropolitanism (i.e., being a global intellectual and cultural hub). (Whyte et al. 2018, 158)

The inscription process involves the erasure of Indigenous presence in the landscape and of Indigenous Peoples themselves through extermination, as the authors further explain in the former case, and, in the case of the latter, imply. One might quibble with the erasure of the capitalist relations underlying settler colonialism, but the trouble already starts with thinking through what ought to be done about it.

Decolonisation entails the opposite of permanence in the settler inscription on Indigenous homelands. A different reading that allows for the permanence of change instead of inscription (incidentally, part of a dialectical materialist worldview) could allow an envisioning of homelands (or landscapes) that can be reinscribed, even if they will be different from what preceded settler colonialism. Yet, given the overwhelming presence of settlers, at least on Great Turtle Island, it seems a rather tall order for Indigenous Peoples alone to undo the harm without having a mass conversion of settlers to Indigenous ways or without mass expulsions (or committing multiple genocides, or both) to get rid of settlers altogether. But then the questions could be which Indigenous ways, especially as they have changed over time, and where such changes should occur.

If the Stockbridge Munsee and the Onyota'a:ka (Oneida) Nations were to decide to, and could, without being threatened with annihilation by settlers (whose majority are mainly Euro-descendent working class), return to their ancestral homelands from what is now Wisconsin State, they would have to reinscribe what used to be their homelands (this includes dealing with some lasting pollution problems) in what is now New York State. It would be useful to know, practically, how to achieve this and work towards it, with much long-term organising and study, if those First Nations would wish to do so.

In the case of the Mamaceqtaw Nation, whose homelands are where they now live or nearby, should settlers simply disappear or adopt seasonal movement, manoomin gathering, fishing, ecologically contextualised logging, and other activities associated with pre-colonial lifeways along with the worldview integral to such a way of life? If one strives for decolonisation, much thinking needs to be done about how to achieve it in concrete terms, not metaphorically, nor by allusions to existing practices, as important as they are, but that have clearly not led to decolonisation.

Perhaps what is meant by decolonisation needs to be specified with greater depth and tied more explicitly to expectations, which should be further clarified (beyond recognition, re-education, and the like), of changes needed in settler communities and in the relations among all communities now inhabiting Mamaceqtaw homelands. Perhaps a conversion to Mamaceqtaw lifeways in Mamaceqtaw ancestral homelands by part of the settler majority of Wisconsin could be workable, but it would necessitate the development of wide alliances and collaborations that are not aleatory, such as those forged out of resistance to mining (Grossman 2017), aside from restoring ancestral traditions among all Mamaceqtawak. Alliances and collaborations would require careful, long-term planning and much coordination among many communities. And there are many communities, Indigenous, such as the Ho-Chunk, Stockbridge Munsee, and the even more numerous non-Indigenous (New Afrikans or Blacks or African Americans, various Euro-descendent ethnicities, Arab, Chinese, Indian, Korean, and many other communities). And all this would have to be achieved by preventing or resolving any major conflict among communities or avoiding conflicts degenerating into wars.

The matter becomes even more complex when considering the reality of wider, beyond human relations that change with or without society changing or as a result of planetary harms, like global warming, which is bringing about major ecological changes to Mamaceqtaw homelands. In fact, these global environmental changes could be a permanent inscription that cannot be addressed by focusing on settler colonialism and self-determination in just one region of the world. This is where an internationalist ecosocialist outlook could be useful in terms of organising to bring about the overcoming of capitalism, even as that alone is insufficient to undo settler colonial damage to Indigenous homelands.

These are issues ignored by ecosocialists of any stripe. It is anyway, and more broadly, unclear how decolonisation or Indigenous self-determination is to be achieved or at the very least how to organise for such ends. What is being attempted in countries like Bolivia by means of taking over mainly the executive part of the state does not curry much favour from at least some Indigenous scholars and ecosocialists, even as it is a strategy developed from within some Aymara and Kichwa communities. But alternative political platforms and concrete strategies from critics of the socialist Bolivian government are rather rare or a bit thin when it comes to addressing military pressures or ensuring everyone will be sufficiently fed or undermining the main mechanisms of capitalist surplus extraction. Critiques are, maybe just as problematically, often framed within national or regional politics, at times even taking settler colonial state boundaries for granted. In contrast to those initiated by, for example, the American Indian Movement or, currently, The Red Nation (2021), many struggles or actions are mainly in response to settler colonial violations or repression instead of being the outcome of organised and planned activities for specific objectives or pre-emptive action.

In the case of ecosocialism, mainly in core-capitalist countries, the situation is much more problematic. As stated above, few take decolonisation as a priority in their politics (for some exceptions, see, for instance, Brownhill et al. 2021, and the 2013 "Bridging Indigenous and Socialist Perspectives" special issue in *Capitalism Nature Socialism*). Nevertheless, there are some major overlaps between First Nation struggles and the mainstream of ecosocialism, at least in the Americas, especially with respect to striving for establishing egalitarian relations and communalistic arrangements free of the state. The already-cited Whyte (Potawatomi), Caldwell (Mamaceqtaw), and Schaefer (Odawa) provide one of the most compelling reasons for ecosocialists to become educated in and prioritise support for First Nations' decolonisation struggles, at least on Great Turtle Island.

> While Indigenous peoples are distinct from one another, they often see themselves in mutual solidarity because they have overlapping political aspirations to continue their own self-determining governance systems in the face of colonial oppression. Planning is an important way in which to exercise collective self-determination. Broadly, we define "planning" as practical activities whereby a collective, such as a society or community or nation, envisions different futures that are more or less desirable for itself and its members, determines what capacities and strategies must be developed today to be prepared for different future scenarios, and revisits and revises its current capacities for preparedness to adjust to current and expected challenges. (Whyte et al. 2018, 155)

With such emphasis on collective self-determination and planning, the overlap should be extendable to those involved in the struggle for ecosocialism. Points of divergence present themselves when one looks at the details of what is meant by collective or planning. Collectives, for First Nations, are inclusive of nonhuman organisms as relations to be heeded, and planning involves ceremonies, oral traditions, lived ecological processes, along with what are typically recognised in conventional understandings of planning (ibid., 156–157). The extent to which First Peoples' planning coincides with ecosocialist planning is bound to be a work in progress, but one that will require ecosocialist commitment to ending settler colonialism (not merely embracing new ideas or definitions), and this could be by starting with prioritising the upholding of treaties and land restitution. How such ends are achieved, though, is not a predictable path and is unlikely to be "democratic" or egalitarian in the sense usually expressed by both of First People representatives or leaders and that tiny fraction of settler colonial society that call themselves ecosocialist, although the notion of egalitarianism may be another and promising point in common. Yet something similarly coercive and filled with strife may be in the offing with planning generally, especially in a capitalist context, which both induces and is nourished by settler colonialism.

Anti-socialist Currents in First People Activism and Scholarship

There are barriers to combining ecosocialism with First Peoples' concerns and hence decolonisation that are not solely due to any inadequacies in ecosocialist viewpoints. Unlike in much of Abya Yala, there is an anti-socialist streak running through First Peoples on Great Turtle Island. It impedes ecosocialist efforts at supporting First Nation self-determination struggles. Prominent influential Indigenous voices, in ways akin to those of some pro-Indigenous proponents of buen vivir,[17] chronically essentialise or caricature Marxism and anything thereto related, sometimes even lumping socialism or communism with capitalism (see, for instance, Barsh 1988; Deloria 1997; Grande 2004).

This is nothing new. For instance, in the 1980s, the evolving tense relationship between the Sandinista government and at least part of the Miskito community was used by detractors to accuse the former of quashing the latter's national autonomy. It is an understandable reaction, given some negative historical or contemporary examples of disrespect, insensitivity, if not violence by at least some leftist governments or parties (Bedford and Irving 2001). At the same time, judgements are often made without due consideration of wider contexts and longer histories. In the case of the Sandinista, the policy was to improve the Miskitos' material conditions and political representation, a community that had been violently oppressed and marginalised by the US-supported Somoza dictatorship. Part of the Miskito community had legitimate grievances towards the new Sandinista government, but these were amplified to belligerence and armed conflict by means of US government covert intervention as well as Sandinista over-reach but also legitimate self-defence from US government aggression. That the Sandinista government managed to get anything done that led to higher living standards in general is a breathtaking achievement, given the siege they suffered (Ballard 1984; Dennis 1993).

One can point to minoritarian Marxist currents agreeing with or actively supporting colonial policies or engaged in disrespectful self-described Marxist interventions (e.g., in Churchill 1983), but accusations of nefariousness or intrinsic colonial attitudes in all Marxism-inspired movements and formations by resorting to the use of unrepresentative

[17] Gudynas (2011) is among many leftist intellectuals who turn what could be a legitimate critique into a totalising anti-socialism. He rightfully underlines the existence of differences among ideas of buen vivir and their unreplicability outside their specific contexts ("There is no room for an essentialist position," he asseverates on page 444), while claiming development policies in countries as diverse as Brazil, China, Venezuela, and Vietnam are all mirror images of "Western development," with the same characterstics and objectives (page 442). Providing public housing, expanding and improving healthcare provisions, measures to reduce poverty in those countries are, to our valiant promoter of alternative Indigenous lifeways, no different than the lack or destruction of public housing, the dearth or undoing of much healthcare, and austerity measures that induce pauperisation in countries like Egypt, India, Israel, Japan, Russia, South Korea, the UK, or the US. As Gudynas aptly puts it (page 446), "Words are not innocent."

samples do little to improve decolonisation prospects. It is entirely rightful to question and challenge settler colonial societies, including their alternative movements, and to demonstrate how Indigenous ways can lay the basis for more constructive ecosocial relations in general. However, as Kanahwake Kenienkeha scholar Taiaiake Alfred himself has pointed out, dichotomies submerge the potentials for all cultural frameworks, including those of Euro-descendent settlers, to express their potentials "to contribute to the achievement of peace and harmony" (Alfred, T. 1999, 21).

With respect to linking First Peoples' anti-colonial movements to socialist or communist movements, the units of contrast most used, such as "Native and white," are much too coarse for much practical political relevance. Furthermore, to put the immensity of First Peoples' worlds on the same plane as a relatively small, diverse, and at times mutually hostile fraction of the political worlds of settler colonial society presents a mismatch in units of analysis with detrimental political repercussions. There is a great diversity of as well as major overlaps in political philosophies and commitments within "Native" and "white" communities. There are minorities (and at times even majorities) within Indigenous communities who embrace capitalist ways and are fine with settler colonial institutions (as intimated by Indigenous currents decrying the loss of traditional ways or inveighing against such institutions as Tribal or Band governments). There are obviously also minorities within settler communities who object to and fight capitalism and colonialism. Such crucial differences and similarities are effaced by viewing the world mainly in terms of Indigenous and settler.

Moreover, there is a broadly strategic usefulness to avoiding untenable generalisations. In the horrific reality of settler colonialism especially on Great Turtle Island, where multiple genocides have reduced First Nations to less than one percent of the total population, the range of political manoeuvre is already much diminished. The range is rendered even tighter by splits, such as on the issue of traditionalism, within and between Indigenous communities and Peoples. Alienating socialist or communist movements among settler colonial peoples—movements currently on the rise—and existing socialist states elsewhere is unlikely to enhance prospects for First Peoples' self-determination in the current conjuncture. An essentialising and dismissive stance towards Marxism (or socialism, communism) narrows an already exiguous range of support from within and outside settler colonial society that thankfully goes beyond mere metaphors, as Unangax̂ scholar Eve Tuck and K. Wayne Yang incisively point out often happens (Tuck and Yang 2012).

Indigenous anti-socialism erases the socialist or communist support that has been important to many Indigenous initiatives. The long-term efforts originating in the 1974 Standing Rock meeting that culminated in the 2007 Declaration on the Rights of Indigenous Peoples (rejected by the settler colonial governments of Australia, Canada, New Zealand, and the US) were facilitated by the support of socialist states and the Non-Aligned Movement. Much earlier, in 1919, the USSR supported a Rotinonhsión:ni Confederacy petition to the League of Nations against the Canadian state's overthrow of a Rotinonhsión:ni self-governing institution (Estes 2019). Communist and socialist parties in Australia played significant roles in helping First Peoples' fight against White Australia policies (Baer 2022). In 1991, the anti-communist socialist Ghaddafi government hosted and donated generously to a Kanien'kehá:ka delegation that had also been part of the first United Nations Conference on Indians in the Americas in 1977 (Hall 2023, 72).

Reducing Marxism, socialism, or communism to some white settler colonial or generic European product means that Indigenous socialist or communist formations are no better than the settler colonial institutions they are fighting. Anti-socialism also implies that the likes of Marxist Frantz Fanon (cited approvingly by Tuck and Yang 2012), or Marxist Amauta José Carlos Mariátegui, or Marxist Claudia Jones, or socialist Evo Morales are enemies of decolonisation, contributors as they would allegedly be to building "[S]ocialist and communist empires" (Tuck and Yang 2012, 4, footnote 2). Or, according to Kanien'kehá:ka scholar Taiaiake Alfred's, such Indigenous movements and both Indigenous and non-Indigenous heroic figures would be read as fighters against "Western capitalism" but not against "Western colonialism" (Alfred 2005, 69).

Ultimately, Indigenous anti-socialism obscures the fact that, worldwide, the overwhelming majority—tens of millions—of communists and socialists are people of colour living in formerly colonised lands. Many of them are the inheritors of successful anti-colonial struggles, if not revolutions, led by Marxists (Prashad 2017). The history of support from socialist or communist institutions, and the involvement of Indigenous people in those institutions, needs to be made widely known.

Similarly, it is unhelpful to frame matters of ecosocial crisis in separatist terms, as in the following statements:

> Indigenous planning, as a way of reflecting on Indigenous sustainability, is about figuring out the planning processes arising from the contexts that we actually live in today, in which our societies are greatly limited and threatened by settler colonialism and other forms of oppression. Reflecting on sustainability in this way—whether regarding climate change, biodiversity conservation, or food sovereignty—keeps us aware of how oppression endures as one of the largest threats to Indigenous peoples and many other groups. Whether settler and other privileged populations ultimately can achieve sustainability in the near- or long-term planning horizons is not so much our concern. Regardless of what happens with them, the odds are that Indigenous peoples will continue to face different forms of colonial oppression and must innovate strategies for protecting our continuance no matter what non-Indigenous populations end up doing. (Whyte et al. 2018, 174–175)

This understandable reaction to settler colonial oppression seems difficult to reconcile with the sort of Indigenous worldview the authors themselves claim, where the world is understood as interconnected, filled with relations. Are settlers not also part of the same world? Is it impossible for a settler to be a relation? The authors' unfortunate framing is a self-defeating dichotomous statement that promotes the delusion, shared by bourgeois ideologues, that what happens to people elsewhere or to nonhuman worlds has no bearing on their own lives. What happens to and within settler populations, whether they arrive at ecological sustainability, will most certainly affect the very viability of Indigenous lives, as the authors must realise. This could not be clearer with respect to long-range air pollution and climate change, but also to class strife within settler colonial society and racist violence from settlers.

An Ecosocialism Worthy of First Peoples' Self-Determination

When combined with inadequacies (if not dismissiveness) in ecosocialist thought regarding First Peoples and decolonisation, the sweeping statements of a few yet prominent Indigenous scholars can undermine or even pre-empt the dialogue needed to create the potential for alliance or collaboration (Baer 2022, 318). For alliances and collaborations to be feasible, non-Indigenous ecosocialists must endeavour to become educated on Indigenous Peoples' specificities, cultural frameworks, and self-determination struggles. But the converse also obtains in that

> for Indigenous peoples to reject or ignore the insights of Marx would be a mistake, especially if this amounts to a refusal … to critically engage his important critique of capitalist exploitation and his extensive writings on the entangled relationship between capitalism and colonialism. (Coulthard 2014, 8)

For Coulthard (Yellowknives Dene), this necessitates transforming Marx's theories relative to Indigenous Peoples' experiences. This can be done by reconceptualising capitalism as a process of ongoing (not just historical) dispossession, understanding that non-capitalist societies are dynamic and have their own trajectories (capitalism is a contingent, not inevitable prospect), and shifting the emphasis to the colonial instead of the capital relation (ibid., 8–11).

As Foster et al. (2020) have shown, based on Marx's writings, these important correctives at most apply to some strands of Marxism (see also Clark and Foster 2021), especially those that would see First Peoples as inevitably proletarianised (Bedford and Irving 2001, especially Chapter 4; see also the critical overview in *New Socialist*, issue 58, published in 2006). Coulthard's ideas could also themselves be interpreted as geographically and historically contingent with respect to the shift in emphasis Coulthard recommends. That is, one could argue that most First Nations' people on Great Turtle Island, given their actually existing wage and market dependence, have long been forcibly absorbed into capitalist relations, more often as racialised and marginalised workers (Rose 2017). Far from denying the need for decolonisation, this acknowledgement should form the basis for building independent and classless First Nations in mutually beneficial relations with classless non-Indigenous Nations across the Americas.

Furthermore, the issue is not confined to ignoring Marx's insights. It is about becoming familiar with the histories of socialism and communism, not only of Marxism, and to do so by looking into the places where most of those histories unfolded and associated theorical frameworks developed, which is outside of Europe and Euro-descendent settler colonies. This could avoid the problem of succumbing to another major constituent of settler colonial ideology, aside from racism, which is anti-communism (Horne 1986, 2019).

Antagonism of the above-described sort or mutual dismissiveness among potentially collaborating communities helps mainly reproduce prevailing relations of settler colonial power. As pointed out in Chapter 1, the net planetary effect of all the constructive actions worldwide to reduce or prevent harmful environmental impacts, including by many First Peoples, is like a drop in the ocean, seeing how, for example, global greenhouse gas emissions continue rising.

It goes without saying that it would be preposterous to expect First Peoples to resolve problems overwhelmingly caused and reinforced by coloniser and imperialist institutions largely from European societies, but also including Japan. If I understand correctly, the Indigenous Peoples who inspire many ecosocialists are first and foremost engaged with self-determination and/or survival/revival projects, involving, at a minimum, regaining sovereignty over territory wrested or being taken over by settler colonial communities. The return of stolen land and the furnishing of reparations tend to coincide with ecologically protective efforts because, in the main, many of the Indigenous Peoples who have survived settler colonial genocides have managed, despite enormous odds, to resist colonialism and retain at least some traditions favouring ecosocially sustainable practices.

But this should not be expected to be always the case. There exist situations when the opportunity for living standards improvements and political and/or economic advancement for a First People or fractions thereof involves ecologically deleterious resource extraction. There are also historical and current attempts of settler colonial regimes at undermining or violently suppressing First Peoples' capacity to transfer their traditional ways across generations. Part of this is the violent uprooting of many Indigenous people from their ancestral lands and from their communities, including through the mass abduction and killing of Indigenous children. The horror of settler colonialism is then exacerbated by the development of tribal businesses and class differentiations within First Peoples, a result of forced assimilation into capitalist relations and at times masked by means of indigenist rhetoric (Rose 2017).

This contradictory process could be one way of interpreting the strides made and setbacks suffered in Ecuador and Bolivia. At bottom, debates and disquisitions over the meaning or uses of sumak kasay and buen vivir conceal the issue of potentials for Indigenous organising to turn into a political force capable of influencing the state and, one could add, wage successful decolonisation campaigns. In fact, the latter is how some have come to understand buen vivir (Lalander and Lembke 2021). Viewed thus, the distinction between Indigenist and socialist currents over buen vivir seems of secondary relevance, especially as they have not always been mutually exclusive in practice (Cuesta-Caza et al. 2020; Lalander and Lembke 2018). As Artaraz et al. (2021, 5) observe, "One key issue that remains to be explored is whether state-sponsored forms of developmentalist modernization can be reconciled with traditional indigenous values." This is certainly among the foremost tensions, if not contradictions, to be confronted and that should resonate with many ecosocialists (Cuestas-Caza et al. 2020; Fernandes 2020).

What seems to be less appreciated, however, is the fact that it is thanks to pressures from well-organised Indigenous movements that there has been any government at all showering such great attention on any First Peoples' worldviews and even attempt to integrate them into policy and practice (for exceptions, see, for instance, Lalander and Lembke 2018, 637; Puente 2011). In fact, it was not until leftists took governments over in Bolivia and Ecuador and started introducing sumak kawsay or suma qamaña into their new constitutions, as diluting and distorting as the process has been, that most leftist intellectuals and movements outside those countries even became aware of those frameworks (Artaraz et al. 2021, 7).

It is one among many constructive effects of leftist, and, in the case of Bolivia, socialist government action that should also be recognised, rather than belittled or denied, or dismissed as yet another case of "Western development." The accomplishments by Indigenous and non-Indigenous movements and parties, especially in Bolivia with the Movimiento al Socialismo Party, are all the more remarkable when considering, for example, neighbouring Peru (never mind Canada or the US, where the political situation for First Nations is at best parlous). Not only are there sizeable Kichwa and Aymara communities living in Peru, but there are proportionately more Indigenous people than in Bolivia and Ecuador. Yet there is hardly any institutional support for Indigenous perspectives, much less any policy or legislative framework inspired by sumak kawsay or suma qamaña. Instead, there is a sustained and violent repression of Indigenous Peoples and suppression of their worldviews, reinforced more recently by the Boluarte coup regime (Espinosa et al. 2023; Merino 2021; see also Blanco 2017; Isla 2019; Löwy 2015). A similarly oppressive system obtains in Chile (Carreño-Calderón 2021).

One should also always keep in mind and never underestimate capitalists' structural unscrupulousness, ruthlessness, and often even viciousness and vindictiveness against any movement that dares hamper their rule and profits. In a context of US and NATO imperialist pressures, with their associated enormously unequal terms of trade and financial flows, and highly uneven power relations among states and peoples, such as in Abya Yala, part of surviving or pre-empting attacks and annihilation from within and without is to adapt, compromise, bring about gradual reforms just to have the potential to engage even in modest wealth redistribution and to concretise improvements in living standards. The massacres and widespread repression of the 2019–2020 Añes coup regime in Bolivia, with their neo-nazi and evangelical contingents, and the support of the Catholic Church, the European Union, the US, and Canada should make this clear (Farthing and Becker 2021). Some scholars, such as Acosta (2017, 88–89), who have been directly involved in government policy making, show awareness of these contradictions and dangers (see also Puente 2011, 347). He recommends continued resource extraction in ways that help eventually overcome it to achieve buen vivir or "biocentric" civilisation. Interestingly, this view resembles what is proposed by the socialist buen vivir current (Le Quang 2020, 106).

Yet the Indigenous lifeways that scholars claim to be central to building buen vivir are rarely if ever described and studied, in contrast to, say, Ecuador's political-economic dynamics relative to the rise of buen vivir politics (Lalander and Lembke 2018; Le Quang 2020, 99–101). Most writings treat Indigenous lifeways and worlds as little more than icons, static, and timeless. Contextualisation and comparative study, especially of the many cosmovisions and practices of Indigenous Peoples (not just those that fit a generic buen vivir perspective), are oddly infrequent in discussions of sumak kawsay or buen vivir.

To state that context and comparisons are important is not to minimise or justify the violence or setbacks emanating from the actions of leftist or socialist governments, as in the instances of ecologically and socially destructive practices (especially in the extractive sector) that undermine some First Peoples lifeways and often also their lives. Rather, it is to understand that there are also interconnected and mutually reinforcing countervailing and resistant forces of oppression at multiple scales that must be faced and for which political strategies and mutually beneficial alliances must be developed to fight such forces of oppression effectively, such as long-standing and diffuse racist and patriarchal colonial social structures, a position of national economic dependency in the capitalist world, threats of and actual US imperialist intervention (including coups), self-aggrandising opportunism even from within Indigenous communities, among other processes. Researchers like Lalander and Lembke

(2018, 637), even if failing to account for imperialism and international processes, constructively identify some fundamental challenges in their appraisal of the new pro-Indigenous and "rights of Nature" laws and policies in Ecuador and Bolivia:

> The contradictory traits of both the constitutional texts and the actual realpolitik signify a seemingly impossible situation in which governments have to choose between the mutually excluding options of ethno-environmental rights versus welfare reforms based on extractive development politics … Other human values and economic aims may be superior vis-à-vis the safeguarding of environmental and ethnic rights in specific political and electoral situations, particularly so in Andean nations still burdened by relatively high indexes of poverty. From the perspective of ecologically concerned indigenous actors, this Catch-22 is articulated in the choice or compromise between development in terms of social welfare conditions on the one hand, and protection of cultural rights and the environment on the other hand.

Contradictory tendencies, which can be and have been deadly, are bound to characterise any decolonisation or ecosocialist project, including one tethered to sumak kawsay or affine perspectives. This is not only because of multiple-scaled oppressive relations, but also because of the multiplicity of alternative forces involved who have distinct material interests, political aims, worldviews, spiritualities, and much else. The struggle to reconcile them, even when their relations sour and degenerate into conflict or when one socialist and/or decolonial formation becomes repressive of another, is just as important as the struggle to overcome the much deadlier oppressive and ecologically destructive forces of capitalism.

At the same time, without specifying the context and scale of struggle, say in Sarayaku Kichwa territory or possibly Ecuador or in Abya Yala more generally, sumak kawsay or suma qamaña, when extended to other social formations as buen vivir, could amount to little more than a truism. Save for the most intransigent sections of capitalist societies, with their profoundly incoherent and delusional politics (where capitalism is perceived as natural and nature and other peoples are regarded as if removed, unrelated, exploitable objects), most any community will strive to be independent or self-determining and to be constructive of or be harmonious with the nonhuman worlds where they live. Buen vivir is an otherwise fine way of broadly understanding ecosocial relations in ways that undermine capitalist notions of nature, but only insofar as it is made appropriate to context and part of a wider set of strategies.

There could even be reasons to abandon buen vivir altogether as a useful politics. Some argue that worldviews like suma qamaña are simply incomparable to "Western" notions of well-being (Beling et al. 2021; see also Gudynas 2011). Perhaps this is so. In any event, it resembles the notion of incommensurability of First Nations' decolonisation projects formulated by Eve Tuck (Unangax̂) and K. Wayne Yang (2012, 28–31). As they have enlighteningly articulated the issue:

> Breaking the settler colonial [settler-native-slave] triad, in direct terms, means repatriating land to sovereign Native tribes and nations, abolition of slavery in its contemporary forms, and the dismantling of the imperial metropole. Decolonization "here" is intimately connected to anti-imperialism elsewhere. However, decolonial struggles here/there are not parallel, not shared equally, nor do they bring neat closure to the concerns of all involved - particularly not for settlers. Decolonization is not equivocal to other anti-colonial struggles. It is incommensurable. (Tuck and Yang 2012, 31)

In this light, more troubling are attempts to extend sumak kawsay as buen vivir—that is, devoid of the context out of which it is born—to other societies, who are the world's majority, characterised by prevailing and diffuse capitalist relations, multiple cultural frameworks and peoples, often in tension and with highly unequal access to or control over even the means of survival, and a central historical role played by the state in their unfolding as the societies they currently are. There really is incommensurability, certainly, but for most of humanity there is by now also a much greater commonality of challenges.

Without joining them to concrete proposals and objectives, as Tuck and Yang at least in part express, appeals to incomparability or incommensurability otherwise shut down alliance or collaboration prospects among different communities and stymie effective action to topple settler colonial regimes. In other words, there may be incommensurability to decolonial struggles, but there is also no net separation. Many people from Indigenous communities inhabit the capitalist world and are subjected to class differentiation processes. They are part of capitalist reproduction by way of such institutions as Tribal government in the US, for example, and by way of upholding private property principles (such as individual home ownership) and wage relations (such as employing other Indigenous peoples in a business under Indigenous ownership). These capitalist processes traverse, are constitutive of, and constituted by the incomparabilities or incommensurabilities that arguably exist regarding First People's decolonisation struggles.

To return to sumak kawsay, what could be useful in terms of developing a practicable alternative way of life—or, more urgently, in the despotic settler colonial situations lived by most First Peoples, a successful self-determination struggle—is to specify the conditions under which suma qamaña, or something like it, could be more widely applicable as part of a coordinated panoply of politics and policies contributing to decolonisation, even if always in tension with other struggles. This, in some ways, regrettably means expecting and planning for deadly contradictions, rather than be surprised by them. Recent struggles in Bolivia and Ecuador are among the cases that can help plan pre-emptive strategies.

Founded in 2014, The Red Nation have articulated yet another and exemplary way to intertwine Indigenous self-determination, decolonisation, ecological concerns, and socialism, or, more precisely, communism. They formed partly as a response to the neglect of First Peoples' struggles in much of social justice organising, offering a ten-point programme stylistically reminiscent, and not coincidentally, of the Black Panther Party for Self-Defence, an influential force in the development of the American Indian Movement, who have served as inspiration for The Red Nation. The programme features issues that will mainly resonate within the context of Great Turtle Island. These include reinstating treaty rights, inclusive of Indigenous people who have been historically forced into living outside reservations or their ancestral lands and currently constitute the majority of members of First Peoples on Great Turtle Island. A corollary demand is the end of violence and discrimination in all their forms (ranging from physical to cultural) against Indigenous people, especially Indigenous women and LGBTQ, as well as the provision or extension of services (education, housing, healthcare, employment, etc.) to improve the material well-being of all Indigenous people. The fulfilment of these demands, including land restitution, necessitate "*an end to all corporate and US control of Native land and resources*" and Tribal government collusion (The Council of the Red Nation 2015, 6; italics in original). These are viewed as preliminary to building Indigenous capacities to treat land and nonhuman relatives ethically, with the ultimate objective being to bring the end of capitalism and colonialism, which, in contrast to the readings of some above-cited other Indigenous scholars, are understood as a single system. In fact, the issue is one of survival for First Peoples: "Capitalism–colonialism means death for Native peoples. For Native peoples to live, capitalism and colonialism must die" (ibid., 7).

The liberation of First Peoples is a universal objective. It implies the construction of "a livable future, not only for Native people but also for all human and nonhuman relatives" (ibid., 1). The technical aspects of building such a future, importantly, involve the combination of conventional science and technologies with long-established Indigenous practices in sectors like farming and forestry. Another way of putting this is to bring scientific work and technological developments in line with an egalitarian world order, whereby mutually respectful relations reign among peoples and among human and nonhuman relatives.

The general political strategy is premised on large alliances that include non-Indigenous communities to pursue objectives of universal benefit yet centred on First Peoples' concerns and locally rooted action. The Red Nation's activities and main concerns have expanded well beyond the US context. They include comprehensive proposals for an Indigenous-led global overcoming of capitalism and colonialism, structures that they view as inseparable and as the primary causes of planetary environmental damage, including and especially climate change (The Red Nation 2021). The coordination of actions and agendas, so far, has been by means of established, mainly First Peoples' networks or by seeking collaborations and mutual support at international levels.

The Red Nation's *Red Deal*, as a response to the Green New Deal, remains mainly within the bounds of US politics but it has much wider application and, because of the global reach of the US empire,[18] a comprehensive vision and set of practical proposals that explicitly ties the feasibility of ecological sustainability with decolonisation and anti-imperialism, drawing from, supporting, and building on, rather than dismissing, revolutionary efforts across the world, successful so far only in the Global South, to overcome colonialism and channel resources primarily to improve the well-being of people (The Red Nation 2021).

[18] To be clear, as The Red Nation underline, the US is an empire by dint of its historical founding as expansionistic colonies, invading and taking over First Peoples' lands and even annihilating First Peoples to form what are too often taken for granted as legitimate political boundaries. The US has always been, since its inception, a prison of First Peoples.

The latter aspect, especially the insistence on the take-over of the state as part of developing people power, may not sit well in in some ecosocialist and Indigenist currents. Yet, overall, in articulating concerns and aims substantively overlapping with ecosocialism, *The Red Deal* should find favour and popularity among ecosocialists especially in core capitalist countries like the US. That so far *The Red Deal* has found little to no resonance with most ecosocialists suggests a disturbing disinterest in First People communism or, perhaps, a tacit fundamental divergence.

Ecosocialism and First Peoples' Self-Determination Struggles

Socialist movements in settler colonial countries, lamentably, have proven to be and still prove to be either irrelevant or, when failing to fight against or at times even supporting colonial projects, an active impediment to Indigenous Peoples' decolonisation struggles. Most ecosocialists have yet to demonstrate any more effectiveness at supporting Indigenous Peoples' struggles, beyond well-meaning speeches and writings. For instance, ecosocialists in North America could put Treaty Rights in Canada and US as central to political platforms and organising instead of including them at most as afterthoughts.

It would be superfluous to state what should be obvious, that First Peoples have their own internally diverse and contrasting political commitments, as in any society. Within ecosocialist writings, there is a tendency to depict First Peoples as fundamentally anti-capitalist and sharing the same general worldview without investigating the specific and historically changing relations within a First People and between First Peoples in relation to shifting modes of settler colonial dictatorship, attempts within settler societies to overturn settler colonialism, and changes in capitalist world-system dynamics.

There are substantial points of agreement in the views and ideas articulated by ecosocialists and many leaders and scholars from First People communities. Regrettably, ecosocialists have tended to ignore Indigenous scholarship and struggles for self-determination or to treat them as secondary in political importance towards building ecosocialism. What is more, there is a paucity of forays into the specificities and histories of First Peoples, which would bring into relief social dynamics, including contradictory movements, that help sharpen strategies for finding substantive commonalities of purpose and making enduring alliances to achieve mutually beneficial ends. On the other hand, there are trends within parts of influential Indigenous leadership that seem intent on preventing such alliances or collaborations by engaging in dismissive and totalising depictions of socialism, communism, or Marxism. Within capitalist countries, this creates additional obstacles to weaving alliances or collaborations with anti-capitalist currents among First Peoples, as a result of having to fight simultaneously against bourgeois anti-socialism and settler colonial ideologies in broader society, when it comes to Europe and settler colonial countries in the Americas, the Pacific Islands, and elsewhere. This state of affairs is especially burdensome for Indigenous communists or socialists, who are compelled to navigate the anti-communist mine fields within their own communities while having to make what should be unnecessary additional efforts to counter the instances of ignorance, superficiality, if not dismissiveness among ecosocialists and socialists generally regarding First Peoples.

The most important problem of the mainstream of ecosocialism is a predominant underestimation, if not neglect of the political ramifications of settler colonialism (along with slavery and other forms of unwaged labour), as a main foundation of the capitalist world. Another problem is with a tendency for using a level of abstraction about Indigenous Peoples that erases their diverse situations, politics, and histories. This relates to a dearth of engagement with cases of ecologically and/or socially deleterious Indigenous practices and with contradictory processes and struggles within and between Indigenous Peoples. Finally, there is a tendency to generalise about Indigenousness from a narrow geographical basis that undermines the credibility of the above-mentioned predominant level of abstraction. These problems debilitate ecosocialist frameworks and can lead to a counterproductive politics.

References

Acosta, Alberto. 2017 "Post-extractivism: From Discourse to Practice—Reflections for Action." In *Alternative Pathways to Sustainable Development: Lessons from Latin America*, edited by Gilles Carbonnier, Humberto Campodónico, and Sergio Tezanos Vázquez, 77–101. Leiden: Brill Nijhoff.

Acosta, Alberto. 2023. "Degrowth and Buen Vivir. Perspectives for a Great Transformation." In *Routledge Handbook of Latin America and the Environment*, edited by Beatriz Bustos, Salvatore Engel-Di Mauro, Gustavo García-López, Felipe Milanez, and Diana Ojeda, 305–314. New York: Taylor & Francis.

Acosta, Alberto, and Mateo Martínez Abarca. 2018. "Buen Vivir: An Alternative Perspective from the Peoples of the Global South to the Crisis of Capitalist Modernity." In *The Climate Crisis: South African and Global Democratic Eco-Socialist Alternatives*, edited by Satgar Vishwas, 131–147. Johannesburg: Wits University Press.

Albó, Xavier. 2011. "Suma qamaña = convivir bien. ¿Cómo medirlo?" In *Vivir bien:¿ Paradigma no capitalista?*, H. Ivonne Farah, and Luciano Vasapollo, 133–144. La Paz: Cides-Umsa.

Albritton, Robert. 2019. *Eco-socialism for Now and the Future*. Geneva: Springer International Publishing.

Alfred, Gerald Robert. 1994. *Heeding the Voices of Our Ancestors: Kahnawake Mohawk Politics and the Rise of Native Nationalism in Canada*. Ithaca: Cornell University.

Alfred, Taiaiake. 1999. *Peace, Power, Righteousness. An Indigenous Manifesto.* Oxford: Oxford University Press.

Alfred, Taiaiake. 2005. *Wasáse: Indigenous Pathways of Action and Freedom.* Toronto: University of Toronto Press.

Altmann, Philipp. 2022. "Buen Vivir." In *Handbook of Critical Environmental Politics*, edited by Pellizzoni, Luigi, Emanuele Leonardi, and Viviana Asara, 104–115. Cheltenham: Edward Elgar Publishing.

Artaraz, Kepa, Melania Calestani, and Mei L. Trueba. 2021. "Introduction: Vivir bien/Buen Vivir and Post-Neoliberal Development Paths in Latin America: Scope, Strategies, and the Realities of Implementation." *Latin American Perspectives* 48 (3): 4–16.

Avilés Irahola, Dennis L. 2020. "The Concept of Buen Vivir: A Decolonial Feminist Perspective." https://radicalecologicaldemocracy.org/decolonial-feminism-and-buen-vivir/.

Baer, Hans A. 2022. "Indigenous Australians and Engaged Anthropologists: How Can They Struggle Together for Social Justice and a Safe Climate?" *AlterNative: An International Journal of Indigenous Peoples* 18 (2): 313–319.

Ballard, Patricia L. 1984. "Toward Indigeneous Liberation: The Sandinistas and the Miskito of Northeastern Nicaragua." *Antipode* 16: 54–64.

Barragán, Luis Alfonso. 2020. "El Buen vivir y el Sumak Kawsay: dos filosofías en disputa. Aproximaciones al caso ecuatoriano y al caso colombiano." *Pacha. Revista de Estudios Contemporáneos Del Sur Global* 1 (3): 9–24.

Barsh, Russel L. 1988. "Contemporary Marxist Theory and Native American Reality." *American Indian Quarterly* 12 (3): 187-211.

Beck, David R.M. 2002. *Siege and Survival. History of the Menominee Indians, 1634–1856*. Lincoln: University of Nebraska Press.

Beck, David R.M. 2005. *The Struggle for Self-Determination. History of the Menominee Indians Since 1854*. Lincoln: University of Nebraska Press.

Becker, Marc. 2021. *The CIA in Ecuador*. Durham, NC: Duke University Press.

Bedford, David, and Danielle Irving. 2001. *The Tragedy of Progress: Marxism, Modernity and the Aboriginal Question*. Halifax: Fernwood Press.

Bedford, David, and W. Thom Workman. 2012. "Whiteness and the Great Law of Peace." In *Working Through Cynthia Working Through Whiteness: International Perspectives*, edited by Levine-Rasky, 25–42. State University of New York Press.

Beling, Adrián E., Ana P. Cubillo-Guevara, Julien Vanhulst, J., António L. Hidalgo-Capitán. 2021. "Buen Vivir (Good Living): A "Glocal" Genealogy of a Latin American Utopia for the World." *Latin American Perspectives* 48 (3): 17–34.

Bieder, R.E. 1995. *Native American Communities in Wisconsin, 1600–1960. A Study of Tradition and Change*. Madison: The University of Wisconsin Press.

Blanco Galdos, Hugo. 2017. *Nosotros los Indios*. Cusco: Centro de Estudios Regionales Andinos Bartolomé.

Brownhill, Leigh, and Terisa E. Turner. 2020. "Ecofeminist Ways, Ecosocialist Means: Life in the Post-capitalist Future." *Capitalism Nature Socialism* 31 (1): 1–14.

Brownhill, Leigh, Salvatore Engel-Di Mauro, Terran Giacomini, Ana Isla, Michael Löwy, and Terisa Turner. 2021. "Preface. An Introduction to Ecoscialism." In *The Routledge Handbook on Ecosocialism*, edited by Leigh Brownhill, Salvatore Engel-Di Mauro, Terran Giacomini, Ana Isla, Michael Löwy, and Terisa Turner, 1–11. New York: Routledge.

Bustamante, Juan P. Vásquez, José Orellana Yáñez, and Juliane Rodrigues Teixeira. 2021. "Del Sumak Kawsay al debate por el Buen Vivir: significados en disputa y disputa por los significantes." *Si Somos Americanos* 21 (2): 120–145.

Carreño-Calderón, Alejandra. 2021. "Living Well and Health Practices Among Aymara People in Northern Chile." *Latin American Perspectives* 48 (3): 69–81.

Chase-Dunn, Christopher, and Thomas D. Hall. 1998. "World-Systems in North America: Networks, Rise and Fall and Pulsations of Trade in Stateless Systems." *American Indian Culture and Research Journal* 22 (1): 23–72.

Chattopadhyay, Sutapa. 2014. "Postcolonial Development State, Appropriation of Nature, and Social Transformation of the Ousted Adivasis in the Narmada Valley, India." *Capitalism Nature Socialism* 25 (4): 65–84.

Churchill, Ward, ed. 1983. *Marxism and Native Americans*. Boston: South End Press.

Clark, Brett, and John B. Foster. 2021. "Marx's Ecology and Metabolic Analysis." In *The Routledge Handbook on Ecosocialism*, edited by Leigh Brownhill, Salvatore Engel-Di Mauro, Terran Giacomini, Ana Isla, Michael Löwy, and Terisa Turner, 89–98. Routledge.

Coral-Guerrero, Carmen Amelia, Fernando García-Quero, and Jorge Guardiola. 2021. "What Is Sumak Kawsay? A Qualitative Study in the Ecuadorian Amazon." *Latin American Perspectives* 48 (3): 35–50.

Coulthard, Glen. 2007. "Subjects of Empire: Indigenous Peoples and the 'Politics of Recognition' in Canada." *Contemporary Political Theory* 6: 437–460.

Coulthard, Glen. 2014. *Red Skin, White Masks: Rejecting the Colonial Politics of Recognition*. Minneapolis: University of Minnesota Press.

Cuestas-Caza, Javier. 2018. "Sumak kawsay Is Not Buen Vivir." *Alternautas* 5 (1). https://journals.warwick.ac.uk/index.php/alternautas/article/view/1070. Accessed 28 July 2023.

Cuestas-Caza, Javier, Rickard Lalander, and Magnus Lembke. 2020. "Andean Intercultural Ecosocialism in Times of Buen-Vivir? A Red-Green-Culturalist Approach." In *Challenging the Right, Augmenting the Left: Recasting Leftist Imagination*, edited by Robert Latham, A.T. Kingsmith, Julian von Bargen and Niko Block, 197–209. Black Point: Fernwood Publishing.

Davis, T. 2000. *Sustaining the Forest, the People, and the Spirit*. Albany, NY: State University of New York Press.

Del Valle Escalante, Emilio. 2014. "Self-Determination: A Perspective from Abya Yala." *e-International Relations*. https://www.e-ir.info/pdf/49610. Accessed 22 July 2023.

Deloria, V. 1985. "The Evolution of Federal Indian Policy Making." In *American Indian Policy in the Twentieth Century*, edited by V. Deloria, 239–256. Norman: University of Oklahoma Press.

Deloria, Vine. 1997. *Red Earth, White Lies: Native Americans and the Myth of Scientific Fact*. Golden, Colorado: Fulcrum Publishers.

Demaria, Federico, Alberto Acosta, Ashish Kothari, Ariel Salleh, and Arturo Escobar. 2020. "El pluriverso, horizontes para una transformación civilizatoria." *Revista de Economía Crítica* 29: 46–66.

Dennis, Philip A. 1993. "The Miskito-Sandinista Conflict in Nicaragua in the 1980s." *Latin American Research Review* 28 (3): 214–234.

Dockry, Michael, and Kyle Whyte. 2021. "Improving on Nature: The Legend Lake Development, Menominee Resistance, and the Ecological Dynamics of Settler Colonialism." *The American Indian Quarterly* 45 (2): 95–120.

Dorney, C.H., and J.R. Dorney. 1990. "An Unusual Oak Savanna in Northeastern Wisconsin: The Effect of Indian-Cause Fire." *The American Midland Naturalist* 122 (1): 103–113.

Dunbar-Ortiz, Roxanne. 2014 *An Indigenous Peoples' History of the United States*. Boston: Beacon Press.

Espinosa, Agustín, Erika Janos, and Martín Mac Kay. 2023. "Indigenous Political Participation in Peru: A History of Racism, Exclusion, and Violence." In *Indigenous Political Representation in Latin America*, 201–232. Cham: Springer International Publishing.

Estes, Nick. 2019. *Our History Is the Future: Standing Rock Versus the Dakota Access Pipeline, and the Long Tradition of Indigenous Resistance*. New York: Verso Books.

Farthing, Linda, and Thomas Becker. 2021. *Coup: A Story of Violence and Resistance in Bolivia*. Haymarket Books.

Fernandes, Sabrina. 2020a. "Ecosocialism from the Margins." *NACLA Report on the Americas* 52 (2): 137–143.

Flader, S.L., ed. 1983. *The Great Lakes Forest: An Environmental and Social History*. Minneapolis: University of Minnesota Press.

Foster, John Bellamy, Brett Clark, and Hannah Holleman. 2020. "Marx and the Indigenous." *Monthly Review* 71 (9): 1–19.

Franco, D. Leandro, and Cintia Giuliana Balaudo. 2019. "El potencial desmercantilizador del derecho en las propuestas del Buen Vivir." *SaberEs* 11 (1): 21–39.

García-García, Aurelio. 2022. "The Limits to Growth of Buen Vivir Socialism: Ecuador's Alternative Development Model from 2007 to 2017." *History of Economic Thought and Policy* (1): 25–59.

Grande, Sandy. 2004. *Red Pedagogy: Native American Social and Political Thought*. Lanham, MD: Rowman & Littlefield Publishers.

Grossman, Zolán. 2017. *Unlikely Alliances: Native Nations and White Communities Join to Defend Rural Lands*. Seattle: University of Washington Press.

Gudynas, Eduardo. 2011. "Buen Vivir: Today's Tomorrow." *Development* 54 (4): 441–447.

Gudynas, Eduardo. 2017. "Value, Growth, Development: South American Lessons for a New Ecopolitics." *Capitalism Nature Socialism* 30 (2): 234–243.

Hall, Louis Karoniaktajeh. 2023. *The Mohawk Warrior Society. A Handbook on Sovereignty and Survival*. Edited by Philippe Blouin, Matt Peterson, Malek Rasamny, and Kahentinetha Rotiskarewake. Oakland: PM Press.

Hidalgo-Capitán, A.L., and A.P. Cubillo-Guevara. 2020. "Buen Vivir: A Reference for Public Policies and Development in Ecuador and the Governments of the Latin American Left." In *Assessing the Left Turn in Ecuador. Studies of the Americas*, edited by F. Sánchez, and S. Pachano. Cham: Palgrave Macmillan.

Holm, T. 1992. "Patriots and Pawns. State Use of American Indians in the Military and the Process of Nativization in the United States." In *The State of Native America: Genocide, Colonization, and Resistance*, edited by M.A. Jaimes, 345–370. Boston: South End Press.

Hook, Andrew. 2019. "Mapping Contention: Mining Property Expansion, Amerindian Land Titling, and Livelihood Hybridity in Guyana's Small-Scale Gold Mining Landscape." Geoforum 106: 48–67.

Horne, Gerald. 1986. *Black and Red. W.E.B. Du Bois and the Afro-American Responseto the Cold War, 1944–1963*. Albany: SUNY Press.

Horne, Gerald. 2019. *White Supremacy Confronted: U.S. Imperialism and Anti-Communism vs. the Liberation of Southern Africa from Rhodes to Mandela*. New York: International Publishers.

Hosmer, B.C. 1997. "Reflections on Indian Cultural 'Brokers': Reginald Oshkosh, Mitchell Oshkenaniew, and the Politics of Menominee Lumbering." *Ethnohistory* 44 (3): 493–509.

Hosmer, B.C. 1999. *American Indians in the Marketplace. Persistence and Innovation Among the Menominees and Metlakatlans, 1870–1920*. Lawrence, KS: University Press of Kansas.

Huber, Matthew T. 2002. *Climate Change as Class War: Building Socialism on a Warming Planet*. New York: Verso Books.

Indian Country Today. 2023. "Navajo Nation Opposes Any Chaco Canyon Buffer Zone." *Indian Country Today*, July 17. https://ictnews.org/news/the-wrap-chaco-buffer-zone. Accessed 19 July 2023.

Isla, Ana. 2006. "Women, Enclosure, and Accumulation: A Rejoinder to Robert Chapman." *Capitalism Nature Socialism* 17 (4): 58–65.

Isla, Ana. 2015. *The "Greening" of Costa Rica. Women, Peasant, Indigenous Peoples, and the Remaking of Nature*. Toronto: University of Toronto Press.

Isla, Ana. 2019. "Indigenous Andoas Uprising: Defending Territorial Integrity and Autonomy in Perù." In *Climate Chaos: Ecofeminism and the Land Question*, edited by Ana Isla, 283–308. Toronto: Inanna Publications & Education Inc.

Jaimes, M.A., and T. Halsey. 1992. "American Indian Women at the Center of Indigenous Resistance in Contemporary North America." In *The State of Native America: Genocide, Colonization, and Resistance*, edited by M.A. Jaimes, 311–344. Boston: South End Press.

Joe, Kianna. 2023. "'Go Home', Haaland's Return Home Derailed by Protest Near Chaco Canyon." *Navajo Times*, July 19. https://navajotimes.com/reznews/go-home-haalands-return-home-derailed-by-protest-near-chaco-canyon/. Accessed 19 July 2023.

Johanson, Bruce. 2004. "The New York Oneidas: A Business Called a Nation." In *Enduring Legacies. Native American Treaties and Contemporary Controversies*, edited by Bruce Johansen and Vine Deloria, 81–94. Westport: Praeger.

Kapp, Amanda. 2019. "The Menominee Indian Tribe of Wisconsin, August, 2019. Climate Change Program." Institute for Tribal Environmental Professionals, Northern Arizona University. Available at: www7.nau.edu/itep/main/tcc/Tribes/gl_menominee.

Kay, J. 1979. "Wisconsin Indian Hunting Patterns, 1634–1836." *Annals of the Association of American Geographers* 69 (3): 402–418.

Kay, J. 1984. "The Fur Trade and Native American Population Growth." *Ethnohistory* 31 (4): 265–287.

Kovel, Joel. 2007. *The Enemy of Nature. The End of Capitalism or the End of the World?* London: Zed Press.

LaDuke, W. 1994. "The Dilemma of Indian Forestry." *Earth Island Journal* 9 (3): 36–38.

Lalander, Rickard, and Magnus Lembke. 2018. "The Andean Catch-22: Ethnicity, Class and Resource Governance in Bolivia and Ecuador." *Globalizations* 15 (5): 636–654.

Lalander, Rickard, and Magnus Lembke. 2021. "Reflexiones Decoloniales sobre Sumak Kawsay y Justicia Social." In *Libro de Acta de Memorias del Congreso Internacional Ciencias Jurídicas, sociales y políticas*, edited by Andrea Aguirre Bermeo; María Beatriz Eguiguren, Jorge Maldonado Ordóñez and Janeth González Malla, 98–124. Loja, Ecuador: Universidad Técnica Particular de Loja/UTPL.

Le Quang, Matthieu. 2015. "Introducción al dossier. Buen Vivir y Ecosocialismo. Enfoques teóricos y políticas públicas." *Theomai* 32: 4–12.

Le Quang, Matthieu. 2020. "Intelectuales del buen vivir y debates sobre el régimen de acumulación en Ecuador." In *Intelectuales, democracia y derechas*, edited by Alfredo Falero, Charles Quevedo y Lorena Soler, 95–114. Buenos Aires: CLACSO.

Loop, W.L., and J. Anderton. 1998. "Human vs. Lightning Ignition of Presettlement Surface Fires in Coastal Pine Forests of the Upper Great Lakes." *The American Midland Naturalist* 140 (2): 206–218.

Löwy, Michael. 2011. *Écosocialisme. L'alternative radicale à la catastrophe écologique capitaliste*. Paris: Mille et Une Nuits.

Löwy, Michael. 2014. "Ecosocial Struggles of Indigenous Peoples." *Capitalism Nature Socialism* 25 (2): 14–24.

Löwy, Michael. 2015. *Ecosocialism: A Radical Alternative to Capitalist Catastrophe*. Chicago: Haymarket Books.

Malm, Andreas. 2021. *How to Blow Up a Pipeline*. New York: Verso Books.

Mantilla, Jorge, and Tamia Vercoutère Quinche. 2021. "Análisis bibliométrico y contextual de la producción científica en torno al Sumak Kawsay en Scopus." *Revista Universidad y Sociedad* 13 (3): 321–332.

Mausel, David L. 2017. "Anthony Waupochick, and Marshall Pecore. 'Menominee Forestry: Past, Present, Future.'" *Journal of Forestry* 115 (5): 366–369.

Medina, Javier. 2011. "Acerca del Suma Qamaña." In *Vivir bien:¿ Paradigma no capitalista?* edited by H. Ivonne Farah, and Luciano Vasapollo, 39–64. La Paz: Cides-Umsa.

Menominee Tribal Enterprises. 1997. "The Menominee Forest-Based Sustainable Development Tradition." https://archive.epa.gov/ecopage/web/pdf/menominee-forest-keepers-1997-25pp.pdf. Accessed 24 July 2023.

Merino, Roger. 2021. "Buen vivir and the Making of Indigenous Territories in the Peruvian Amazon." *Latin American Perspectives* 48 (3): 136–151.

Mohawk, John C. 2004. "The Iroquois Land Claims: A Legacy of Fraud, Politics, and Dispossession." In *Enduring Legacies. Native American Treaties and Contemporary Controversies*, edited by Bruce Johansen and Vine Deloria, 81–94. Westport: Praeger.

Olson, Stacey K., Erica A. H. Smithwick, Melissa S. Lucash, Robert M. Scheller, Robert E. Nicholas, Kelsey L. Ruckert, and Christopher M. Caldwell. 2021. "Landscape-Scale Forest Reorganization Following Insect Invasion and Harvest Under Future Climate Change Scenarios." *Ecosystems*, 1–19.

Ourada, P.K. 1990. *The Menominee*. New York: Chelsea House Publishers.

Owens Howard B. 2023. "Genesee County Will Benefit from $50 Billion CHIPS and Science Bill Passed by Senate, Schumer Says." *The Batavian*, July 27. https://www.thebatavian.com/howard-b-owens/genesee-county-will-benefit-from-50-billion-chips-and-science-bill-passed-by-senate. Accessed 1 August 2023.

Pecore, M. 1992. "Menominee Sustained-Yield Management. A Successful Land Ethic in Practice." *Journal of Forestry* 90 (7): 12–16.

Peroff, N.C. 1982. *Menominee DRUMS. Tribal Termination and Restoration, 1954–1974*. Norman: University of Oklahoma Press.

Peters Michael A., and Carl T. Mika. 2017. "Aborigine, Indian, Indigenous or First Nations?" *Educational Philosophy and Theory* 49 (13): 1229–1234.

Pietrorazio, Gabriel. 2023. "Onondaga Nation General Counsel Joe Heath Debriefs on Latest Decision for Nearly Decade-Long Land Theft Petition." Central Current, July 7. https://centralcurrent.org/onondaga-nation-general-counsel-joe-heath-debriefs-on-latest-decision-for-nearly-decade-long-land-theft-petition/. Accessed 1 August 2023.

Planas, Roque. 2023. "How Tribal Hunters Became the Scapegoat for Yellowstone's Bison 'Slaughter'." *Huffpost*, May 21. https://www.huffpost.com/entry/yellowstone-tribal-bison-hunts_n_64668149e4b0005c6057df4a. Accessed 19 July 2023.

Prashad, Vijay. 2017. *Red Star over the Third World*. New Delhi: Leftword Books.
Puente, Rafael. 2011. "'Vivir Bien' y descolonización." In *Vivir bien:¿ Paradigma no capitalista?* edited by Ivonne Farah, H., and Luciano Vasapollo, 345–364. La Paz: Cides-Umsa.
Ranta, Eija. 2018. *Vivir Bien as an Alternative to Neoliberal Globalization: Can Indigenous Terminologies Decolonize the State?* Abingdon: Routledge.
Rengifo Arias, Carlos G., Sonia E. López Pulgarín, Diana M. Lopera Montoya, and Diana Mayerly Díaz Tapasco. 2022. "El Buen Vivir en comunidades andinas: un reto ecosistémico." *Naveg@mérica. Revista electrónica editada por la Asociación Española de Americanistas*, n. 29, http://revistas.um.es/navegamerica. Accessed 25 July 2023.
Rivers, Tom. 2023. "Tonawanda Nation, Environmentalists Speak Out Against STAMP." OrleansHub, May 12. https://orleanshub.com/tonawanda-nation-environmentalists-speak-out-against-stamp/. Accessed 1 August 2023.
Rose, Samuel W. 2017. "Marxism, Indigenism, and the Anthropology of Native North America: Divergence and a Possible Future." *Dialectical Anthropolology* 41: 13–31.
Robbins, R. 1992. "Self-Determination and Subordination: The Past, Present and Future of American Indian Governance." In The *State of Native America: Genocide, Colonization, and Resistance*, edited by M.A. Jaimes, 87–122. Boston: South End Press.
Runstrom, A., R.M. Bruch, D. Reiter, and D. Cox. 2002. "Lake Sturgeon (Acipenser fulvescens) on the Menominee Indian Reservation: An Effort Toward Co-management and Population Restoration." *Journal of Applied Ichthyology* 18: 481–485.
Salleh, Ariel. 2009. "Ecological Debt: Embodied Debt." In *Eco-sufficiency & Global Justice: Women Write Political Ecology*, edited by Ariel Salleh, 1–40. London: Pluto Press.
Salleh, Ariel. 2017. *Ecofeminism as Politics: Nature, Marx and the Postmodern*. London: Zed Books.
Saito, Kohei. 2023. *Marx in the Anthropocene: Towards the Idea of Degrowth Communism*. Cambridge: Cambridge University Press.
Sands, Benjamin A., and Marc D. Abrams. 2011. "A 183-Year History of Fire and Recent Fire Suppression Impacts in Select Pine and Oak Forest Stands of the Menominee Indian Reservation, Wisconsin." *American Midland Naturalist* 166 (2): 325–338.
Schwartzman, Peter, and David Schwartzman. 2019. *The Earth Is Not for Sale: A Path Out of Fossil Capitalism to the Other World That Is Still Possible*. New York: World Scientific.

Sedrez, Lise F. 2014. "Rubber, Trees and Communities: Rubber Tapers in the Brazilian Amazon in the Twentieth Century." In *A History of Environmentalism. Local Struggles, Global Histories*, Marco Armiero and Lise Sedrez, 147–166. London: Bloomsbury.

Simpson, Leanne Betasamosake. 2017. *As We Have Always Done: Indigenous Freedom Through Radical Resistance.* Minneapolis: University of Minnesota Press.

Tanuro, Daniel. 2015. *L'impossible capitalisme vert.* Paris: la Découverte.

The Council of the Red Nation. 2015. "Native Liberation Struggles in North America: The Red Nation 10-Point Program." *Capitalism Nature Socialism* 26 (2): 1–7.

The Red Nation. 2021. *The Red Deal: Indigenous Action to Save Our Earth.* Brooklyn: Common Notions Press.

Thiri, May Aye, Sergio Villamayor-Tomás, Arnim Scheidel, and Federico Demaria. 2022. "How Social Movements Contribute to Staying Within the Global Carbon Budget: Evidence from a Qualitative Meta-Analysis of Case Studies." *Ecological Economics* 195: 107356.

Tronnes. Libby R. 2002. "'Where Is John Wayne?': The Menominee Warriors Society, Indian Militancy, and Social Unrest During the Alexian Brothers Novitiate Takeover." *American Indian Quarterly* 26 (4): 526–558.

Trosper, Ronald L. 2007. "Indigenous Influence on Forest Management on the Menominee Indian Reservation." *Forest Ecology and Management* 249 (1–2): 134–139.

Tuck, Eve, and K. Wayne Yang. 2012. "Decolonization Is Not a Metaphor." *Decolonization: Indigeneity, Education & Society* 1 (1): 1–40.

Vanhulst, Julien, Ana P. Cubillo-Guevara, Adrian Beling, and Antonio L. Hidalgo-Capitán. 2020. "Los discursos políticos latinoamericanos sobre el buen vivir (1992–2017)." *Ayer. Revista de Historia Contemporánea* 119 (3): 163–193.

Velitchkova, Jackie Smith, and Austin Choi-Fitzpatrick. 2010. "Windows on the Ninth World Social Forum in Belém." *Societies Without Borders* 4 (2): 193–208.

Viteri Gualinga, Carlos. 2002. "Visión indígena del desarrollo en la Amazonía." *Polis* 3. https://journals.openedition.org/polis/7678?lang=pt. Accessed 25 July 2023.

Wall, Derek, and Hugo Blanco. 2010. *The Rise of the Green Left: Inside the Worldwide Ecosocialist Movement.* London: Pluto Press.

Wallis, Victor. 2015. "Intersectionality's Binding Agent." *New Political Science* 4: 604–619.

Wallis, Victor. 2018. *Red-Green Revolution: The Politics and Technology of Ecosocialism.* Toronto: Political Animal Press.

Walshe, Rory, and Alejandro Argumedo. 2016. "Ayni, Ayllu, Yanantin and Chanincha: the Cultural Values Enabling Adaptation to Climate Change in Communities of the Potato Park, in the Peruvian Andes." *GAIA-Ecological Perspectives for Science and Society* 25 (3): 166–173.

Warrior, Robert, ed. 2015. *The World of Indigenous North America*. New York and London: Routledge.

Whitaker, James A. 2020. "Strategic Alliance and the Plantationocene Among the Makushi in Guyana." *Social Anthropology/Anthropologie sociale* 28 (4): 881–896.

Whitney, G.G. 1996. *From Coastal Wilderness to Fruited Plain. A History of Environmental Change in Temperate North America from 1500 to the Present*. Cambridge: Cambridge University Press.

Whyte, Kyle, Chris Caldwell, and Marie Schaefer. 2018. "Indigenous Lessons About Sustainability Are Not Just for 'All Humanity'". In *Sustainability: Approaches to Environmental Justice and Social Power*, edited by Julie Sze, 149–179. New York, USA: New York University Press.

Wikstrom, Lindsey. 2023. *Designing the Forest and Other Mass Timber Futures*. London: Taylor & Francis.

Wisconsin State Journal. 2023. "Counties with the Shortest Life Expectancy in Wisconsin." *Wisconsin State Journal*, January 31. https://madison.com/news/state-and-regional/counties-with-the-shortest-life-expectancy-in-wisconsin/collection_90ccc395-9cd6-5447-ab38-e3e662074ae0.html#38. Accessed 24 July 2023.

Wolfley, J. 1998. "Ecological Risk Assessment and Management: Their Failure to Value Indigenous Traditional Ecological Knowledge and Protect Tribal homelands." *American Indian Culture and Research Journal* 22 (2): 151–169.

Yampara, Simon, ed. 1993. *Naciones autóctonas-originarias: Vivir-convivir en tolerancia y diferencia*. La Paz: CADA.

Yampara, Simon. 2001. *El ayllu y la territorialidad en los Andes: Una aproximación a chambi grande*. La Paz: UPEA/Inti-Andino/CADA.

CHAPTER 8

Ecosocialism and Ecological Civilisation

The People's Republic of China as Challenge to Ecosocialists

Among the more contentious topics within ecosocialism are the political status and potentials of the People's Republic of China (PRC), including the notions of socialism with Chinese characteristics and, more recently, Ecological Civilisation (生态文明, shēngtài wénmíng) (see Boer 2021; Boer and Yan 2021; Huan 2021). The latter presents a potentially decisive, even if unrecognised challenge to ecosocialists. Were an Ecological Civilisation transition to materialise in the PRC, major components of the aims of ecosocialism would arguably be on the way to being fulfilled in one of the largest countries on Earth. It would imply that ecosocialism would be of marginal relevance to the struggle for a worldwide shift away from capitalism and towards global ecological sustainability. Given the calibre of most ecosocialist discourse on the topic, it would seem that ecosocialists have yet to appreciate the stakes involved.

The challenge is much wider than for a tiny political movement. The last half century witnessed in the PRC a world-consequential gradual then rapid intensification of industrialisation and capitalist relations leading to much environmental damage, some of long-term consequence, and widening social inequalities. Just as important to emphasise are the tremendous improvements in the living conditions and life chances for most in the PRC, including the elimination of extreme poverty for nearly

S. Engel-Di Mauro, *Ecosocialism*, Marx, Engels, and Marxisms,
https://doi.org/10.1007/978-3-031-74288-0_8

850 million people within the span of roughly four decades (World Bank 2022). Worthy of ecosocialist attention is that the latter, unprecedented development in human history coincides with a large and ecologically destructive increase in throughputs, in cases of ecosystem degradation, and in pollution levels.

The widening inequality and exacerbation of class differentiation were doubtless unwanted corollaries for most, but arguably part of the long-term challenges of establishing the material conditions for socialism. The road to a brighter future cannot but be tortuous and filled with setbacks for a country first under imperialists' siege right after surviving decades of war, then, having undergone multiple convulsions, in partial subordination to economically ravenous imperialist liberal democracies, and finally again increasingly under economic attack and threatened militarily by the same insatiable imperialist bunch. The move towards capitalist relations to be contained through communist party rule is a point of vehement contention and everyday social struggle. It is, to say the least, a primary contradiction ushered in by such historically momentous and system defining processes, a contradiction with which the Communist Party of China (CPC) and movements within wider society have been grappling for decades (Hammond 2023; Lau 2018; Li 2017; Lin 2006; Xu 2018).

The nature and trajectory of the PRC are vexing issues hardly confined to ecosocialist disputes. What happens in the PRC is of enormous global consequence, for social and ecological reasons (Li 2016). In fact, the PRC could be headed towards assuming a leadership position that would be useful towards establishing the global and likely coercive—unless capitalist ruling classes selflessly oblige—redistribution system required for a worldwide transition to ecosocialism, especially in its degrowth-promoting guise (Gare 2021; Li 2017; cf. Baer 2018, 172). Most ecosocialists (and degrowth promoters) living in the highest per capita consumption countries, mainly liberal democracies, may find such conclusions disagreeable or incorrect. A discussion of prospects for worldwide ecosocialism is beyond the scope of this chapter (see instead Chapter 9), but it can be useful to keep such issues in mind while considering the current striving for Ecological Civilisation in the PRC (a more comprehensive ecosocial view on the PRC is offered in Engel-Di Mauro 2021).

Ecosocialist Contentions over the People's Republic of China

Ecosocialists have had mixed reactions to the PRC generally, both over the kind of politics and economy that has been developing there as well as the ecological and social repercussions of such an economy within and beyond the PRC. A main disagreement is whether the PRC is a socialist country. Viewpoints generally converge on the capitalist character of the PRC's economy and range between hostile and critical. Sympathetic or supportive, even critically supportive evaluations tend to be rare.

Positive depictions are mainly from those promoting ecosocialism within the PRC. According to their understanding, the PRC remains socialist because "Key economic sectors continue to be dominated by state-owned enterprises, and they tend to serve the political interests of the state, whereas commercial interests are secondary" (Zhu and Lo 2022, 2; see also Huan 2010, 7).[1] They point to concrete evidence of what amounts to ecosocialism in practice. For instance, state forestry institutions, with their successful reforestation programme throughout the country since the late 1990s,

> serve dual roles as the implementer of central policy and as a buffer between the central government's policy goals and local communities' interests. As such, forest conservation in China can be understood as a form of eco-socialism, where environmental-friendly policy is delivered by reconstructing the mode of production and living under a socialist system. (Zhu and Lo 2022, 2)

The development of ecosocialism in the PRC is also grounded in Marxist theory relative to the different and complementary ecological aspects of Marx's and Engels' writings (Huan 2022, 339–340). Regarding the meaning and application of Ecological Civilisation, there are different tendencies within the CPC, ranging from reformist to transformational. Some advocate for the continuation of industrialisation until the economic resources are plentiful enough to tackle environmental aspects, while others press for a deep reconstitution of social relations to reorient towards ecologically sustainable relations in all facets of human

[1] The authors describe socialism "as an economic system in which the principal mechanism of resource allocation is central planning and control, rather than market" (Zhu and Lo 2022, 2).

life (Gare 2021). Further, the interpretation of Ecological Civilisation's history is divided into two major schools of thought. One views it as a novel framework formulated by Xi Jinping and another points to the collective basis of its production and in continuity with the PRC's history (Huan 2022, 342–343).

Zhang (2019) surmises that all societies exhibit different forms of ecological civilisations. The question is whether their development is in harmony with the material conditions for their existence. In this light, the current trend in the PRC of exceeding ecological capacities will lead to self-destruction. One possible solution is to combine Marxism with the ecological sensitivity intrinsic to pre-existing traditional cultural norms (see also Gare 2012; Wang et al. 2014, 54).

In contrast, to Wang et al. (2013, 51), concurring in some respects with the predominant ecosocialist position outside the PRC, both capitalism and socialism (read "first epoch" socialism) are forms of "modernity" giving rise to the separation of people from the rest of nature. According to this viewpoint, the ruling modernity ideology in the PRC explains the presence of much environmental destruction. The combination of "constructive postmodernism" and Eco-Marxism is claimed to furnish the necessary basis for a solution by overcoming modernity.

Taking a wider social lens, it would be premature to argue for the existence of an ecosocialist movement in the PRC as a social force of political consequence. A potential convergence, though, exists between environmentalist organisations and unions through governmental and CPC guidance. The reasons for a lack of red-green movements are due to differing material interests, relative dependence for both on institutional support, and international pressures restraining the influence of environmentalism (Huan 2014, 56–58). As shown in the geographical survey in Chapter 4, one could arrive at similar conclusions regarding ecosocialist movements in the rest of the world, even if involving different causal mechanisms.

Few ecosocialists elsewhere recognise the existence of ecosocialism in the PRC (e.g., Baer 2021b, 159; Gare 2021). Most ecosocialists, mainly in NATO and like countries, view the PRC as essentially capitalist (e.g., Baer 2021a, 53), with any form of socialism having disappeared, including in Cuba, by the early 1990s (e.g., Löwy and Tanuro 2021, 27). This is implicit in the mainstream of Eco-Marxism. Since the socio-ecological crisis is explained by capitalist dynamics, it follows that the PRC, being traversed by much environmental destruction, must be capitalist (Wang et al. 2014, 43). Such an argument is made explicitly by Tanuro (2020, 74–75). Many ecosocialists, implicitly or explicitly, regard the PRC as no different in its exploitative foundations as other capitalist countries but with an authoritarian political system with allegedly worse environmental effects than liberal democracies (Flanagan and Raphael 2022). Within this anti-PRC camp, there inhabits an alarmist position as well, accusing the CPC of practising a form of environmentally devastating "police-state capitalism" that will bring the downfall of us all (Smith 2020).

To some who are ecosocialism adjacent, the civilisational trope would coincide with "a discourse of competition between civilisations that is increasingly nationalist, racialized, and illiberal" (Webber 2022, 66). In this perspective, Ecological Civilisation is little more than a "techno-market solution of ecological modernisation" that is "supported by increasing government-led investment and that "only portends a state-led, more environmentally friendly form of capitalist production" (Webber 2022, 72).

Other ecosocialists have less dismissive things to say about the PRC, but their critique is often blinkered by expectations of post-revolutionary society involving immediately or in a historically linear manner the application of participatory democracy and workers' control of the economy. The late Saral Sarkar (2010), for instance, understood the importance of the CPC's historical accomplishments (but especially the one-child policy) and control over substantive aspects of the national economy. Such control, he surmised, will have to be used towards ecosocialist transformation because it is no longer feasible to build the material conditions for socialism without incurring collapse (cf. Smith 2020).[2] Thus, decades of CPC experience and theoretical development are to be summarily swept

[2] Sarkar's fatalistic attitude about resource availability, mirrored in his fixation with population control, obstructed the possibility for another option in the struggle against imperialism and for global wealth redistribution. Disturbingly, as a long-time resident in Germany, he seemed to have internalised the rampant racism still imbuing European societies:

> The prospects for eco-socialism, I guess, are better among the peoples of the poor Third World countries. The distance between an ideal sustainable way of life and their present real way of life is not as great as that among the peoples of Western Europe and North America. In the Third World, many of us still remember having lived without electricity and motor vehicles. In India, even today about two thirds of the population do not have access to electricity. But one very depressing aspect of the situation there is the unabated population growth. (Sarkar 2010, 220)

aside in favour of adopting a presumably more enlightened and fresh perspective by way of an Indian male who lived in Germany.

Foster offers a more constructive view (Foster 2022). As he has put the matter,

> the necessary forging of an ecological civilization ... has been recognized most fundamentally in China, which ... draws on the theoretical legacy of historical materialism—adding to its long civilizational–cultural dynamic a materialist revolutionary outlook. (Foster 2017, 440)

Many Marxist thinkers in the CPC have also borrowed from other intellectual currents in other countries on their own intellectual terms and have ascribed their own differing meanings to such formally borrowed ideas (cf. Gare 2021).

Sometimes, however, it seems that appraisals rely on little to no evidence. For instance, Baer (2018, 49) regards the current PRC as "the largest postrevolutionary society in the world," but with a less Marxist-Leninist and increasingly Confucian and nationalistic ruling communist party (ibid., 65). This would be a credible statement if followed by, for example, estimates of institutions dedicated to Confucianism relative to the hundreds of Schools of Marxism. More empirically sound is Baer's observation that while environmental impacts have been mainly destructive, there are signs of betterment. He, like Albritton, recognises that, unlike the US, "China, the world's largest source of green-house gases (GHG) is making serious efforts to reduce its dependency on fossil fuels" (Albritton 2019, 100). Data-driven comparisons can be useful towards preventing rash judgement.

Ecosocialists at times consider historical factors and the current conjuncture to qualify their appraisals. Wallis (2020, 14), for instance, stresses the internal and external capitalist pressures characterising the development of socialist states like the PRC. However, according to that view, Deng is alleged, without supporting evidence, to have been

Such lamentable statements, where "true" ecosocialism (ibid., 221) is to be found in Malthus rather than Marx and Engels, hardly flatter the late author's intellect.

anti-Marxist[3] (ibid., 19) and, while said pressures eventually have ended up winning the day for capitalists (capitalist restoration), "there remain certain beneficial legacies of China's socialist revolution" (ibid., 20). Among them are advances in bettering environmental impacts (reforestation, developing solar energy infrastructure) and resisting US imperialism under Xi Jinping's leadership, but also worker exploitation within and outside the PRC and repression against Marxists supporting working-class organising. These are examples of critical ecosocialist perspectives on the PRC more broadly that do not impinge directly on the issue of Ecological Civilisation. However, they give an indication of the range of ecosocialist views on the PRC.

Ecosocialist Critiques of the Ecological Civilisation Framework

Ecological Civilisation has been approached from different perspectives aside from ecosocialist ones. Only those pertaining to ecosocialism or coming directly from ecosocialists are entertained here. Of these, a useful systematic and thorough treatment has been offered by Xue Jin, who is concerned with the successful spread of ecosocialism. Unlike most ecosocialists, Xue regards the PRC as socialist, though a sort of socialist formation that contradicts ecosocialism. In Xue's reading, the Ecological Civilisation framework amounts to a PRC version of ecological modernisation because economic growth and market mechanisms retain primacy of place, even explicitly in the very policies' official formulation (Xue 2016, 196; see also Gare 2021). What is more, using economic policies to tackle social problems like wealth maldistribution is reductionistic in that it is social relations that constitute economic activity. This is why social inequalities persist. Wealth redistribution involves policies directed at society as a whole, not one of its components, that is, the economy.

Attempts to raise levels of affluence in the PRC will also bump against biophysical limits. As Xue reasons, on the basis of data from the Global Footprint Network, that

[3] One can consult the archival material available in English to quickly dishabituate oneself of such notion: https://www.marxists.org/reference/archive/deng-xiaoping/index.htm (see also Boer and Yan 2021). The evaluation of interpretations and applications of Marxist method should be met with scientific evidence, rather than buried by one's political ideals and predilections.

> Due to limited resources within the borders of China, 57 percent of the total demand would have to be provided by the rest of the world. This means that, even though China may obtain a high eco-efficiency by following the ecological modernization model, the environmental gains from eco-efficiency would be counteracted by the high consumption level, resulting in huge pressure on the global environment. (Xue 2016, 198)

The pattern so far observed in any case points to reducing the rates of negative environmental impacts per unit of GDP, while the total resource consumption increases. Overall environmental damage may also rise through lasting or irreversible impacts building up over time.

These processes and the flaws of ecological modernisation, Xue intimates, are sufficient reasons to regard the PRC as a form of "productivist socialism" lacking in and necessitating a turn to ecosocialism. This implies that an ecological rationality must be introduced in the PRC, "founded on the collective ownership of the means of production and democratic planning through which the society decides investment and production priorities" (Xue 2016, 200). Furthermore, democratic planning decisions would have to be guided by degrowth principles, first to be applied in the wealthiest countries, where socially equitable reductions in production and consumption levels are put into action so as to ensure that economic activities are within the limits of biocapacity.

To Xue, ecosocialism would be an advance over existing policies in the PRC because, as laid out by critical realist Roy Bhaskar in terms of four planes of being, it is a perspective that is mindful of ecological embeddedness, inter- and intra-subjective relations, the importance of social institutions and structures, and individual subjectivity (e.g., the collective and personal transformations needed for developing ecocentric production). This view overlaps substantively with that of Pan Yue (former Vice-minister of China's State Environmental Protection Administration), who sees the matter as one of transforming society to overcome all forms of capitalist relations and rule of capital (Gare 2021; see also Baer 2021b, 99).

According to Xue, prevailing reforms in institutions and practices and cultural frameworks in the PRC, embodied in the concept of Ecological Civilisation, are taking the PRC further away from ecosocialism. While harmony is a cardinal aspect of Ecological Civilisation, development strategies undertaken contradict such a worldview by violating limits and intrinsic values of nature. Decentralising planning and devolving decision-making, fiscal systems, investments, and other economic and political instruments to local authorities and increasingly to the private sector at the expense of state enterprises are undermining the socialist nature of the PRC and reducing the coordinating role of the state. The prerogative given to market-based and business decision-making processes and the continuing primacy given to economic growth compared to ecological sustainability are salient examples.

Cultivating consumerism and promoting the imitation of lifestyles in core capitalist countries exemplify another process impeding the development of ecosocialist possibilities. The latter may be justified by the need to fulfil the goal of zero poverty, but there seem to be few signs for economic processes to be eventually subsumed under democratic planning or social control, with markets subordinated to ecocentric production coordinated through the state and wider society. Thus, the prospects for any ecocentric production and a socially controlled economy are being pushed farther away in the political horizon, if not directly impeded through reforms that are progressively eroding "the necessary institutional settings for the ecosocialist restructuring," that is, social control over the means of production enabling democratic planning, which was the direction the PRC was undergoing prior to the 1978 reforms (Xue 2016, 203).

At the same time, Xue argues that the CPC, given their capacity to induce radical social transformations within even a single generation, could steer the PRC towards ecosocialism and degrowth by means of instruments like the national planning system, by expanding or intensifying existing socialist features, such as state ownership of key economic sectors, and by revitalising aspects of Chinese traditions that exalt frugality and harmony with the rest of nature.

The Context of the People's Republic of China

The above kind of more constructive critique tends to be rare among ecosocialists. Xue's perspective intimates that matters are much more complex, challenging, and contradictory than most ecosocialists allow. This is important to realise in a country characterised by and rooted in a couple of millennia of historical change, and now hundreds of millions of people, diversified in at least 56 nationalities or ethnicities with many speaking mutually unintelligible languages. The governing structures feature more than 90 million CPC members, with at least seven internal political currents whose range of substantive and mutually incompatible differences easily exceeds that of any liberal democracy (Cheng 2012), and where there exist several independent parties allied to the CPC (much like the Democratic Socialists of America with respect to the officially anti-socialist US Democratic Party). Economically, a socialist-oriented and regulated capitalist mode of production—characterised by state-run, multinational, and national firms and myriad small businesses—partly sponges off from and sits uneasily with much subsistence-oriented economic activity, mainly at household levels. These dynamics make for highly heterogeneous political economic pulls that are made opaque by the coarse CPC-capitalists dyad typically used by most ecosocialists (e.g., Webber 2022).

That major and negative environmental impacts have occurred in the PRC is not in dispute, certainly not within the PRC. The timing for much of the increase in ecologically destructive practices is traceable to political shifts during the 1970s. Introduced since 1978, the policies of 改革开放 (Gǎigé kāifàng, "reform and opening up") have enabled increasing numbers of capitalist firms to enter the country, first by means of special economic zones, and for private enterprises to form within the PRC. They have been allowed to exploit workers and ecosystems, leading to harmful ecosocial consequences. Various PRC administrations have thereby facilitated the expansion and intensification of the industrialisation process, transforming the PRC into the world's main manufacturing centre, rapidly rising in terms of conventional economic indicators to become among the most economically influential countries. Among the salient impacts have been rising inequalities, mass displacements (including for projects like the Three Gorges Dam), environmentally destructive urban expansion, large tracts of land and surface water (lakes, rivers) polluted through industrial emissions and resource extraction (mining, agrochemicals-heavy farming), air and water quality decline in many cities, various kinds of soil degradation, problems of deforestation, and spiralling greenhouse gas emissions (Baer 2018, 64–65, 95–96).

However, if one understands striving to improve the lot of working classes as part of the struggle for socialism, one should not be remiss about improvements and decisive turns in countering the above-described processes. Social conflicts and struggles over wages and working conditions may persist, yet standards of living have greatly improved since the 1970s (Baer 2018, 65; cf. Baer 2021b, 97–192). Substantial strides have been made in improving the well-being of most people in the PRC. It bears reiterating that freeing about 850 million people from abject poverty within the span of just 40 years is an unprecedented historical feat.

This is not to conceal that this most impressive result was not also reached by means of enormous social upheaval and environmental degradation. Large and expanding social inequalities, mass displacement and unemployment, highly uneven health care and other socially constructive provisions, and repression have been exacerbated by insufficient environmental policy frameworks and enforcement, major pollution problems, biodiversity decline, soil erosion, and several cases of ecosystem destruction, among other destructive effects (Li et al. 2021; Lu et al. 2020; Xue 2016; Zheng and Cao 2015). What is more, the PRC has surpassed the US as the country with the highest total CO_2 emissions.

The Development of Ecological Civilisation as Principle and Policy

A challenge for many ecosocialists appears to be conceptualising socialism and developing strategies in terms of multi-generational struggles and planning (Huan 2021, 9–10). In some respects, the multi-generational aspect is in continuity with the CPC's New Democracy principles, which are infused with Marxist method. Successive CPC administrations have recognised the destructive nature of their policies and have responded to the challenges in proactive ways, even if not necessarily in the most effective manner. Effectiveness, it should also be kept in mind, is not a feature solely under the control of those striving to be more effective. Economic sabotage and military pressures from countries like the US should not be hidden from view, for example. Progress has nevertheless been made even within a largely inimical global context.

Interest in developing an ecologically sustainable economy can be dated to at least the early 1970s (Yan and Bocharnikov 2022). The concept of Ecological Civilisation was instead introduced in the early 1980s through scholarly exchanges with the USSR, where the matter was framed as a cultural struggle as well (Gare 2021). However, influences may be traceable to much earlier philosophical currents even as far back as the 1920s. In some ways, similar ideas were reintroduced with the Rural Reconstruction movements of the 1990s and were inscribed in the 11th Five-Year Plan in the form of the New Rural Reconstruction programme (Wen et al. 2012). Hence, conceptualisations leading to the current constitutional modifications emerged through existing practices and legal frameworks that addressed to different degrees, even if sometimes inadequately, the ways that post-revolutionary society is to interact with the rest of the biophysical world. Environmental protection legislation existing since the 1950s was further refined between 1978 and 1991 to meet the environmental problems expanding or rising from the application of capitalist-oriented reforms (Huan 2013; Wang et al. 2014).

This was followed by a sustainable development model not unlike the "green" capitalist perspective described in Chapter 1, but with the crucial socialist difference of devising and implementing social policies aimed at eradicating poverty. Since 2002, a science-based development approach has informed environmental policy. The CPC have attempted thereby to reduce wealth inequalities, reduce environmental impact, expand and improve on environmental remediation, and strengthen enforcement. This even included attempts to curb economic growth, something that in the wealthiest capitalist countries would be simply inconceivable (Huan 2013).

Even the reformist pro-capitalist Club of Rome is recognising such feats. In a mere four years, the PRC

> went from a capacity of 0.3 GW in 2009 to 13 GW by 2013, adding 30.5 GW of renewable energy in 2015, 16.5 GW of that solar. China still burns a lot of coal, but its Green Horizons programme has committed to clean air in its cities and cut carbon intensity by 40-45% from 2005 levels within the next 5 years. In 2016, the total recorded CO2 emissions were reduced by 5%despite a 7% economic growth rate. (von Weizsäcker and Wijkman 2018, 122)

The irony is that this and other accomplishments, which are in accord with much of what the Club of Rome and similar organisations have been advocating for decades, are not happening under multi-party democracy, nor by way of states under the command of capitalist ruling classes, but in a country where the state is controlled by a single communist party.

It is undeniably thanks to the line followed first by those in the CPC led more recently by Hu Jintao and then by Xi Jinping that a turn occurred towards intensifying measures for environmental remediation and protection, as well as degradation prevention. The environmental turn was not exactly a break in policy, but to some extent a long-term strategy despite convulsive episodes like the Cultural Revolution. First the CPC strove for rapid industrialisation, in part started and developed under Mao Zedong, and then, under Deng Xiaoping, capitalist export-led expansion of manufacturing (first in designated regions), which created massive social and environmental problems but also formed a basis for improving people's lives.

Once major strides were made in getting most people out of poverty, a renewed and much more concerted shift towards attending to environmental issues and disasters appeared by the 2006–2010 guideline (the 11th and formerly known as Five-Year Plan). The 17th CPC Congress in 2007 saw the first explicit proposal to build an Ecological Civilisation (Yan and Bocharnikov 2022; Huan 2014). By that point, policies had been geared mainly to improving the state of environments at local levels. Lowering greenhouse gas emissions became the priority of the 2011–2015 or 12th guideline, focusing on expanding renewable energy sectors and resource use efficiency (von Weizsäcker and Wijkman 2018, 192–195). With the 2011 introduction of the National Functional Zones Planning, a legally enforceable framework, government institutions at every level have been coordinated to optimise spatial, economic, demographic, and land-use strategies to reach scientific development objectives of reducing social inequalities and improve environmental conditions (Xue 2016).

In 2012, the CPC ratified the inclusion of the concept and objective of achieving Ecological Civilisation in the constitution. This is a shift of major consequence, at least with respect to legal frameworks, policies, and potential for more effective enforcement. The 13th guideline (2015–2020) evolved substantively from this new constitutional modification. The guideline included following UNEP green economy criteria (overlapping, but not coextensive with solely pro-capitalist ones), improving on food safety, and establishing new and enlarging existing ecological preserves (von Weizsäcker and Wijkman 2018, 192). It must be emphasised that Ecological Civilisation is grounded in Marxist understandings harkening back to the original texts of Marx and Engels. It is thereby a socialist paradigm or a socialist form of Ecological Civilisation, as articulated explicitly in official pronouncements at the topmost leadership levels (Huan 2022; Zhang 2019).

Ecological Civilisation in Practice

The potential for the dire consequences that Xue and others have ably illustrated has not gone unnoticed to the CPC leadership (Wang et al. 2014, 49–50). The scientific development perspective incorporated into policy has been arguably in effect until the 2017 19th CPC Congress, when a major change was announced prioritising the struggle for Ecological Civilisation. The CPC have been responding to concerns over widespread ecological damage, which they themselves cannot escape. A major reason is traceable to political dynamics since the CPC, in charge of key economic sectors, can be held directly responsible for much of environmental harms. There is also widespread dissatisfaction, including unionised worker protests, in increasing numbers of communities (Huan 2014; Zhou 2021). The new direction has borne some fruit over even the short time that it has been in effect, notably building on historical achievements briefly illustrated below (for a more detailed overview, see Engel-Di Mauro 2021).

Ecological Civilisation not only involves defining locality-specific baselines (drawn up at district levels, which is scientifically a dubious undertaking relative to ecological processes), but also the remediation of damaged ecosystems, all reached on the basis of scientific expertise and discussions with local residents. This is part of the recently instituted Whole-Process People's Democracy, whereby public input is mobilised and conditionally channelled through local administrations (Boer 2021; Hu and Wu 2023; see also case studies in Mao and Mei 2014; Zhu and Lo 2002). Discussions may lead to set aside areas for protection or to build up a green space to offset degradation in another part of an area. Four model cities have been designated to develop quantifiable balance sheets relative to how much "natural capital" is preserved or undermined through economic activities (von Weizsäcker and Wijkman 2018, 193).

Some successful results have been shown in such cities in terms of coupled economic growth and ecological improvements, but these may mask damage incurred elsewhere (Xue 2016). To clarify, this is not the sort of pro-business "natural capital" accounting known in countries like the US, limited to still a marginal number of firms voluntarily taking up process redesigning to reduce costs and look environmentally friendly as long as profit margins are kept. In the PRC, the greening of businesses is part of a mandate for an entire city, which means legally enforceable reductions of impacts that are not necessarily bound to profitability.

It is more akin to the enforcement of environmental regulation that is seldom introduced and carried out successfully and in a sustained fashion in core capitalist countries to tackle degradation problems like air and water pollution.

Another outcome of Ecological Civilisation policies so far was the 2014 logging ban in areas under Natural Forest Conservation Programme (introduced in 2000). It has been through socialist state forestry, joining political and economic control over forested lands, that the above-cited forest protection successes have been achieved while contributing to improving living conditions for some locals by means of ecologically constructive employment and attendant welfare provisions. However, the process undermined the livelihoods of sacked forestry workers and service provisions like public transport have also been undercut by under-investment. This is a result of a failure of political mechanisms responsive to local voices. Still, this mixed ecosocial result contrasts favourably with the long-term forestry budget cuts, relaxation of logging regulation, and much more extensive damage to local inhabitants' livelihoods in the forestry sectors of the wealthiest capitalist countries (Zhu and Lo 2022).

Nevertheless, there remains much to be undone and more to be done. As stated above, the PRC have reached the highest CO_2 emissions in the world, even as attempts are being made to reduce them drastically and as rapidly as possible. Xu (2020) demonstrates that the introduction of profit motives in farming leads to rising inorganic fertiliser consumption and undermines prospects for ecologically sustainable agriculture. Soil and water pollution from the intensification and expansion of mining, industrial production (including agrochemicals-intensive farming), and waste generation, among other sources, has created lasting damage that will entail much time and effort to neutralise or mitigate to safe levels for people and many other organisms. This is to enumerate only some of the challenges, including air pollution, common to any industrialising or, in terms of pollution legacies, industrialised countries.

However, it would be premature to conclude that the PRC is therefore no better or worse than core capitalist countries, as many ecosocialists living in such countries tend to believe. Context-sensitivity and prevailing political objectives should be considered prior to expressing judgement. In capitalist countries, the primary purpose of the political process is to ensure the application of capitalist and landowner prerogatives, while in socialist countries it is to provide the conditions for the betterment of living conditions for the majority as a prerequisite to build socialism.

It must be underlined that the PRC's per capita CO_2 emissions, according to 2022 data, are about 61% of those of the US (second ranked in total emissions behind the PRC; see also Baer 2021b, 98). Moreover, CO_2 emissions have been on a downturn and increasingly less tied to economic development indicators thanks to multiple, concerted technical interventions that have simultaneously led to marked reductions in air pollution levels, though O_3 and $PM_{2.5}$ levels remain problematic (Chen et al. 2023; Li et al. 2023; Liu et al. 2023). Basic and applied scientific research is being mobilised towards identifying problems and concretising Ecological Civilisation policies (Lu et al. 2020; Zeng et al. 2019).

Many local government initiatives have also emerged to induce practices that contribute to emission reductions. For instance, Li Qiang (2021) reports that in Fuzhou progress is being made to incentivise inhabitants financially to reduce consumption levels and to reorient their economic activities away from those solely concerned with exchange-value. Thus, unlike in the US, there is much reason for optimism in the 2060 carbon neutrality goal set by the CPC.

Ecological Civilisation policies' relative effectiveness relies on advances already reached by previous governments as well, as adumbrated above. Over the past quarter century, environmental improvement efforts in the PRC have come to account for 25% of vegetation cover expansion worldwide, 42% of which is due to reforestation through the state-run Natural Forest Conservation Programme (Zhu and Lo 2022). Moreover, investments in renewable energy have since the 2010s far outpaced those in any of the wealthiest capitalist countries (Baer 2018, 96; Wang et al. 2014, 50). There have been major improvements in the expansion of ecological preserves (Liu and Diamond 2005) and successes in biodiversity protection (Mao and Mei 2014). Similarly, strides have been made in reforestation and afforestation, especially useful in turning back desert encroachment and attenuating carbon emissions (Shi et al. 2011; Zhu and Lo 2022). Notably, these achievements have occurred often thanks to freely established and mutually beneficial collaboration among grassroots organisations and government institutions, broadly in line with New Democracy principles and perhaps prefiguring Whole-Process Democracy policies.

Despite the major challenges ahead in reaching the goal of Ecological Civilisation, the outcomes so far are no small feat for a country that in 1949 inherited extensive environmental disasters and that had been under foreign imperial oppression for at least a century, suffered through several decades of virtually uninterrupted warfare with long-term socially and environmental devastating consequences, underwent two separate rapid phases of industrialisation, and for a good 50 years has been the main manufacturing centre of the capitalist world economy. One should wonder what possible excuse liberal democracies like the United Kingdom, Canada, Australia, or the United States could have relative to the social and environmental disasters meted out through settler colonial genocidal conquest and imposed on many other parts of the world while never subjected to anything even remotely as awful as what peoples in the PRC have gone through.

Recognising the Different Lengths of Revolutionary Paths

Ecosocialists, mainly outside of the PRC, are divided over the characteristics of the PRC and its role in the promotion of ecosocialism. Most of those who regard the PRC as a socially and environmentally worse variety of capitalism live, like the author, in countries with governments set on schemes to recolonise if not dismember the PRC, namely the US and satellite states. An alternative ecosocialist perspective could be to consider the disadvantageous position of a capitalist context necessitating much struggle just to establish a system ruled through a socialist politics and framework, compared to striving to effect changes towards ecological sustainability in an already socialist country. In this manner, it becomes clear that Ecological Civilisation is a possible model for a post-revolutionary society and ecosocialist movements would instead be more appropriate to pre-socialist contexts (see also Huan 2021, 11–14). The latter is implied by the socialist roots of the presently constitutionally mandated efforts in the PRC at achieving Ecological Civilisation.

This is a minoritarian ecosocialist view within a minoritarian leftist view in capitalist countries, where substantive checks on capitalists are deemed "authoritarian" and where many leftists believe dissent within the PRC entails progressive opposition or proof of CPC "authoritarianism" (see, e.g., Webber 2022, 68–69).[4] Whether the CPC will identify with ecosocialism at all, even implicitly, remains to be seen, but most ecosocialists are sceptical, if not hostile to characterising the PRC even as socialist, much less ecosocialist. Consequently, any improvements in living conditions and achievement in ecological sustainability in the PRC, as there have been, will have to be chalked to the overall benefits of an "authoritarian capitalist regime" by such ecosocialists. Ultimately, any successes by the CPC in bringing about Ecological Civilisation will be globally beneficial, whatever attitudes ecosocialists may reserve for the PRC (Huan 2010).

Negative judgement about the PRC, in any case, is probably extended to Laos and Vietnam, although they are rarely even mentioned, unlike Cuba, which paradoxically receives mostly positive marks from ecosocialists. The DPRK is overwhelmingly rejected entirely as anything but an authoritarian system devoid of any socialism, a viewpoint that the present author had mistakenly held in the past (Engel-Di Mauro 2021). It is often unclear whether ecosocialists understand that such countries are called socialist by most people living in them, not just by governing authorities. Even so, socialism in such countries as the PRC is regarded as fraudulent. Apparently, in these "purported socialist regimes" (Wallis 2020, 139), tens of millions of communist party members are feigning to be concerned with socialism, unlike a handful of enlightened far away ecosocialists true to principles. The courtesy could easily be reciprocated by calling ecosocialism a sham, given the failure so far to bring about any social change in countries from where ecosocialists chide socialist states as being fake. Perhaps it should not be surprising that ecosocialism exerts no social or formal institutional influence in any socialist country, if ecosocialism is even considered at all seriously. It would be more constructive to develop an ecosocialist strategy for a country like the PRC or DPRK than simply characterise that country as this or that sort of failure (see, e.g.,

[4] In this connection, terms like "authoritarian capitalism" to describe a possible future for the PRC (Webber 2022, 71) imply an otherwise "democratic capitalism," which only makes sense from a classical liberal point of view. It is in such instances that leftists betray a visceral liberalism. Alternatively, one must imagine a "localised, non-capitalist, sustainable lifestyles [that] can wriggle free of the CCP's [CPC's] dominance" (Webber 2022, 72). But without any specificity about the "non-capitalist" part (for example, it could be feudal or slave-based) and modes of resource of procurement for what is unavailable locally, one merely returns to another version of liberalism, the negatively romantic petit bourgeois sort rampant in the mainstream environmentalism.

Witherspoon et al. 2022; Xu 2024). The fate of the DPRK, in particular, but also of communist parties suppressed in liberal democracies (e.g., the Black Panther Party) may be similar to what would be forthcoming to any ecosocialist formations who reach national or regional levels of influence, or even before they even get close. This is the case whether this is achieved by parliamentary roads and/or mass movements, the so far preponderant kinds of strategies in the current ecosocialist imaginary.

Ecosocialism will likely remain irrelevant to post-revolutionary societies until ecosocialists contribute more than critique or dismissiveness, offering ideas for practicable alternatives or, more helpfully, dedicating themselves to eradicating regimes attacking socialist countries and organising for socialist revolutions where they live. Short of that, any critique of and recommendation by ecosocialists should be informed at the very least by a study of local history, sensitive to international relations of power, and free of moralistic judgements (especially when coming from imperialist countries, like those within NATO).

With respect to the international dimension, it could be useful to discern which forces are to be fought as priority and, due to ecosocialists' political insignificance so far, to develop alliances with globally influential formations that have potential to create openings for ecosocialist prospects. The latter may be too controversial or unacceptable, as it often implies bending at least some principles (e.g., anti-authoritarianism) to arrive at the possibility for the concrete applicability of the same transgressed principles (e.g., increasing workers' control over the economy). Making means and ends coincident is a rare possibility in practice, if it ever is, because social conditions generally and relations of power in particular are far beyond self-described revolutionaries' capacities to influence or change wider society.

Ecosocialists' ideas will lack substance without proving them to be implementable at the very least in the countries where ecosocialists live. Without also successfully organising to defeat ruling class forces in their respective countries (not coincidentally the same forces constantly attacking socialist countries), ecosocialists will remain as irrelevant to socialist countries as their critiques. Hence, it should not be too surprising to find ecosocialism having little resonance so far in socialist countries. However, ecosocialism may find interest in some academic and policy circles, as attested by ecosocialists who have been invited in the PRC for speaking tours or collaborative research projects (e.g., Salleh 2008; Wallis 2020).

Whether one agrees or not with this reasoning, it remains clear that so far ecosocialism, in any guise, has brought none of the achievements of often much-disparaged socialist states like the PRC. This should be a troubling contrast to the achievements of "first epoch" socialism within sometimes less than a few decades of existence. Sometimes in even less time. In Burkina Faso, a "first epoch" socialist formation led by Thomas Sankara, who gained state power by means of a military coup, in a handful of years achieved what no ecosocialist formation has so far, such as greater women's empowerment, mass literacy and vaccination campaigns, and successful reforestation programmes, among other incredible strides. These were undone by a murderous French government backed coup, which the French left of whatever stripe failed to prevent. The alternatives cogitated by ecosocialists will become convincing when ecosocialists will organise into a politically coherent force and accomplish concrete ecosocial advancements, rather than confine themselves predominantly to theorisation and activism. In this, ecosocialists should regard "first epoch" socialism as their more likely future, in terms of the challenges to be confronted in a post-revolutionary society.

For those who ground their ecosocialism on abjuring (e.g., Flanagan and Raphael 2022, 11) instead of building with actually existing socialism, the world-impacting capitalism-overcoming potentials of the PRC will be as irrelevant as they are to the hundreds of millions taking part of the gigantic Ecological Civilisation effort. A multi-generational perspective would be more appropriate when attempting to contribute to overturning centennial time-scale processes such as the transformation of a mode of production and, even more challenging, at the spatial scale of the world. There are, on the one hand, the ultimate objectives of overcoming capitalism once and for all by attaining ecosocialism (or why not communism?). On the other hand, the roads are long and arduous to get even to the point of revolutionary change and then to face and resolve decades of subsequent challenges that may or may not lead to the any durable form of ecosocialism. The historical shifts from feudalism and other kinds of society to some sort of capitalism, for example, have taken centuries and were never entirely complete. As Wallis has articulated the issue, "The revolutionary process is a long one, and it would be presumptuous to insist that a particular moment of it—especially in its early stages—should yield a clear-cut and definitive result" (Wallis 2020, 142). This important insight should be a source of critical self-reflection to ecosocialists who continue to describe socialist states as "productivist" or authoritarian variants of capitalism, among other kinds of premature verdicts. It should be expected that post-revolutionary ecosocialist communities, if ever they were to arise, will inevitably experience setbacks or involutions, with constant potentials for or actual incidences of reactionary reversals or further revolutionary change. This is where ruling communist parties, especially in the post-revolutionary societies of East Asia and in Cuba, but also anti-colonial movements among First Peoples, have much to teach ecosocialists.

References

Albritton, Robert. 2019 *Eco-Socialism for Now and the Future*. Geneva: Springer International Publishing.

Baer, Hans A. 2018. *Democratic Eco-Socialism as a Real Utopia: Transitioning to an Alternative World System*. New York: Berghahn Books.

Baer, Hans A. 2021a. *Global Capitalism and Climate Change. The Need for an Alternative World System*. Lanham: Lexington Books.

Baer, Hans A. 2021b. "Toward Democratic Eco-socialism as the Next World System: A Vision for the Future." In *The New Systems Reader: Alternatives to a Failed Economy*, edited by James Gustav Speth, and Kathleen Courrier, 127–144. New York and London: Routledge.

Boer, Roland. 2021. *Socialism with Chinese Characteristics. A Guide for Foreigners*. Singapore: Springer.

Boer, Roland, and Ping Yan. 2021. "'Not Some Other-ism'—On Some Western Marxist Misrepresentations of Chinese Socialism." *International Critical Thought* 11 (2): 171–189.

Chen, Yang, Ziheng Zhao, Wen Yi, Jingke Hong, and Bo Zhang. 2023. "Has China Achieved Synergistic Reduction of Carbon Emissions and Air Pollution? Evidence from 283 Chinese Cities." *Environmental Impact Assessment Review* 103: 107277.

Cheng, Enfu. 2012. "Seven Currents of Social Thoughts and Their Development in Contemporary China with a Focus on Innovative Marxism." *The Marxist* XXVIII (4). https://wapescholar.elsevierpure.com/ws/portalfiles/portal/28100106/Seven_Currents_of_Thoughts_and_Their_Development_in_Contemporary_China_with_A_Focus_on_Innovative_Marxism.pdf. Accessed 26 April 2024.

Engel-Di Mauro, Salvatore. 2021. *Socialist States and the Environment: Lessons for Ecosocialist Futures*. London: Pluto Press.

Flanagan, Erin, and Dennis Raphael. 2022. "From Personal Responsibility to an Eco-socialist State: Political Economy, Popular Discourses and the Climate Crisis." *Human Geography* 16 (3): 244–259.
Foster, John B. 2017. "The Earth-System Crisis and Ecological Civilization: A Marxian View." *International Critical Thought* 7 (4): 439–458.
Foster, John B. 2022. "Ecological Civilization, Ecological Revolution: An Ecological Marxist Perspective." *Monthly Review* 74 (5). https://monthlyreview.org/2022/10/01/ecological-civilization-ecological-revolution/. Accessed 26 April 2024.
Gare, Arran. 2012. "China and the Struggle for Ecological Civilization." *Capitalism Nature Socialism* 23 (4): 10–26.
Gare, Arran. 2021. "The Eco-Socialist Roots of Ecological Civilisation." *Capitalism Nature Socialism* 32 (1): 37–55.
Hammond, Ken. 2023. *China's Revolution and the Quest for a Socialist Future*. New York: 1804 Books.
Hu, Jieren, and Tong Wu. 2023. "'Whole-Process People's Democracy' in China: Evidence from Shanghai." *Chinese Political Science Review*, 1–28.
Huan, Qingzhi. 2010. "Eco-socialism in an Era of Capitalist Globalisation: Bridging the West and the East." In *Eco-Socialism as Politics: Rebuilding the Basis of Our Modern Civilisation*, edited by Qingzhi Huan, 3–12. Dordrecht: Springer.
Huan, Qingzhi. 2013. "Growth Economy and Its Ecological Impacts Upon China: An Eco-Socialist Analysis.' In *Eco-socialism as Politics*, edited by Qingzhi Huan, 191–203. Springer, Dordrecht.
Huan, Qingzhi. 2014. "Development of the Red–Green Environmental Movement in China: A Preliminary Analysis." *Capitalism Nature Socialism* 25 (3): 45–60.
Huan, Qingzhi. 2021. "Socialist Eco-Civilization as a Transformative Politics." *Capitalism Nature Socialism* 32 (3): 65–83.
Huan, Qingzhi. 2022. "Ecosocialism and Socialism with Chinese Characteristics." In *The Routledge Handbook on Ecosocialism*, edited by Leigh Brownhill, Salvatore Engel-Di Mauro, and Terran Giacomini, Ana Isla, Michael Löwy, and Turner Terisa, 339–345. Oxon and New York: Routledge.
Lau, Kin Chi. 2018. "A Subaltern Perspective on China's Ecological Crisis." *Monthly Review* 70 (5): 45–57.
Li, Minqi. 2016. *China and the 21st Century Crisis*. London: Pluto Press.
Li, Minqi. 2017. "Political Economy of China from Socialism to Capitalism, and to Eco-Socialism?" In *The Political Economy of Emerging Markets Varieties of BRICS in the Age of Global Crises and Austerity*, edited by Robert Westra, 79–100. London: Routledge.

Li, Qiang. 2021. "Ecological Valuation and Promoting Green Behavior in the Process of Eco-Civilization Construction in Fuzhou City, China." *Capitalism Nature Socialism* 32 (3): 115–126.

Li, Shengyue, Shuxiao Wang, Qingru Wu, Yanning Zhang, Daiwei Ouyang, Haotian Zheng, Licong Han, Xionghui Qiu, Yifan Wen, Min Liu, Yueqi Jiang, Dejia Yin, Kaiyun Liu, Bin Zhao, Shaojun Zhang, Ye Wu, and Jiming Hao. 2023. "Emission Trends of Air Pollutants and CO_2 in China from 2005 to 2021." *Earth System Science Data* 15 (6): 2279–2294.

Li, Xiushan, Yu Wang, Youqing Luo, Junbao Wen, Houhun Li, Eckhard Gottschalk, Josef Settele and Oliver Schweiger. 2021. "Opportunities to Improve China's Biodiversity Protection Laws." *Nature Ecology & Evolution* 5: 726–732.

Lin, Chun. 2006. *The Transformation of Chinese Socialism*. Durham: Duke University Press.

Liu, Xinyuan, Chaoyi Guo, Yazhen Wu, Chen Huang, Keding Lu, Yuanhang Zhang, Lei Duan, Miaomiao Cheng, Fahe Chai, and Fengqiao Mei. 2023. "Evaluating Cost and Benefit of Air Pollution Control Policies in China: A Systematic Review." *Journal of Environmental Sciences* 123: 140–155.

Liu, Jianguo, and Jared Diamond. 2005. "China's Environment in a Globalizing World." *Nature* 435: 1179–1186.

Löwy, Michael, and Daniel Tanuro. 2021. "Introduction. Le XXIe Siècle Sera le Siècle de la Question Écosociale." In *Luttes Écologiques et Sociales dans le Monde. Allier le Vert et le Rouge*, edited by Michael Löwy and Daniel Tanuro, 7–35. Paris: Les Éditions Textuel.

Lu, Xi, Shaojun Zhang, Jia Xing, Yunjie Wang, Wenhui Chen, Dian Ding, Ye Wu, Shuxiao Wang, Lei Duan, and Jiming Hao. 2020. "Progress of Air Pollution Control in China and Its Challenges and Opportunities in the Ecological Civilization Era." *Engineering* 6 (12): 1423–1431.

Mao, Da, and Xueqin Mei. 2014. "Protecting the Tibetan Antelope: A Historical Narrative and Missing Stories." In *A History of Environmentalism. Local Struggles, Global Histories*, edited by Marco Armiero and Lise Sedrez, 83–105. London: Bloomsbury.

Salleh, Ariel. 2008. "Eco-Socialism and "Ecological Civilization" in China." *Capitalism Nature Socialism* 19 (3): 123-129.

Sarkar, Saral. 2010. "Prospects for Eco-Socialism." In *Eco-Socialism as Politics: Rebuilding the Basis of Our Modern Civilisation*, edited by Qingzhi Huan, 207–222. Dordrecht: Springer.

Shi, Lei, Shuqing Zhao, Zhiyao Tang, and Jingyun Fang. 2011. "The Changes in China's Forests: An Analysis Using the Forest Identity." *PLOS ONE* 6 (6): e20778.

Smith, Richard. 2020. *China's Engine of Environmental Collapse*. London: Pluto Press.

Tanuro, Daniel. 2020. *Trop Tard our Être Pessimiste! Écosocialisme ou Efondrement*. Paris: Textuel.
von Weizsäcker, Ernst Ulrich, and Anders Wijkman. 2018. *Come on! Capitalism, Short-Termism, Population and the Destruction of the Planet*. New York: Springer.
Wallis, Victor (2020) *Socialist Practice: Histories and Theories*. London: Palgrave Macmillan.
Wang, Zhihe, Meijun Fan, Hui Dong, Dezhong Sun, and Lichun Li. 2013. "'What Does Ecological Marxism Mean for China?" Questions and Challenges for John Bellamy Foster." *Monthly Review* 64 (Fasc. 9): 47–53.
Wang, Zhihe, Huili He, and Meijun Fan. 2014. "The Ecological Civilization Debate in China: The Role of Ecological Marxism and Constructive Postmodernism-Beyond the Predicament of Legislation." *Monthly Review* 66 (6): 37–59.
Webber, M. 2022. "China: Capitalism and Change?" In *Post-Capitalist Futures: Paradigms, Politics, and Prospects*, 65–74. Singapore: Springer Singapore.
Wen Tiejun, Lau Kinchi, Cheng Cunwang, He Huili, and Qiu Jiansheng. 2012. "Ecological Civilization, Indigenous Culture, and Rural Reconstruction in China." *Monthly Review* 63 (9): 29–44.
World Bank. 2022. *Four Decades of Poverty Reduction in China. Drivers, Insights for the World, and the Way Ahead*. Washington, DC: The World Bank.
Xu, Zhun. 2018. *From Commune to Capitalism: How China's Peasants Lost Collective Farming and Gained Urban Poverty*. New York: Monthly Review Press.
Xu, Zhun. 2020. "Farm Size, Capitalism, and Overuse of Agricultural Chemicals in China." *Capitalism Nature Socialism* 31 (3): 59–74.
Xu, Xhun. 2024. "Industrial Agriculture: Lessons from North Korea." *Monthly 1511 Review: An Independent Socialist Magazine* 75 (10): 30–47.
Xue, Jin. 2016. "China at the Crossroad. Ecological Modernization or Ecosocialism?" In *Crisis System. A Critical Realist and Environmental Critique of Economics and the Economy*, edited by Petter Næss and Leigh Price, 192–208. London and New York: Routledge.
Yan, Jun, and Vladimir Bocharnikov. 2022. "Knowledge and Understanding of Ecological Civilization: A Chinese Perspective." *BRICS Journal of Economics* 3 (4): 231–247.
Zeng, Yingying, Yuanfei Cao, Xue Qiao, Barnabas C. Seyler, and Ya Tang. 2019. "Air Pollution Reduction in China: Recent Success But Great Challenge for the Future." *Science of the Total Environment* 663: 329–337.
Zhang, Yunfei. 2019. "On the Historical Position of Ecological Civilization." *Capitalism Nature Socialism* 30 (1): 11–25.
Zheng, Heran, and Shixiong Cao. 2015. "Threats to China's Biodiversity by Contradictions Policy." *Ambio* 44 (1): 23–33.

Zhou, Xin. 2021. "Ecological Civilization in China: Challenges and Strategies." *Capitalism Nature Socialism* 32(3): 84–99.

Zhu, Liyuan, and Kevin Lo. 2022. "Eco-socialism and the Political Ecology of Forest Conservation in the Greater Khingan Range, China." *Political Geography* 93: 102533.

CHAPTER 9

Strategies and Achievements of Actually Existing Ecosocialism

Bountiful Speculation and Meagre Reality

Ideas abound, often of a speculative character, on desired ends or on what ecosocialism should look like. The reader is directed to cited sources to gain a sense of the breadth of coverage, from the character of decision-making processes to the content of environmental policies to the shaping of suitable technologies (e.g., Adaman and Devine 2022; Schwartzman and Saul 2015; Wallis 2018). It is common to find a list of desired ends and policies, along with historically untested principles. These are often presumed to be applicable to national or even global scales, such as the notion of fully participatory or democratic planning (e.g., Albritton 2019; Devine 2022; Löwy 2007; Magdoff and Williams 2017, 285; Smith 2021), an ecosocialist world-system (Baer 2018), a "global regime of peace and cooperation" (Schwartzman and Saul 2015, 16), or a Global Green New Deal (Schwartzman and Schwartzman 2018), a care-oriented paradigm shift (Tanuro 2020, 37), and so on.

Otherwise, ecosocialists will point to relatively successful cases in an urban neighbourhood or in small areas where practices akin to ecosocialism (or potentially ecosocialist) illustrate what is already in existence and feasible (e.g., Akuno and Hall 2021; Löwy and Tanuro 2022; Nelson 2022a; Wall 2010). Sometimes, in attempts to demonstrate the practicability of ecosocialism, authors draw from a few historically delimited examples, like a single case of biological conservation planning in the

S. Engel-Di Mauro, *Ecosocialism*, Marx, Engels, and Marxisms,
https://doi.org/10.1007/978-3-031-74288-0_9

USSR or the cybersyn project in Chile under the short-lived Allende government (Gare 1993; Vettese and Pendergrass 2022).

Rarely, there may even be recognition that a socialist country, such as Cuba, is "on the ecosocialist path," at least "with respect to agroecologies and renewable-energy development in rural areas" (Schwartzman 2016, 103; see also Schwartzman 2022, 286; Wallis 2022, 294). As this suggests that "first epoch" socialism provides thus far the most viable model, one might want to explain how ecosocialism would be relevant to socialist countries (cf. Suing 2018).[1] It is a similar quandary for Eco-Marxists who acknowledge strides made in ecological thought in the USSR during the 1970s (Foster 2015a), while claiming "Soviet" Marxism stifled the development of ecological thought originating in the writings of Marx and Engels (Foster 2023; a similar argument is made in Saito 2022, 7).

Rather than linger on ecosocialists' prescriptions and projections or on case descriptions of "green left" projects, the emphasis here is on ecosocialists' development of ideas for practicable political strategy and, even more importantly, on ecosocialists' own concrete achievements at the time of writing, as much as discernible from the literature available and the author's direct experiences in managing the ecosocialist journal *Capitalism Nature Socialism*, presiding over the operations of Ecosocialist Horizons, and co-founding the First Ecosocialist International. Taking stock of accomplishments is especially pertinent to discussions of strategy because concrete results, not just speculation, vision, or critique, should guide the development and implementation of ecosocialist politics. To sort of give away the punch line, if it is admitted that ecosocialism has so far had its greatest influence and achievements in Bolivarian Venezuela, it should follow that the strategies that have led to such achievements should be at a minimum closely studied by ecosocialists the world over.

Though largely underdeveloped, political strategies to arrive at ecosocialism have been considered arguably since the inception of ecosocialist thought. The matter is rarely taken up directly but can be deduced from what is taken for granted or from an author's activities. Put another way, ecosocialist strategies, as political strategy generally, exist by dint of actions taken (or of omissions). They are rarely articulated outwardly, much less expressed in systematic fashion. For example, the approach advocated by SERA (1980) is of building strength and gaining greater social reach by

[1] Perhaps this explains, in some measure, the largely negative view of socialist countries propagated by most ecosocialists.

acting from within an existing party, in their case the UK Labour Party, while linking with or taking part of mainly labour and environmental movements. As evinced from the history of ecosocialism (Chapter 3), one strategy has been to use institutions where one is already an active member (a political party, an environmental organisation, a trade union, etc.) in order to introduce and spread ecosocialist ideas and policies. Alternatively, the default is promoting ecosocialism by means of affinity groups or loosely organised networks whose adherents may be involved in existing, perhaps leftist institutions and/or grassroots organisations, and thereby promote ecosocialism. Sometimes, the unstated strategy is an unaccountable individual promulgating ecosocialist ideas by various kinds of intervention, often in published form.

Rather than attempt to excavate every instance of strategy formulation ensconced in writings that do not aim to address the issue, the task here is to discuss what has been written expressly to address the means whereby ecosocialism is to be achieved. More importantly, what ecosocialists have accomplished also need to be assessed, beyond the production and spread of ideas. This is one way to understand the degree to which ecosocialist objectives are being met by ecosocialists themselves, including at the scale—the world scale—they themselves invoke when arguing for the necessity or urgency of ecosocialism. For the most part, it has been through a few parties and leftist organisations or affinity groups that ecosocialist projects are being attempted and ecosocialist thought is being conveyed. The influence of ecosocialist thought and action may be globally insignificant when compared with the Ecological Civilisation paradigm, but it is not negligible. There is much potential that is not to be underestimated of ecosocialist ideas and ecosocialism itself becoming part of mainstream political discourse in the imperialist countries, even if much of the reach into wider politics is more vicarious than direct. Given the global reach and mainly destructive ecosocial effects of those countries, a political result of that calibre could have global repercussions of the sort conforming to ecosocialists' recommendations. Prospects for that degree of political relevance remain little more than faint. Some of the impediments to making direct inroads into the mainstream are discussed in the concluding section of this chapter. Special attention is devoted to Bolivarian Venezuela, the only country in the world where ecosocialism has been embraced within both state and grassroots institutions.

Political Parties, Social Movements, and the State

Delineations of political strategy or platforms have been relatively rare, but they are valuable in revealing what is considered viable as well as in bringing forth aspects of the sort of future being imagined. To some degree, the preferred strategy framework is a function of how ecosocialism is defined or envisioned (see Somerville 2022). As already shown in chapter two, socialism has been diverse from the start. And just as there has never been any single socialist movement or any single set of socialist ideas or theories, so has it been the case for ecosocialism to encompass different kinds of politics that may also be at odds with each other. The targets of strategy formulations oscillate between party and social movement building. At times, ideas are proposed about how to join the two kinds of political organising and structures, but such ideas often lack concreteness or omit existing examples. The time frame for political action is often left unspecified or articulated as urgent, but it usually recognised that that there are shorter- and longer-term processes involved. Some understand the change to be brought about as more gradual and others as a radical rupture (e.g., Albert 2023, 26; Löwy 2007; Schwartzman 2013; Tanuro 2020, 251).

At times, it is difficult to discern the difference between ecosocialist and reformed social democratic politics. Somerville (2022), for instance, offers little more than a call to concentrate on increasing government regulation and basic universal services in the context of the UK (not too dissimilar from the implications of the discourse study of Flanagan and Raphael 2022). Arguably, from an internationalist perspective, such as one that other ecosocialists adopt, much more should be expected of ecosocialists working from within an imperialist power. The scale of politics assumed by ecosocialists can vary considerably and affects the formulation of strategies. The debate over the usefulness of the Green New Deal, discussed below, in part exemplifies a lack of internationalism and attentiveness to imperialism, in great contrast to the works of other ecosocialists like Derek Wall (2010).

Many ecosocialists would agree with Derek Wall that “to get to … ecosocialist society, the state must be transformed; it cannot simply be ignored” (Wall 2010, 121; see also Devine 2022; Flanagan and Raphael 2022; Löwy 2007; Ryle 1988, 60; Schwartzman and Engel-Di Mauro 2022; Tanuro 2020, 147). At the very least, this necessitates parties, or like forms of organising, as main vehicles to effect the transformation of the state (Wallis 2020, 33: Wolf and Löwy 2007; see also O'Connor 1998, 306–310). The sort of organisations to form for such purpose is seldom made clear. The default appears to be the classical Menshevik or social democratic model, where membership requirements amount to little more than agreeing to broad principles and paying dues.

Democratic eco-socialists, who largely draw from "New Left" socialist democracy, hold the sanguine view (see below on addressing capitalist violence) that the shorter-term and primary strategy is the take-over of capitalist states by means of "new left parties." There are to be "transitional radical reforms," described as "loose guidelines" (Baer and Singer 2022, 88; Tanuro 2020, 262). These substantially overlap with other ecosocialists' ideas, where there is at times the acknowledgement for the need of a vanguard "with organic grounding in the masses" to mount effective political struggles, but as a complement to "autonomous self-managed associations," or a vanguard accountable to a "mass movement" (Wallis 2020, 34–35).[2] In democratic eco-socialist terminology, this roughly corresponds to a "blend of centralism and decentralism" (Baer and Singer 2022, 86; see also Satgar 2018; Satgar and Cock 2022).

For Baer, democratic eco-socialism is achievable through "much theorizing and social experimentation" (Baer 2016, 6), though experimentation is so far largely missing, especially at regional and national scales (the exception being Venezuela, as discussed below). To him, though, the process must be global. It can be witnessed emerging in different ways in different places and by means of many different anti-systemic movements. How to arrive at the global is not specified, but the political struggle also includes fighting for transitional changes. These are the kinds of revolutionary reforms associated with socialists like André Gorz, like laws mandating more energy efficiency, easily repairable consumer items, and more workplace democracy.

[2] Wallis is referring to a Leninist view of the political vanguard and claims that, after Lenin, vanguard parties "overlooked" the need to be attentive to "the actual evolution of workers' outlooks" (Wallis 2020, 34). This is surprising, since Wallis elsewhere demonstrates an appreciation for capitalist powers' "military and economic aggression" and capitalism's widespread social or cultural influence as a further long-term impediment to developing socialism (Wallis 2020, 40). Given the enormity of the task and the constant attacks from within and without, one should be amazed just how far communist parties were able to go historically in facilitating any lasting changes in their respective societies where successful revolutions were had.

The overall strategy for more centralist ecosocialists is to form leftist parties capable of taking over the state and merging with other like leftist parties in power in different countries to reach global-scale coordination or capacity of intervention. Policies include devising and enforcing effective greenhouse emissions taxes and the nationalisation of the means of production to reduce business power. To arrive at such leverage within a country or worldwide will likely entail a mixture of peaceful and violent means, depending on the situation (Baer 2021b, 162). Much more needs to be studied about past socialist revolutions and post-revolutionary conditions, arguably, if one is to prepare for that mixture.

Another issue is to reconcile the painstaking process of party formation and state take-over worldwide with the realisation that "history tells us that social changes can occur quickly once economic, political, social structural and environmental conditions have reached a tipping point" (Baer and Singer 2022, 88). Given the exiguous progress made so far even in terms of popularising ecosocialism, let alone the glacial and fitful pace of ecosocialist organising generally, the more likely prospect is for ecosocialists to be caught unprepared even if they identify what constitutes a final threshold.

The late Joel Kovel also advocated for the formation of ecosocialist parties that, similarly to democratic eco-socialist renditions, are to be internally democratic, unlike "Leninist parties" (Kovel 2007, 263). However, it is not by parliamentary means that ecosocialism will prevail, in his view. Hence, ecosocialist parties must be in constant linkage with communities struggling against capitalism, funded by members, assembly-based and transparent, with a rotational structure and with delegates subject to recall. In other words, means and end must be made to cohere and delegation entails administering and coordinating, rather than dictating actions. Much else was imagined by Kovel with respect to transitioning to an ecosocialist future, but the main point is that the party form would be a conduit for coordinating diffuse actions and projects, rather than an organising force propelling anti-capitalist communities forward (ibid., 264–266). This third type of political party has an affinity with ecosocialists calling for joint centralist and decentralist methods (e.g., Akuno and Nangwaya 2023; O'Connor 1998; Schwartzman and Engel-Di Mauro 2022, 572–573; Tanuro 2020, 253–254; Wallis 2020). It may or may not differ substantially from what other more centralist ecosocialists have in mind regarding parties. An inventory and comparative analysis could be useful towards systematising knowledge about this issue.

In any case, an example of practical achievements through a similar ecosocialist party framework, involving institutional linkages and grassroots organising, is the much-celebrated Cooperation Jackson Project, already discussed in Chapter 3. The project's outcome, however, may leave one with some sobering thoughts. After a decade of the project's existence and its background in "over forty years of community organizing and base building" (Akuno and Nangwaya 2023, 59),[3] one of the principal organisers and founders is forced to concede that

> ...we still haven't stimulated the self-organization of the Black working class on a mass scale. More work—profoundly more—must be done to accomplish the main tasks in this regard, which are to elevate and strengthen the class consciousness of the community, foster and cultivate new relationships of social solidarity among the working class, and co-construct and advance new social norms and values rooted in radical ecological and humanitarian principles. In effect, what we are aiming to do is to develop a new transformative culture (Akuno 2023, 15)

Such aspirations are perhaps too ambitious for a relatively small group living under one of the most oppressive systems in the US, where many in Black communities are denied, among other basic needs, potable water as well as heating during extreme winter weather.

In fact, a lot has been accomplished quite above the potential for a small cooperative by securing land (by way of securing funds to establish a land trust, since communal ownership is not recognised by the capitalist state) and setting up urban cultivation areas, a café and catering cooperative, and a landscaping cooperative focused on composting and other such services (ibid., 31–32). Many other projects are in gestation, if ever the conditions arise to fulfil them. This is all impressive considering the context of an underserved neighbourhood in a city subject to the whims of white supremacists dominating state-level administrations. Still, it is questionable whether the above-described plans of consciousness-raising and cultural transformation would bring about the desired outcomes when the material basis in the control over the means of

[3] The much earlier sacrifices and accomplishments made by Black radicals in the Communist Party USA are perplexingly omitted (Campbell 1994).

production (including the most influential media) and in healthy ecosystems is missing. Perhaps needless to add, this kind of situation is hardly confined to Cooperation Jackson.

Another main alternative to party politics is a variously described autonomous, grassroots, or bottom-up approach, among other terms (e.g., Giacomini 2022; Haro García 2022; Nelson 2022a; Pepper 2010; Saul 2023; Turner 2022). Herein, it is generically labelled decentralist. The cases cited for this form of ecosocialism have not actually involved ecosocialists in founding the organisation or community. Nor would such movements or organisations identify as ecosocialist. They may have in their ranks people identifying as ecosocialist, but such cases are rare. A current example is Cooperative Jackson, already discussed in chapter three, where a few of the founders espouse ecosocialism.

Otherwise, there is a disturbing trend to try and lay claim to the fame of movements that are not ecosocialist but may have some features in common with ecosocialism. For instance, a celebrated early instance alleged to be ecosocialist, the Alliance of Forest Peoples (Löwy 2011; Wall 2010),[4] points to ways of unifying workers, First People and peasant communities, religious organisations, and others concerned with deforestation. There is no evidence that they are based on or promoting ecosocialism. One of its founders, the late Chico Mendes (assassinated by landlord henchmen), was a unionist who was involved in founding the *Partido dos Trabalhadores* (Workers' Party), with Luiz Inácio Lula da Silva, the president of Brazil at the time of writing. It is true that some ecosocialists derive inspiration from the Alliance of Forest Peoples, but inspiration should not be conflated with identity of political conviction. More credible and constructive would be to refer to Bolivarian Venezuela, a bastion of ecosocialism.

[4] Wall (2010) even describes Thomas Sankara, the Marxist-Leninist Burkinabé revolutionary, as an ecosocialist. This is yet another unhelpful example, quite common among ecosocialists, of resorting to superimposing one's favoured ideology on past heroes to compensate for a persistent failure in the present. If one were to extend the same logic to other, a typically reviled figure like Stalin, whose leadership led to major ecologically constructive shifts in the USSR (see, e.g., Brain 2011), then even Stalin must be regarded as an ecosocialist.

Bolivarian Ecosocialist Venezuela

The Bolivarian government of Venezuela and the comunas movement (self-governing communal councils) have as their stated objective the construction of an ecosocialist and bottom-up state (Alvarado and Bastidas 2022; Burbach and Piñeiro 2007; Ciccariello-Maher 2016). The Bolivarian epithet alludes to the liberation process symbolised by Simón Bolívar, a historical political and military figure who led the independence struggles against Spain and who originated what became Venezuela. In Venezuela, ecosocialism has attained the most institutional influence anywhere in the world. This is because the Bolivarian government officially adopted ecosocialism as the basis of "XXI Century Socialism," under the leadership of Hugo Chávez Frías (1999–2013). In fact, the Chávez and the succeeding Maduro governments have repeatedly and explicitly called for the collective construction and consolidation of ecosocialism throughout society. This was formally expressed and divulged with the *Ley del Plan de la Patria or National Planning Law of 2013–2019* (Rodríguez Cota 2018).

The Bolivarian government was established in 1999 ostensibly by electoral means, led by the socialist Fifth Republic Movement party. It was not until 2006 that the socialist content of the new administration would be made explicit, while the ecosocialist objectives were introduced officially at least by 2012. Eventually, many of the leftist parties that had lent support to the effort fused into the United Socialist Party of Venezuela (PSUV) in 2007. The PSUV remains the ruling party in the government of Venezuela at the time of writing. It cannot be sufficiently emphasised that the electoral success was mainly the result of previous decades of radicalisation and transformation from within the military establishment combined with pressures from anti-capitalist leftist movements (Wall 2010; Wallis 2020). This feat has been named the Bolivarian Revolution, which is an ongoing process. As already pointed out, this should be of major interest to ecosocialists advocating for a more centralist or mixed approach to politics, one that involves state institutions (e.g., the military) in tune with and supported by social movements.

When the Chávez government embarked on redirecting for social and environmental ends part of the profits made by private US and other oil firms, the US government swiftly responded by attempting to overthrow the Venezuelan government. The main method was by lavishly funding opponents and supporting military coups and assassination attempts, which have so far failed (including against President Maduro). Military pressure and political interference by successive Republican and Democratic Party US administrations were accompanied by increasingly harsher economic sanctions. They were imposed by 2003 on key financial transactions that, among other effects, impeded foreign trade (the US had been the largest trading country). They became generalised to almost all economic activities by 2015 and intensified still more since then. At the time of writing, the US imposed sanctions remain in place.

Sharp economic contraction followed, tied to the collapse of the oil industry sector, which had constituted the largest share of the economy. This was a result of decades of policies privileging fossil-fuel production linked to US corporate and local oligarch interests, creating conditions of dependence on imports for almost all consumer items, including food, medicine, and the products enabling oil drilling and extraction. The results of the sanctions have been greater and more frequent oil spills, malnutrition problems, and tens of thousands of preventable deaths from otherwise curable diseases. The US government have thereby curtailed the Bolivarian government's efforts to diversify the economy through investments in other sectors and technologies and undermined state capacity to divert oil proceeds and other funding streams to socially and ecologically constructive ends (Wilpert 2022).

Despite relentless US imperialist attack, aided by NATO satellite states, the embattled Bolivarian government have contributed to noteworthy achievements in combination with the work of grassroots collectives organised through the communal councils. For the first time, First Peoples and Afro-Descendent communities are officially recognised and given the resources to and encouraged to preserve their cultures and languages. Health and housing provisions for the first time reached people until then cut off from such state services. The extension of medical services was aided by Cuban medical personnel in exchange for oil and other resources for Cuba, who have been under a US economic blockade since 1960. Food supplement packages and other services to help raise farming productivity have alleviated the terrible conditions resulting from US sanctions. At the international level, aside from furnishing crucial energy supplies to countries like Cuba and Haiti, and even disadvantaged communities in the US, the government established Bolivarian Alliance for the Peoples of Our America (ALBA) in 2004 that presently includes ten countries with socialist or affine governments, mainly in the Caribbean islands arc. It is an organisation aiming to strengthen mutual support among member countries. ALBA became an important force that has pushed for international climate change policies opposing imperialist powers' impositions and thereby much closer to ecosocialist positions (Barreto 2020; Ciccariello-Maher 2016; Gilbert 2023; Marquina and Gilbert 2020; Raymundo 2022; Schwartzman and Saul 2015; Wall 2010, 122–123; Wallis 2020, 121–122).

The Bolivarian Revolution introduced a mixed political system, the Bolivarian process, that should be the envy of ecosocialists elsewhere in the world. Developing policies and legislation involves a consultation process with the widest social inclusion feasible (Burbach and Piñeiro 2007; Marquina and Gilbert 2020). It is not too dissimilar from the national consultation processes in Cuba and in the People's Republic of China (PRC) (see Chapter 8), but in a context of multi-party parliamentarism and capitalists' and landlords' economic predominance. Under sanctions that hamper the development of transportation and communication networks, in a vast country with a history of inadequate investment in such public infrastructure, the consultation process can also be quite challenging, however.

The first major outcome of ecosocialist importance resulting from the Bolivarian process was the introduction of anti-capitalist components to the constitution (e.g., the promotion of use- over exchange-values) and the expansion and prioritisation of environmental legislation and policies. The latter includes commitment to reducing greenhouse gas emissions by a fifth, even though Venezuela contribute a mere 0.5% or less of global totals, and efforts have been under way since then with the implementation of several interventions, such as expanding public transport and wind energy (Barreto 2020; Schwartzman and Saul 2015; see also Wall 2010).

The results have been a decline, for example, in per capita CO_2 emissions (Fig. 9.1). For comparison, Germany also exhibited a decline from about 10 to 8 tonnes CO_2 per capita from 2012 to 2020, while in Venezuela it was from about 7 to roughly 3 within the same time-period (the latter being a similar level as those of México and neighbouring Guyana). The US figures declined from 17 to close to 14, while the world average oscillates about the value of 4.6 (Ritchie et al. 2023). Notably, the downward shift in emissions has occurred irrespective of any ecosocialist influence. The precipitous drop in CO_2 emissions in Venezuela is certainly related to the terrible hardships imposed by US sanctions. This does not negate the more positive impacts of government investment in greenhouse gas reduction technologies and infrastructure. It would take a specialised study to determine the proportion of the massive fall in CO_2 emissions in Venezuela that are ascribable to ecosocialist policies.

Another way of gauging developments at the national scale is by way of following ecological footprint[5] and sustainable development index (SDI)[6] fluctuations over time (Hickel 2020; Wackernagel and Beyers 2019). The first shows a similar initial and general increase followed by rapid decline (Fig. 9.2). Patterns are not synchronous with CO_2 emissions, which are embedded in ecological footprint estimates. These point to resource consumption relative to biocapacity. The 2012–2013 period does nevertheless show rough correspondence in ecological footprint and emissions.

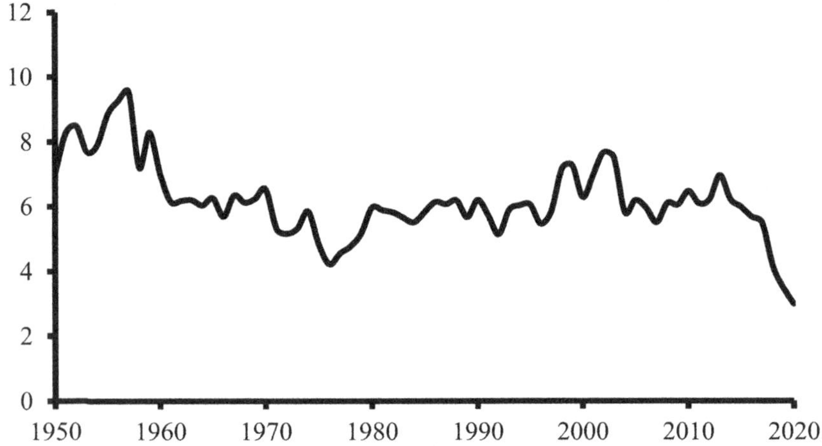

Fig. 9.1 Venezuela per capita CO_2 Emissions (tonnes), 1950–2020 (*Source* Ritchie et al. [2020])

[5] An estimate of how much of the regenerative capacity of the biosphere is occupied by human activities. The footprint is compared to the biocapacity within a country's territory to determine whether the country is in ecological deficit or surplus (reserve). The unit of estimation is the global hectare (gha). It is an area with a biological regeneration rate equal to that of a world-average biologically productive hectare. Every gha has the same amount of bioproductivity (Global Footprint Network 2024).

[6] This is defined by dividing a development index (the cube root of the multiplication product of the indices of life expectancy, education, and income) by an ecological impact index (a measure of over- or undershoot of material use and CO_2 emissions, determined according to baseline thresholds).

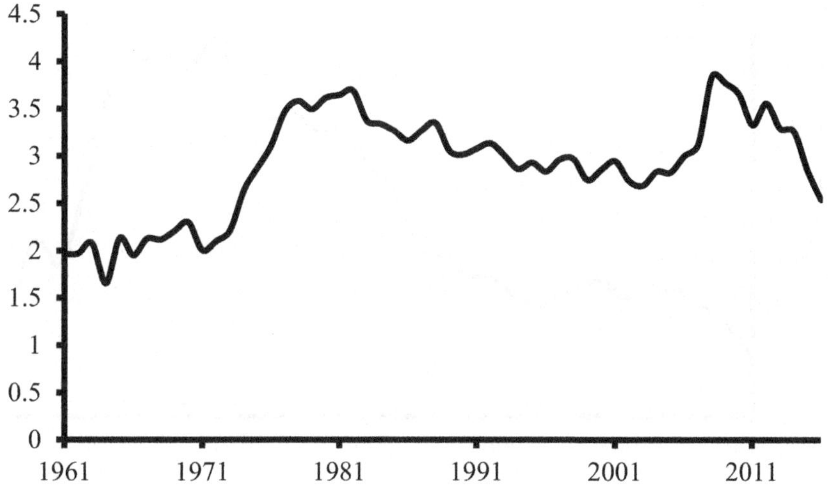

Fig. 9.2 Venezuela's ecological footprint (global hectares), 1961–2016 (*Source* Global Footprint Network, https://www.footprintnetwork.org/)

SDI is an estimate of ecological efficiency of human development. If one considers SDI (Fig. 9.3), it becomes more evident that gains in environmental indicators since 2013 (the sharp drop in per capita CO_2 emissions and ecological footprint) reflect worsening conditions in material well-being for many people in Venezuela. In other words, the patterns are likely explainable by the combined effects of improvements in environmental practices and the negative social repercussions of US sanctions.

The mixed effects of government environmental policies notwithstanding, another noteworthy change ushered in by the Bolivarian Revolution was the formal recognition of and support for the *comunas*. This came about with the passage of the 2006 Law of Communal Councils. This resulted from popular socialist demands, after thousands of councils had already been established. The process is rooted in the deeply collective spirit developed over many generations within much of the countryside, but state support was crucial, especially to fend against landlords, who continue to claim land that should be communal property and to assassinate comuneros and comuneras to achieve their aims of restoring their private control over land (Ciccariello-Maher 2016; Gilbert 2023;

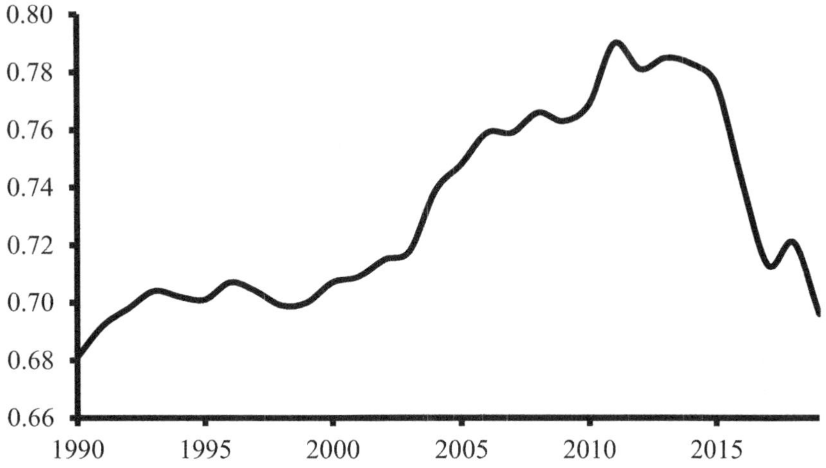

Fig. 9.3 Venezuela's Sustainable Development Index*, 1990–2019 (*Source* Sustainable Development Index, https://www.sustainabledevelopmentindex.org/)

Marquina and Gilbert 2020). State support extended to elected neighbourhood or village councils, who are those involved in initiating and overseeing local community development projects. *Comunas* are especially influential in developing, implementing, and spreading ecosocialism as well as environmental and agroecological policies. As of 2022, there are an estimated 19,500 *comunas* and a National Communard Union has been introduced to improve inter-communal and communes-government coordination (Marquina and Gilbert 2020; Gilbert 2023).

Other outcomes of pertinence to ecosocialism have flowed from combining centralist and decentralist institutions under the aegis of the periodically revised Plan Patria (National Plan). One is the Law for Integrated Agricultural Health (2008). Agroecology was officially adopted as the scientific basis for sustainable agriculture in conjunction with the wider perspective and knowledge of konucos (small farming plots; Núñez 2021). The policy includes a mandate for phasing out agrochemicals and much collaborative pedagogical work has been devoted to that end (Núñez 2022b), in part facilitated by the scarcity due to US sanctions. The 2010 Biodiversity Policy established a binding framework to inscribe ecosocialist principles into biodiversity conservation measures.

The process was put into place in the 2010–2020 National Strategy for the Conservation of Biological Diversity, where it was reiterated that ecosocialist ethics guide biodiversity and general environmental policy.

Ecosocialists were a main force behind the formulation of the Monte Carmelo Declaration (2012), which involved thousands of konucos. The document underlined the need to protect and promote seed and crop commons as well as sovereignty for konucos and First Peoples. The movement's efforts to bring the issues to parliament led to the National Assembly's ratification of the Seed Law of 2015. Large investments into agricultural research and development, mainly for smallholders, farming cooperatives, and *comunas*, focusing on developing and spreading agroecological methods. Major advances in food self-sufficiency at the national scale were subsequently reached by 2008 with some staple crops and livestock and 75% of food produced being consumed domestically (Alvarado and Bastidas 2022; Núñez 2021).

The same year when the Seed Law was introduced saw the establishment of an ecosocialist Ministry, the Ministerio del Poder Popular para el Ecosocialismo (the Ministry of People's Power for Ecosocialism; http://www.minec.gob.ve). It was charged, among other responsibilities, with coordinating and implementing environmental conservation policies. There are also projects aimed at streamlining tourism with ecosocialist objectives (Mendoza Suárez 2023). In May 2022, the Maduro government elevated the presidential climate adaptation commission to vice-presidential status, further empowering the Ministry. They have decreed that the Ministry be charged with coordinating climate change research and funds for climate adaptation development and implementation. Thus far, this is the only Ministry in the world that is dedicated to promoting and developing ecosocialism.

Efforts to promote self-sufficiency, agroecology, cooperatives, worker/konuco-managed *comunas*, First Peoples' and Afro-Descendent communities' autonomy, among many others, have been hindered by US-supported national bourgeois reaction (including violence and infrastructure sabotage), the US blockade, and internal strife within the governing party and among leftists. Ecosocialist policies and technical interventions to improve on ecological practices have been undercut directly by US sanctions by making even basic lab, monitoring, and field equipment unavailable or scarce. Many social inequalities persist, mainly inherited from centuries of oppressive relations but also exacerbated by US belligerence, including greater centralisation of power in contradiction with the Bolivarian process and communal council autonomy and the re-introduction of some neoliberal policies. There have also been not a few occasions of abuse of power and officials being unhelpful, if not impediments towards communal councils' struggles against landlords to keep or secure land and towards projects led by First Peoples and Afro-Descendent communities. At times, government actions have also contradicted stated ecosocialist goals and have generated conflicts with environmentalists. For the most part, such aims have had to be sacrificed mainly in an effort to improve social conditions or prevent catastrophic drops in living standards (Gilbert 2023; Marquina and Gilbert 2020; Schwartzman and Saul 2015; cf. García-Guadilla 2014, who confuses dilemmas and contradictions with some unstated "extractivism" policy).

External Ecosocialist Support for Bolivarian Ecosocialism

In the imperialist countries, ecosocialist support for the Bolivarian Revolution has been underwhelming, akin to most leftist responses (Baer 2018, 171–172), with a smattering of exceptions (e.g., Wall 2020; Wallis 2018). Schwartzman and Saul (2015) are the only ecosocialists who have gone beyond description to suggest how to transition away from fossil fuels, but without examining why prior attempts to do so during the early 2000s have been stymied.

The main message is otherwise defined by a prevailing silence on the ecosocialist strategies developed and strides made in Venezuela (e.g., Albritton 2019; Brownhill and Turner 2020; Pepper 2010; Schwartzman and Engel-Di Mauro 2022). In Michael Albert's discussion on political strategy (Albert 2023), Venezuela is not even worthy of a mention. Kovel (2007, 258) dedicated a mere couple of sentences to the Bolivarian project, mainly to express the fear that, as alleged in the case of Cuba, "the socialistically-oriented state … eventually becomes too strong and stifles the emergence of freely associated labor, causing the movement toward ecosocialism to stall." Such a view would have one worry more about an incipient result ("freely associated labor") than the means to keep and expand on it (the survival of the socialist government).

More positively, Hans Baer recognises the Bolivarian government's successes in bettering conditions for the majority (Baer 2018, 22, 157), their role as a challenge to world capitalism (ibid., 153), and their ecosocialist programme, while under a US siege (see also Baer 2021a, 53–54).

Notwithstanding this appreciation for the Bolivarian government and the threats they face, Baer fails to recognise the ecosocialist character of the Bolivarian movement and relativises the socially positive outcomes by citing similar trends occurring elsewhere in South and Central America. The following remarks sum up how the Bolivarian Revolution is treated, at once vaguely sympathetic and gratuitously critical:

> While still alive, Chavez developed Plan Pa[t]ria, an eco-socialist program, which included training thousands of cadres in agroecology, bioremediation, sustainable land and water management, and urban agriculture. Despite these efforts, Venezuela has the highest per capita murder rate in Latin America. (Baer 2018: 156)

The author provides no logical connection between the two faces of the Venezuelan situation. It is unclear how murder rates are addressed by, say, bioremediation. A more plausible explanation would relate the ghastly statistic with the activities of US-aided bourgeois forces in running the economy, creating conditions of diffuse strife, and fomenting social unrest. Furthermore, the impression of an ecosocialist programme contingent on Chávez's presence is contradicted by the Maduro government's reinforcing and expanding on their commitment to ecosocialism.

Baer's statements are quite puzzling because of his understanding of the Venezuelan government's precarious situation as a country targeted by US-sponsored coup attempts that is also a semi-peripheral economy largely in capitalist hands (ibid., 157–158). The Bolivarian government's achievements are therefore much greater than those of governments elsewhere in South America, who are not savaged by US imperialism and constant sabotage from within by capitalist forces. They are certainly much greater accomplishments than those of any ecosocialist formation elsewhere in the world.

Another major ecosocialist exponent, Löwy (2014, 22), has engaged in a similar kind of equivocation. After excoriating centre-left governments like those of Lula in Brazil for their "*social-liberalism*" (italics in original), he highlights a basic contradiction in the leftist governments of Venezuela, Bolivia, and Ecuador, who try to overcome neoliberalism while their budgets are mainly tied to fossil fuel. He acknowledges that they

> have attempted to break with neoliberal policies and have confronted the interests of the oligarchy and multinationals. All these anti-imperialist and anti-oligarchic governments recognize the importance of ecological challenges and are disposed to take measures to safeguard the environment. However, all three of their governmental budgets remain totally dependent on returns from fossil fuels (gas and petroleum), i.e., the fuels responsible for climate change. (ibid.)

He claims the Bolivarian government to have "hardly examined this issue," with "few efforts to develop alternative energies." He points to "the absence of a sizeable or organized indigenous population at the sites of exploitation" as one reason for this situation. At least Löwy recognises the proscription of industrial fishing as a step forward in Venezuela. And he is aware that

> it would be unrealistic to ask the governments of Venezuela, Bolivia, and Ecuador to immediately stop oil (or gas) production, since this is, at the moment, the main source for the financing of their significant social programs. But they could take some partial initiative concerning specific areas, ... to be a positive example for other parts of the world on how to fight climate change (by leaving the fossil energies under the soil), putting pressure on the rich countries of the North, with the demand of indemnities for the unexploited oil. (ibid., 23)

To overcome such pressures as economically crippling sanctions, attempted coups, economic sabotage by national business owners, bribery of government officials, and threats of military invasion, Löwy's political strategy is to force a shift in governments like the Bolivarian one in Venezuela by applying "sufficient pressure from autonomous social movements and public opinion on the leftist governments." This is to occur with the aid of unsubordinated, independent Indigenous communities acting as political auxiliaries, although this is only possible if, according to Löwy, they could just get beyond their parochialism (ibid.).

Aside from failing to recognise the Bolivarian government's attempts at diversifying the economy away from fossil-fuel export dependence, there is in such assertions an unhelpful paternalism towards and a lack of analysis of First Peoples' political histories and current situation, as well as the diversity of interests (within as well as between communities). It could be useful to explain how such presumably "autonomous" movements and public opinion are to be developed—especially given the overwhelming influence of capitalist media, something Löwy recognises as a huge obstacle—and how they could be organised and mobilised to change the direction of the Bolivarian government while pre-empting a US military invasion and/or a return to neocolonial conditions under US dominance.

The main challenges to Bolivarian Venezuela's progress towards ecosocialism are rather to be found in the imperialism of US and satellite NATO states, continued Venezuelan oligarchs' control over most of the economy, and the persistence of bourgeois political influence within government and state institutions. For ecosocialists outside of Venezuela, a major task should then be, at a minimum, to organise and concentrate efforts in their respective countries towards the defence of ecosocialist Bolivarian Venezuela (Schwartzman and Saul 2015, 18). Ecosocialists not concurring with this conclusion have yet to demonstrate an alternative politics that creates at least the same level of social influence and institutional presence as achieved in Venezuela.

Disadvantages to Existing Ecosocialist Strategies Outside of Venezuela

There are challenges to both centralist and decentralist approaches, or to attempts at their combination. Regarding the former, it is unclear how such parties or vanguards or mass movements are to be put together, what accountability entails, or how vanguards and "free associations" are to develop in complementarity. A recent overview of social and environmental struggles makes it clear that even putting grassroots organisations together is yet to prove feasible or lasting in many instances (Löwy and Tanuro 2021). This is an underwhelming result after decades of red-green initiatives and efforts. The only viable model so far is Bolivarian ecosocialism, but most ecosocialists seem uninterested in studying and pursuing the strategies that have resulted in the only successful form of ecosocialism at a national scale. As briefly described above, the Bolivarian Revolution came about through mass dissatisfaction and movements (since the late 1980s), organised opposition within military rank and file, and leftist parties able to channel mass protests into effective political action.

This is not entirely unknown to non-Venezuelan ecosocialists. The contours of this are adumbrated by Devine (2022, 276), who "would begin with Gramsci's concepts of wars of position, organic crises, and wars of manoeuvre," but has yet to contribute more than this abstract allusion.[7] A war of position refers to struggles to gain influence throughout society, while a war of manoeuvre entails open conflict, which may include military clashes. An organic crisis describes a state of waning legitimacy for ruling classes. It is a period when their hegemony (political leadership through both coercion and consent mechanisms) starts to dissipate thanks to a successful war of position. The case of ecosocialism in Bolivarian Venezuela could be described accordingly. The outcome of the war of position remains indeterminate, in suspension, and constantly on the verge of wars of manoeuvre whenever open conflict is not occurring. There is a prolonged organic crisis with ruling class hegemony undermined yet potentially restored at any time. The situation is a result of, among other interrelated factors, a persistent capitalist and landowner control over much of the economy (including communication infrastructure), US interventionism (with NATO satellite states' assistance), strife among socialists, and contradictions within and between diverse communities.

What is more, the emphasis on communal councils and the attempt their coordination and coherence with state institutions provides an illustration, if not a model, of how to join political strategies. A feasible way to connect and coordinate decentralist and centralist modes of organising therefore already exists. Yet, to Wallis (2020, 117), the "effective synthesis between the self-management impulse and a political strategy has yet to be worked out," even while discussing the successful case of Venezuela at length. The trouble is that, as the same author recognises, social change is envisaged in major ecosocialist currents as only possible "from below" (Wallis 2018), thereby precluding strategies involving some form of centralism and coercive means. Another problem is circumscribing his musings over strategy mainly to historical examples where socialist revolutions have not happened, such as the US.

[7] Devine is not the only one to cite Gramsci without explaining that the communist luminary's politics were firmly grounded in Bolshevik-type or vanguard party organising, which most ecosocialists in the imperialist countries reject, as shown above. However, concepts like war of position and manoeuvre are deprived of meaning without understanding that centralised decision-making processes and disciplined, coordinated, and even military activities were taken for granted by Gramsci as means to succeed in gaining hegemony or in cases of open conflict with ruling class forces. In a Gramscian reading, the Bolivarian Revolutionaries could even be faulted for having been insufficiently centralist and disciplined in their handling of enemy forces bent on annihilating any socialist current.

Almost as if trying to answer Wallis' quandary, Patrick Bond views a sort of ecosocialist potential emerging from the contradictions produced by "the radicalization of ecological modernization by environmental justice advocates" (Bond 2022, 511), where the former includes "Corporations, states, global agencies" (ibid., 507, Table 1).[8] It is doubtful, however, that decentralist ecosocialists will find palatable the resulting "red-green hybrid networks" Bond envisions. There is a fundamental divergence of principles that is bound to lead to conflict among ecosocialists, if they ever reach any appreciable social influence. The conflict is for now restrained to words, but in Venezuela, while it persists as an unresolved (likely unresolvable) tension, the challenges of combining and coordinating grassroots and state efforts are recognised and addressed, even if under tremendous strain, contradictory movements, and at times openly clashing.

The difficulties in designing viable or consistent political strategies are hardly confined to centralist ecosocialists. There is also inconsistency among avowedly anti-state or decentralist ecosocialists. The characterisation of intrinsically repressive and violent structures they heap on parties and states, could be argued as characterising the Rojava Revolution, which includes coercive measures and killing people for self-defence or to enable the project's existence. Yet the Rojava Revolution is touted as decentralist and entails some of the most advanced forms of ecological and socialist feminist thought among social movements worldwide (e.g., Brownhill et al. 2022, 5; Nelson 2022b). Likewise, ecosocialists supporting the EZLN (e.g., Brownhill and Turner 2020, 2fn3; Nelson 2022a, 38; see also Baer 2018, 181–182) should explain how the latter's practices of military organisation (the EZLN is, after all, an army) and exclusion of rival political formations align with decentralist frameworks. It is in such instances that it could be helpful to raise awareness about the tension between the two processes (centralist and decentralist) not being reducible to a matter of principles (which leads to self-undermining moralism),[9] but being also shaped by external interventions, including

[8] This may perhaps align with Devine's allusion to Gramsci's concept of organic crisis, but Bond does not specify the contradiction this way.

[9] By this I mean insisting on a pre-defined ethics, sets of morals, or political tenets, etc. to be applied regardless of situation. This undermines ecosocialist politics in several ways. One is by restricting the applicability of ecosocialism to a narrow set of contexts where, for instance, one's principles are not put under strain. Another is the incrustation

military invasions.[10] In a less charitable reading, ecosocialists insisting on such distinctions tend not to be directly affected by the violence inflicted on the formations they support and tend to express ideas from positions of relative material luxury compared to most of humanity. This can cloud judgement and narrow the range of politics to be considered as legitimate.

There are also potential problems in so-called bottom-up or autonomy-based approaches on matters of principle and that are yet to be addressed. For instance, logging by a community upstream could generate greater potential erosion and higher magnitude flooding for communities living downstream. This challenge is rarely explored in ecosocialist writings. For instance, the "tensions and contradictions between the self-managed associations themselves and between them and the representative assembly," as envisioned by Devine (2022, 277), is presumably resolvable by means of "negotiated coordination." This presumes that a sufficiently universal

of judgementalist politics, which prevents collaboration with otherwise affine forces on account of their violation of certain principles. The reasons for horrific acts are buried under abstract universalism, so that practicable preventative measures cannot be developed to address future or other circumstances that can lead to similarly regrettable actions. This impedes the spread of ecosocialism as well. Those ecosocialists who hold dear the notion of diversity of perspectives should be among the first to be sensitive to the possibility of having to accept a relaxation or temporary suspension of at least some principles, by considering events or processes beyond the control or influence of judged parties. Ethics, morals, principles, etc. are the product of social relations and material circumstances, and therefore not homogenous in content across time and space. Ecosocialists' frequent disparagement of socialist states (or "first epoch" socialism) is a main example of an ultimately self-undermining moralism (see also Albert 2023).

[10] The Rojava system of government in Northern Syria is called democratic confederalism and it is not evident how well it coheres with ecosocialism. Even if one were to include Rojava revolutionaries in the ecosocialist camp, one would have to explain the existence of centralised decision-making authorities in the militias (the Syrian Democratic Forces), local police (the Asayish), and political parties like the Kurdish national Congress and TEV-DEM (Movement for a Democratic Society) coalition, all institutional frameworks that deviate from democratic socialist principles. One would also have to account for there being effectively a ruling party called the PYD (Democratic Union Party), who are associated with a very much centralist party called the Kurdish Workers' Party and operate as part of the Kurdistan Communities Union, through which decisions are made relative to the implementation of democratic conferedalist policies. Similar arguments could me made of the EZLN, who do identify as ecosocialist and do not allow multi-party elections or state (public) property within areas where "they learn as they walk" with the people, or where they practise local autonomy. This is evident as well in the latest restructuring of the system for local self-governing in response to rising "defense and security" concerns (see https://enlacezapatista.ezln.org.mx/2023/11/13/ninth-part-the-new-structure-of-zapastista-autonomy/, accessed 22 March 2024).

form of ecosocialism already prevails worldwide. There is no sense of how such "negotiated coordination" could be made applicable prior to any global ecosocialist hegemony being reached.

The challenge becomes even greater when the issue is of global environmental significance, such as cumulative land cover changes from converting forests to arable land (altering overall landmass reflectance, with climate change implications). Planetary interconnections are not necessarily self-evident. If community autonomy results in carelessness towards other autonomous communities, some coercive coordination measures are also necessary to pre-empt the harms on some communities resulting from parochial understandings of self-management. This is part of what communist parties have repeatedly faced in post-revolutionary situations and have often had to resort to draconian methods to address and, in the case of recurring famines, to resolve historically once and for all (the USSR, Mongolia, and PRC being among salient examples).

Just establishing coordination mechanisms, especially at a planetary scale, requires the attainment of sufficient political influence or control to disempower ruling classes and to impose rules and standards on communities that would otherwise interpret autonomy to mean doing whatever they want without any consultation with other communities. Moreover, determining what constitutes environmental harm and who may be affected by what sort of environmental practice necessitates coordination and enforcement of mutually agreed rules and standards. Ecosocialists, a relatively tiny group, already bicker among themselves about principles, strategies, and the best course of action. To imagine that violence, coercion, repression, authoritarianism can be avoided while struggling to alter the very basis of existence for hundreds of millions of people is to engage in a most unhelpful evasiveness.

In any case, if achieving the complementarity of strategies advocated by ecosocialists like Baer and Wallis is by means of the ballot box (or "majority vote"; Devine 2022, 277) and/or instituting parallel power structures, ecosocialists should not be surprised at being rapidly outmanoeuvred by parties and media well-financed by capitalists or arrested on fictitious charges or assassinated when they start succeeding in the electoral system or in establishing viable alternative institutions. These have been among the standard treatments for socialist forces whenever they gain enough political clout to threaten ruling classes' interests or the capitalist order itself. At the very least, ecosocialists should plan for acts of capitalist resistance and violent repression.

Supposing success in gaining state power or effective social influence, more state-centred ecosocialists would introduce high taxation on emissions, except for those with low income, abbreviated work weeks, public ownership of the means of production, legislation mandating "sustainable" travel, trade, town layouts, and buildings, and policies promoting social equality within and between countries. The latter would feed into the aim of global population stabilisation, alongside "overall degrowth for countries in the Global North and development for countries in the Global South" (Baer and Singer 2022, 88).

Those are instruments that decentralist ecosocialists forego in favour of building alternatives within the interstices of capitalist society. As many of those advocating for decentralist ecosocialism also find the present state of the world alarming, it would be useful to know how the even slower process of building parallel institutions, compared to party and state take-over strategies, can address fast unfolding ecosocially destructive processes. The urgency often expressed by decentralist ecosocialists is at odds with the strategies they envision. The strategy is also rather vacuous so far as a result of what they have achieved concretely as ecosocialists. Instead, such ecosocialists have tended to contribute to struggles as supportive individuals for a movement or community organised by other political or social forces, such as La Vía Campesina or Lakota communities (in the ultimately unsuccessful protests to halt a major pipeline; see Chapter 6). It is laudable in itself for decentralist ecosocialists to lend such support, but it is inadequate to the task they themselves have set for themselves.

Even if there are major differences about strategies and convergence relative to rarely significant results, the general striving for economic equalisation meets the agreement of any ecosocialist. It is nevertheless unlikely that a unification is feasible among the main ecosocialist currents. Democratic eco-socialists, in contrast to decentralist ecosocialists, for example, would have public ownership of the means of production, as well as representative and participatory democracy, to arrive at environmental sustainability. The aim is nothing short of establishing "an eco-socialist world system" (Baer and Singer 2022, 87). It will be based on "common ownership" which combines "centralism and decentralism" to manage a "steady-state economy" for the world. Along with this, ultimate aims include socialist planning, workers' democracy, global population stabilisation (as stated above), meaningful work, ending consumerism, and the like. It is understood that there is no guarantee that environmental disasters will be prevented, but such a world-system would have "the potential to place constraints on resource depletion" (ibid., 86).

Other policies would involve ways to develop "sustainable" buildings and town layouts, while undermining consumerism and draw down the military and nuclear sectors. Energy policies would include renewable energy and raising energy efficiency, based on "green jobs" and "sustainable technology," inclusive of transport. Food production and forestry systems would be overhauled for ecological sustainability. For this, there are plenty of examples even within the current capitalist world economy (see also Mamaceqtaw forestry in Chapter 5), but also and especially in socialist countries like Cuba, as some ecosocialists recognise (e.g., Schwartzman 2022; Wallis 2022).

In the realm of ideological struggles, a salient characteristic of ecosocialism is the absence of strategy regarding the means of communication. There are preoccupations with advertisement industries, which are rightfully identified as pernicious organs of mass deception and fomenters of environmentally destructive consumption (Löwy 2015; Baer 2018; Wallis 2022). There is very little discussed about the infrastructure that is in the hands of the private sector or pro-capitalist public sector (e.g., submarine fibre-optic cables, satellites, etc.) and about policies to be devised and implemented regarding media content and diffusion.

A main strategy regarding political campaigning or messaging, if there is one at all, rests on the already discussed disavowal of "first-epoch" socialism. Wallis has articulated well the rudiments of the sort of political platform that emerges from this disavowal. In discussing the aftermath of the overturning of socialism in the USSR and Eastern Europe:

> ... socialism stands to benefit in the long run from the complete collapse of its sham versions. Better to start again from scratch than to entrench even further people's identification of socialism with its siege expression; better to bring pre-socialist prejudices out into the open than to pretend that decrees and territorial arrangements had somehow abolished them. (Wallis 2020, 52)

With this logic, an ecosocialist platform will have difficulty explaining any support for still existing socialist countries, unless ecosocialists are supposed to jettison them as well. This option would have to be explained further, given the largely positive assessment at least some of the same ecosocialists have for "first epoch" socialist Cuba (e.g., Schwartzman 2022, 286; Wallis 2020 120–121).[11] There would also need to be some clarification regarding the discrepancy between, on the one hand, praising Cuban accomplishments and, on the other hand, failing to make the struggle to overturn the US government's policies on Cuba into an ecosocialist priority. More troubling is the call to efface, rather than build on historical achievements (rather than just focus on failures), and for promoting distortions about the efforts, lasting successes, as well as setbacks in socialist countries to overcome "pre-socialist prejudices" (see, for instance, Bartels and Bartels 1995).

In Wallis' more specific case, but this is not unusual in ecosocialist circles, there is also a parochial reading of politics that constrains the ecosocialist imaginary to what is deemed acceptable within liberal democracies, where anti-communism has been pounded into most people's brains across generations. One could just as legitimately argue that the opposite strategy should be promoted, which is to dedicate more ecosocialist energies on showing and stressing what socialist countries managed to accomplish under relentless threat of obliteration by reactionary forces especially from within liberal democracies.

Furthermore, it makes little sense to "start again from scratch" in countries like the US, the main if not the only political reference point for Wallis and many other ecosocialists, since there has never been even a remote chance for socialist revolution in that settler colonial dictatorship. Given the rising interest in socialism within the US that Wallis himself underlines, for example, it seems more reasonable to explain, without covering up any of the terrible aspects, the net superiority of socialism in any form compared to capitalism in any form. Selectivity about what constitutes socialism may also backfire politically. As distortions on socialism preponderate in capitalist countries while ecosocialists still have a relatively negligible social reach, it seems inappropriate to expect from most people an ability to discern among different forms of socialism.

[11] Elsewhere, Wallis calls Cuba "the closest current approximation to a socialist society" (Wallis 2022, 294). No criteria are offered to determine how a country qualifies as socialist. This is a recurring problem of arbitrariness in the literature on socialism.

Addressing Ruling Class Violence

A curious dimension of ecosocialist thought is a penchant for evading the crucial question of ruling class reaction. Regrettably, this is considered, often implicitly, as separable from the multiple forms of violence, especially rape culture, that undergirds capital accumulation (Brownhill and Turner 2020). Ruling class reaction encompasses all forms of oppression, but there must also be greater appreciation of its connection to the international expression of anti-socialist violence. The "siege expression" of socialism to which Wallis alludes is not some unfortunate state of affairs that can be pre-empted, overcome, or pretended away. It is something for which ecosocialists must prepare if substantive social change is to be had. Derek Wall has put the question in the following manner:

> Ecosocialism cannot be created in just one part of the world. If it were, it would soon be crushed by hostile corporate forces and would be, even if left alone, too limited to create the international change needed to protect the planet. However, it is unlikely that ecosocialism will win simultaneous victories in many countries across the planet at the same time. (Wall 2010, 122)

Wall sees the international blocs, such as the above-described ALBA, as one pre-emptive strategy in defending against "hostile corporate forces." The improbability of worldwide simultaneous ecosocialist gains should be addressed by those seeking global change primarily by parliamentary means (e.g., Baer's notion of establishing an ecosocialist world-system). Yet Wall goes no further with respect to the question of capitalist reaction and violence.

Michael Albert has been another of the few who have given any substantive thought to this aspect of transitioning to ecosocialism. He surmises that

> ecosocialists should reflect more systematically on the domestic and international threats that ecosocialisms-in-transition would confront – including capital flight, of course, but also militant far-right resistance, cyberattacks, social media-enhanced disinformation operations from capitalists and counterrevolutionary states, and other forms of sabotage – and how they might respond. ... such problems must be anticipated and systematically explored in advance in order to work out creative solutions. (Albert 2023, 23)

To him, this means not shying away from state power and resort to it as needed for self-defence. The danger is therefore much more than a "rightwing reaction" to a hypothetical crisis ushered in by reforms like the Green New Deal in the US (Albert 2023, 21). This is anyway a curious argument coming out of a US context where such reaction to any threat on profit rates is the norm, whichever political party is in power.

Ecosocialists rarely discuss the violent repression typifying the regular workings of capitalist rule. Warfare, including civil war, is a consequence of seeking any control of the means of production, or state take-over, or even the "transitional radical reforms" articulated by democratic ecosocialists. Arguably, any power even to implement "degrowth," an idea becoming dear to increasing numbers of ecosocialists, will be met with very well-armed violence and well-oiled cultural suppression mechanisms on behalf of the ruling classes. Keeping to the above-described principles, ecological sustainability, and all the socially positive policies listed while being attacked militarily is quite the challenge. The much-touted example of the EZLN and especially of Rojava, with its anti-patriarchal foundations, should make it clear how difficult it is to keep to principles.

It is when ecosocialists attempt to address capitalist repression that an inadequate grasp of socialist histories is revealed. Joel Kovel has provided a rare attempt to discuss the question. He envisioned the issue is in terms of a "heroic phase" characterised by "much sacrifice" (cf. Tanuro 2020, 263) following the prior establishment of "numerous" and diffuse, well-organised communities capable of low-input self-sustenance and of "shutting down normal production," coordinated and supported by means underground mutual aid networks. These developments, buttressed by the "spiritually superior" behaviour of revolutionaries, are supposed to suffice in making capitalist forces lose confidence and leading militaries and police to lay down their weapons (Kovel 2007, 267).

As Michael Albert has pointed out, "This is quite an idealistic scenario, and Kovel only weakly attempts to describe conditions in which a global-scale revolution of this sort might occur" (Albert 2023, 15). The consequent is being placed in front of the antecedent. The resources to be used can easily be made scarce or denied in countries where nearly the entire economy is in private hands, like most farmland, forests, and much else. Even the most basic inputs would be difficult to secure without already being well-organised and armed to confront extreme forms of violence. The loosely organised, self-funded, always transparent, assembly-based ecosocialist parties Kovel described will be infiltrated and run aground as soon as or even before they gain any substantive social influence. So-called Leninist parties arose for such a reason and will doubtless remain necessary, regrettably, just to embark even on beginning the process Kovel envisioned. One could go on in critiquing such fanciful ideas and assumptions. Capitalist forces' capabilities and ruthlessness must never be underestimated.

Unfortunately, the field of view offered by Albert and most ecosocialists is quite restricted. Usually, politics, whether centralist and/or decentralist, are assumed to be pursued within a parliamentary framework or in any situation by means of dissident grassroots movements. A single-party centralism and/or state institutions answerable to diffuse pressures and decision-making processes organised at workplace and neighbourhood levels, as in Cuba, is not to be fathomed or is implicitly considered a contradiction. The alternative to multi-party systems or community-based organising is only authoritarianism. In Albert's case, the way to think about the political options is reduced to preoccupations with potential (and undefined) "authoritarian" turns in attempts to establish an ecosocialist order. Longer-term political strategies to be developed are presented in the following dichotomous form:

> A trade-off might ... be unavoidable, as noted earlier: do we follow the path of rich-world degrowth and flirt with the heightened risks of authoritarian statism, or advocate a GND/socialist politics of "more" that would threaten biospheric tipping points (while producing its own authoritarian risks, as I'll expand on below)? (Albert 2023, 24)

It could be argued, even in keeping with Albert's framing, that without some initial intensification of environmental degradation (not necessarily reaching thresholds), the transition to degrowth is not feasible. This can be argued for an expansion of renewable energy systems requiring further, temporary fossil-fuel use, while at the same time avoiding displacement of environmental degradation to the most disadvantaged or politically least empowered communities. Continued, but mitigated destructive effects can lead to their opposite of ecological sustainability. This would be a more dialectical understanding in lieu of an unreconcilable binary opposition.

More importantly, with respect to political strategy and in particular imperialism, the question posed already presupposes that "authoritarian statism" is as worrisome as the transgression of "biospheric tipping points." This betrays a view of the world that denies the importance of defeating the actually existing dictatorships on which "rich-world" societies depend for the resource base enabling such obscenely maldistributed wealth. Furthermore, the framing amounts to denying other options more appropriate to the North American context about which Albert refers. Salient among these is to give primacy to the struggle for the dismantling of NATO countries' militaries, starting with the US. This would already undermine the ecosocially destructive industrial infrastructure therewith associated and enable the redirection of resources to more constructive ends (Schwartzman 2013; Schwartzman and Schwartzman 2018).

Another aspect that should be worthy of consideration is the ecosocial strides made through "first epoch" socialism, with all its contradictions (Engel-Di Mauro 2021; O'Connor 1998, 256–265). There is plenty of experience from the histories of successful and unsuccessful socialist revolutions from which ecosocialists can and should draw to understand the enormity of the task. The issue of repression, suppression, and conflict (including among ecosocialists) cannot be addressed by radical reforms, nor by worrying about potential authoritarianism,[12] and least of all by presuming peaceful transitions, which assume such challenges and horrors away. As soon as property relations are called into question (the "male deal," settler colonialism, racist exclusions all pertaining to and complicating the issue), which would be exemplified by such necessities as taking over land to produce food for ecosocialist ends, ruling classes will respond and have responded violently. Countries where successful ecosocialist revolutions occur will otherwise be militarily invaded or otherwise attacked, including by means of enrolling oppressed groups (see Chapter 2). This is a thread that runs through all histories of "first epoch" socialism. The continuing blockades on Cuba and the DPRK, as well as military threats to and encirclement of the PRC. These are current instances that often meet with indifference among

[12] The matter is contingent on what precedes social change and what it takes to mount effective self-defence, which likely involves possibly violent suppression of enemy forces or dissenters who prefer debacle over compromise with what is objectively achievable. There is not a single historical example of successful revolutions where such terrible circumstances and decisions have not had to be faced. Rather than worrying about a generic authoritarian turn, it would be more constructive to be concerned about what institutional structures are feasible in a given set of conditions, as much as can be expected, that could at least help curb negative turns.

ecosocialists fixating on authoritarianism or instances of repression or oppression within socialist countries, instead of paying more attention to working-class majorities' health conditions and standards of living and longer-term, cross-generational improvements to people's material well-being in socialist countries.

Often, ecosocialists will invoke peaceful means to achieve their ends without incurring conflagrations. A recurring trope is one of hoping for the least amount of violence. According to Wallis (2018, 60), for example,

> The best hope for keeping violence to a minimum lies in building the most solidly based, coherent, and effective movement that we can. To the degree that we (i.e., all those who share this vision) are successful, the resulting network of mobilized communities will assure that a violent response from the ruling class can only backfire and multiply our support.

These remarks stand at odds with the recognition that "first epoch" socialism developed under siege conditions (Wallis 2018, 76) and the "disastrous impact on the [US] Left" of multiple bouts of repression (ibid., 168). There is little reason to expect that capitalist powers will refrain from trying to annihilate or putting a siege on any future socialist project. Mass mobilisation is unlikely to provide the capacity to restrain ruling class violence. Since all examples so far of successful mass mobilisations have not deterred extreme violence and successful repression by the ruling classes affected, such claims of assurance are at best sanguine and at worst delusional.

One might also want to consider how to defend against the combined forces of police, military, and secret services all converging to smash any ecosocialist movement before it reaches such a degree of mobilisation. It is even more fanciful to believe that even if such backfiring would occur that there would not be a military invasion from other capitalist countries, as happened with virtually all socialist revolutions, from the Paris Commune onwards. It appears that ecosocialists tend to underestimate the effectiveness of outfits like NATO and their underground international networks (e.g., Operation Gladio) in suppressing any socialist project. Aside from contradictions within and between countries that can spawn bloody conflicts, there are internal repressive apparatuses as well as geopolitical and international coordination processes that ecosocialists should study, rather than merely hope for the best. There are objective historical processes shaping the development of post-revolutionary societies. Internal conflicts and repression of dissent, contradictory policies undermining aspects of socialism over the short-term, environmentally destructive practices to enable survival or develop capacities to improve people's material conditions are not deviations from socialist objectives. They are the multi-generational ordeals to surmount just to have the chance for any socialist or ecosocialist future.

Technical Feasibility Issues

Even in a most unlikely situation of state capture and transitional radical reforms unfolding peacefully, some dense specks of technical dust will blemish the ecosocialist ointment. There is a plethora of ecosocialist thought devoted to general recommendations on how technology should be developed and for what purpose, as well as what types of technological complex to be favoured (e.g., Devine 2022; Schwartzman and Schwartzman 2018; Wallis 2018). Much of what is proposed hinges on the development and spread of ecologically sustainable or ecosocialist technology related to climate change and at times farming, with not much dedicated to industrial production, forestry, mining, and much else. The definition and feasibility of transitioning to a different technological complex literally engulfs almost if not all societies at once, as there is no community entirely self-sufficient on locally available resources. The issue seems much more daunting than ecosocialists' pronouncements and analyses would lead one to believe. There are myriad materials, parts, and processes involved even in producing what are now basic consumer items like telephones or computers. There are also enormous numbers of interlinkages across the globe that would have to be reconfigured to avert interruptions leading to sudden mass shortages, such as by worsening already poor food distribution systems or impeding the manufacture of reagents for hospital use. Interrupting existing linkages, especially those requiring large amounts of initial investment and resources (for instance, underwater cables, container ships), could also be disastrous, such as with sudden mass unemployment and insufficient resources for the upkeep of capacities for hazardous waste disposal and neutralisation of nuclear bombs.

As ecosocialists provide no details on how to arrive at "sustainable technology," nor any plans on how to replace existing production systems and interconnections affected by developing and applying ecosocialist technology (even a couple of examples could be enlightening), the technological aspect could prove to be the keystone that, once removed, would make an ecosocialist edifice crumble before it is even at the beginning stages of construction (see also Huber 2022, 560). This is not to suggest that ecologically sustainable technologies are unachievable, and in any case many already exist (Schwartzman and Schwartzman 2018), but that the physical forces of production cannot be as easily transformed as most ecosocialists imagine. To address the issue of transforming technologies, the very least that ecosocialists should do is to make of enlisting technical experts to the cause a priority, or to gain expertise themselves so as to avoid argumentative vacuity (e.g., Smith 2021). The few ecosocialist works that raise and discuss this question do not yet provide the detailed blueprints necessary to embark on recruiting technical experts and building ecosocialist technical education programmes and alternative production systems (see also Schwartzman and Schwartzman 2018; Wallis 2018; White 2020).

The technical challenges are many but await the depth of consideration they deserve. Somerville (2022) finds recurring flaws in ecosocialist discussions, the few that exist, on both technology and finance,[13] though one could also add in the realm of the biophysical sciences broadly. There continues to be an assumption that fossil fuels can be dispensed with and replaced with renewable energy systems as if it were a seamless process and without investigating and discussing aspects of technical feasibility, timing, and much else related to an energy transition. The work of Schwartzman and Schwartzman (2018) is unfortunately still rare (and curiously omitted by Somerville and most other ecosocialists). The less than stellar grasp of technical matters among most ecosocialists, anyhow, reflects not so much a lack of ideas among ecosocialists as an evident lack of technical expertise (and perhaps a lack of priority given to gaining such expertise; see Schwartzman and Engel-Di Mauro 2022). It is perhaps one factor that contributes to a tendency for ecosocialist thought to veer towards

[13] A similar problem persists with the ways of using financial means to buttress and establish ecosocialist policies in lasting ways. It is unclear, however, how a money-free economy, as advocated by Nelson (2022b), would enable the global economic coordination needed to address environmental problems at the planetary or just continental scale. Another problem is that all the examples cited of practising nonmonetary systems rely on money-based transactions for the most basic necessities that are not locally available or producible, such as most clothing, most building materials, energy provisioning systems for heating, laboratories to develop vaccines, etc.

alarmism and/or utopianism as well as levels of abstraction of doubtful usefulness to concrete projects. In this, ecosocialism resembles "non-scientific ecology" or "green" thought (Chen 2015; Hayward 1990, 2; see also Suing 2018).

Reckoning with Under-Achievement

The bourgeois electoral road is the default strategy for many ecosocialists, represented by statements that take for granted the election of "leftist parties following ecosocialist platforms" (e.g., Albert 2023, 22; see also Baer 2018).[14] This reflects a parochial view of political systems, as in many countries electoral systems systematically exclude leftist anti-capitalist parties (such as Singapore, the UK, and the US, for example), involve pre-emptive mechanisms to deny power to any such parties (all liberal democracies) or exist to rubber-stamp a monarchical or like order (for example, Bhutan, Morocco, Saudi Arabia, etc.). Otherwise, there are those preferring the creation and diffusion of parallel institutions or communities that somehow will overtake capitalist society and usher in ecosocialism. Some envision this to be compatible with electoral politics (as in the case of Venezuela), with a major task becoming how to make the two kinds of strategies cohere.

There are two ways that progress can be generally gauged according to the aims of any form of ecosocialism. One is social and another is ecological. The former includes the spread of ecosocialist ideas and political organising as well as actions leading to improvements in the living conditions for the communities ecosocialists purport to support. Social improvements are difficult to evaluate, since ecosocialists have rarely gained enough social influence to make much of a dent anywhere. Perhaps at most some localities, likely in core capitalist countries, may have featured some positive impacts from explicitly ecosocialist projects for a handful of people. This is hardly anything to shout about compared to the millions who are immediately affected by state intervention. The

[14] Just in case this is unclear, in socialist countries, there is no need to elect leftists to office, since state institutions are run by leftist anti-capitalist or communist parties (often not just one, as in the case of the PRC and Vietnam, or in the former German Democratic Republic). Of course, most ecosocialists will disagree about the socialist character or political legitimacy of such parties, but that is part of the problem with such forms of ecosocialism.

same unimpressive outcome extends to the ecological aspects of ecosocialist struggle. This is mainly because of having so little political clout to change policies at any administrative or to facilitate major changes in environmental practices at any societal level.

Regardless of strategy, with about four decades of efforts at developing ecosocialist movements, accomplishments have been scarce so far with respect to building political forces influential enough to address the socio-ecological crisis at any scale. The notable exception is Venezuela, where Bolivarian ecosocialism consists of combining and striving to coordinate a collectivist and decentralist communal council framework with a political party leading the take-over of key government and state institutions. This is the only successful case so far of ecosocialist politics gaining wide institutional presence and mass support and contributing to substantive positive ecosocial change at national and international levels. Yet most ecosocialists outside of Venezuela largely ignore the political strategies that have enabled the Bolivarian Revolution resulting in such ecosocialist achievements.

This is peculiar for a movement defined by the purpose—a purpose at times expressed as urgency—of changing the world towards ecosocialism. It is just as peculiar to find a scarcity of detailed political strategy and even just a list of accomplishments in much ecosocialist writings. Instead, much ink is spilled over grim appraisals of what exists and what could be in store and over celebratory depictions of bygone or current political projects that may in part resemble but are not manifestations of ecosocialism. Exaggerating ecosocialism' status will hardly draw the gulf between self-defined purpose and ultimate aims actual capacities and achievements. A recent example is Albert's claim that

> Ecosocialism is one of the foremost among [post-capitalist alternatives], which has become worldwide movement that challenges the hegemonic project of "green capitalism" while also addressing the ecological weaknesses of earlier forms of socialist politics. (Albert 2023, 11)

This is a dubious assertion for a movement whose number of mostly uncoordinated adherents unlikely exceeds the tens of thousands worldwide (e.g., compare tens of millions of CPC members), that is virtually absent in the most populated regions of the world, and whose achievements are at most rather poor. Even inflating figures, as other ecosocialists do (e.g., Schwartzman 2016; Wall 2010), through the appropriation of movements or countries characterised by "first epoch" (e.g., Marxist-Leninist) socialist politics that ecosocialists reject, such as the Burkinabé Revolution or Cuba, will scarcely support the contention that ecosocialism is "worldwide" and "challenges" capitalism at that scale.

Critical evaluations of ecosocialist actions or achievements are as exceptional as their sheer recognition or description. Typically, as the previous chapters have also in part illustrated, a plethora of mainly Marxian theoretical overviews, often introduced by descriptions of the socio-ecological crisis (mainly climate change) generated by capitalism, accompany lists of desired changes or envisioned futures (e.g., Vettese and Pendergrass 2022; Wallis 2020). This very book to some extent contributes to this typical formula, but then the aim here is to describe and explain a socialist current, rather than exhort readers to join it (even if the present author would recommend such a possibility to be at least considered, precisely to work on overcoming the drawbacks discussed). As shown above, with few exceptions of far from global consequence, ecosocialism's direct and concrete achievements thus far fall quite short of its proponents' wishes or depictions of what a future world could or should look like.

References

Adaman, Fikret, and Pat Devine. 2022. "La coordination négociée, un modèle écosocialiste." *Nouveaux Cahiers du socialisme* 28 (2022): 41–46.

Akuno, Kali. 2023. "Build and Fight: The Program and Strategy of Coopera789 tion Jackson." In *Jackson Rising Redux. Lessons on Building the Future in the 790 Present*, edited by Kali Akuno and Matt Meyer, 12–55. Oakland: PM Press.

Akuno, Kali, and Ajamu Nangwaya. 2023. "Toward Economic Democracy, Labor Self-Management, and Self-Determination." In *Jackson Rising Redux. Lessons on Building the Future in the Present*, edited by Kali Akuno and Matt Meyer, 56–90. Oakland: PM Press.

Akuno, Kali, and Sacajawea Hall. 2021. "Building Economic Democracy to Construct Eco-socialism from Below." In *The New Systems Reader: Alternatives to a Failed Economy*, edited by James Gustav Speth, and Kathleen Courrier, 433–440. New York and London: Routledge.

Albert, Michael J. 2023. "Ecosocialism for Realists: Transitions, Trade-Offs, and Authoritarian Dangers." *Capitalism Nature Socialism* 34 (1): 11–30.

Albritton, Robert. 2019. *Eco-socialism for Now and the Future*. Geneva: Springer International Publishing.

Alvarado, Rómulo and José Daniel Romero Bastidas. 2022. "La Dinámica Agroecológica para la Construcción del Nuevo Estado Ecosocialista y Autogestionario en Venezuela. Perspectiva Evolutiva de la Política Agroecológica y su Relación con las Iniciativas Populares." In Agroecologías Insurgents en Venezuela. *Territorios, Luchas y Pedagogías en Revolución*, edited by Francisco F. Herrera and Olga Domené-Painenao, 251–282. Caracas: Fonacit.

Baer, Hans A. 2016. "Toward Democratic Eco-Socialism as the Next World System." https://seed.uw.edu/wp-content/uploads/baer-eco-socialism-2016.pdf. Last accessed 4 September 2022.

Baer, Hans A. 2018. *Democratic Eco-Socialism as a Real Utopia: Transitioning to an Alternative World System*. New York: Berghahn Books.

Baer, Hans A. 2021a. *Global Capitalism and Climate Change. The Need for an Alternative World System*. Lanham: Lexington Books.

Baer, Hans A. 2021b. "Toward Democratic Eco-socialism as the Next World System: A Vision for the Future." In *The New Systems Reader: Alternatives to a Failed Economy*, edited by James Gustav Speth, and Kathleen Courrier, 127–144. New York and London: Routledge.

Baer, H.A., and M. Singer. 2022. "Can Ecological Modernisation Contain Climate Change: An Eco-Socialist Perspective." *Journal of Australian Political Economy* 90: 75–91.

Barreto, Guillermo R. 2020. "Ecosocialism in Venezuela and the North-South Struggle." Progressive International, https://progressive.international/blueprint/18974adb-867d-4d55-87e2-6a24ee81fed5-barreto-ecosocialism-in-venezuela-and-the-north-south-struggle/en. Accessed 16 April 2023.

Bartels, Dennis A., and Alice A. Bartles. 1995. *When the North Was Red. Aboriginal Education in Soviet Siberia*. Montréal: McGill-Queen's University Press.

Bond, Patrick. 2022. "The Case for Ecosocialism in the Face of the Worsening Climate Crisis." *Science & Society* 86 (4): 485–515.

Boykoff, Jules. 2013. *The Suppression of Dissent: How the State and Mass Media Squelch USAmerican Social Movements*. New York: Routledge.

Brain, Stephen. 2011. *Song of the Forest: Russian Forestry and Stalinist Environmentalism, 1905–1953*. University of Pittsburgh Press.

Brownhill, Leigh. 2022. "The Ecology of Misogyny." In *The Routledge Handbook on Ecosocialism*, edited by Leigh Brownhill, Salvatore Engel-Di Mauro, and Terran Giacomini, Ana Isla, Michael Löwy, and Turner Terisa, 109–116. Oxon and New York: Routledge.

Brownhill, Leigh, Salvatore Engel-Di Mauro, and Terran Giacomini, Ana Isla, Michael Löwy, and Turner Terisa. 2022. "Preface: An Introduction to Ecosocialism." In *The Routledge Handbook on Ecosocialism*, edited by Leigh Brownhill, Salvatore Engel-Di Mauro, and Terran Giacomini, Ana Isla, Michael Löwy, and Turner Terisa, 1–11. Oxon and New York: Routledge.

Brownhill, Leigh and Terisa E. Turner. 2020. "Ecofeminist Ways, Ecosocialist Means: Life in the Post-capitalist Future." *Capitalism Nature Socialism* 31 (1): 1–14.

Burbach, Roger, and Camila Piñeiro. 2007. "Venezuela's Participatory Socialism." *Socialism and Democracy* 21 (3): 181–200.

Campbell, Susan. 1994. "'Black Bolsheviks' and Recognition of African-America's Right to Self-Determination by the Communist Party USA." *Science & Society*, 440–470.

Chen, Hong-Rui. 2015. "The Influence of David Pepper's Ecological Theory of Socialism on China's Era." In *Energy, Environment and Green Building Materials: Proceedings of the 2014 International Conference on Energy, Environment and Green Building Materials (EEGBM 2014), November 28–30, 2014, Guilin, Guangxi, China*, edited by Ai Sheng, 447–450. Boca Raton: CRC Press.

Ciccariello-Maher, George. 2016. *Building the Commune: Radical Democracy in Venezuela*. New York: Verso Books.

Devine, Pat. 2017. "Ecosocialism for a New Era." In *Varieties of Alternative Economic Systems. Practical Utopias for an Age of Global Crisis and Austerity*, edited by Richard Westra, Robert Albritton and Seongjin Jeong, 33–51. Oxon: Routledge.

Devine, Pat. 2022. "Ecosocilist Economics." In *The Routledge Handbook on Ecosocialism*, edited by Leigh Brownhill, Salvatore Engel-Di Mauro, Terran Giacomini, Ana Isla, Michal Löwy, and Terissa Turner, 271–279. New York: Routledge.

Engel-Di Mauro, Salvatore. 2021. *Socialist States and the Environment: Lessons for Ecosocialist Futures*. London: Pluto Press.

Engel-Di Mauro, Salvatore, Danny Faber, and Christina Schlegel. 2023. "Social Struggles in Post-Bolsonaro Brazil: An Interview with Cassia Bechara of the Movimento dos Trabalhadores Rurais Sem Terra (MST)." *Capitalism Nature Socialism* 34 (2): 57–68. https://doi.org/10.1080/10455752.2023.2212513.

Flanagan, Erin, and Dennis Raphael. 2022. "From Personal Responsibility to an Eco-Socialist State: Political Economy, Popular Discourses and the Climate Crisis." *Human Geography* 16 (3): 244–259.

Foster, John B. 2015. "Late Soviet Ecology and the Planetary Crisis." *Monthly Review* 67 (2): 1–20.

Foster, John B. 2023. "Planned Degrowth: Ecosocialism and Sustainable Human Development." *Monthly Review* 75 (3): 1–29.

García-Guadilla, María-Pilar. 2014. "Neo-extractivism, Neo-rentierism and Social Movements in Venezuela's 21st Century Ecosocialism: Discourses, Conflicts and Resistance." In *Environmental Movements around the World: Shades of Green in Politics and Culture*, vol. 1, edited by Sherilyn MacGregor, and Timothy Doyle, 77–104. Santa Barbara: Praeger.

Gare, Arran. 1993. "Soviet Environmentalism: The Path Not Taken." *Capitalism Nature Socialism* 4 (4): 69–88.
Giacomini, Terran. 2021. "Agroecology and Ecofeminist Activism." In *The Routledge Handbook on Ecosocialism*, edited by Leigh Brownhill, Salvatore Engel-Di Mauro, Terran Giacomini, Ana Isla, Michal Löwy, and Terissa Turner, 189–200. New York: Routledge.
Gilbert, Chris. 2023. *Commune Or Nothing!: Venezuela's Communal Movement and Its Socialist Project*. New York: Monthly Review Press.
Global Footprint Network. 2024. "What the Ecological Footprint Measures." https://www.footprintnetwork.org/what-ecological-footprints-measure. Accessed 21 April 2024.
Haro García, José Luis. 2022. "Ecosocialism and Workplace Democracy." In *The Routledge Handbook on Ecosocialism*, edited by Leigh Brownhill, Salvatore Engel-Di Mauro, Terran Giacomini, Ana Isla, Michal Löwy, and Terissa Turner, 214–222. New York: Routledge.
Hayward, Tim. 1990. "Ecosocialism—Utopian and Scientific." *Radical Philosophy* 56: 2–14.
Hickel, Jason. 2020. "The Sustainable Development Index: Measuring the Ecological Efficiency of Human Development in the Anthropocene." *Ecological Economics* 167: 106331.
Huber, Matthew T. 2022. "Ecosocialism and the Problem of Industrialism." *Science & Society* 86 (4): 546–563.
Kovel Joel. 2005. "The Ecofeminist Ground of Ecosocialism." *Capitalism Nature Socialism* 16 (2): 1–8.
Kovel, Joel. 2007. *The Enemy of Nature. The End of Capitalism or the End of the World?* London: Zed Press.
Kovel, Joel. 2008. "The EIN." *Capitalism Nature Socialism* 19 (1): 1–4.
Kovel, Joel. 2009. "The EIN: Chapter Two What Is to be Done?" *Capitalism Nature Socialism* 20 (2): 1–12.
Levidow, Les. 2022. "Green New Deals: What Shapes Green and Deal?" *Capitalism Nature Socialism* 33 (3): 76–97.
Li, Minqi. 2009. *The Rise of China and the Demise of the Capitalist World Economy*. New York: Monthly Review Press.
Li, Minqi. 2016. *China and the 21st Century Crisis*. London: Pluto Press.
Li, Minqi. 2017. "Political Economy of China from Socialism to Capitalism, and to Eco-Socialism?" In *The Political Economy of Emerging Markets Varieties of BRICS in the Age of Global Crises and Austerity*, edited by Robert Westra, 79–100. London: Routledge.
Löwy, Michael. 2007. "Eco-Socialism and Democratic Planning." In *Coming to Terms with Nature: Socialist Register*, edited by Leo Panitch and Colin Leys, 294–309. London: The Merlin Press.

Löwy, Michael. 2014. "Ecosocialism: Putting on the Brakes Before Going Over the Cliff." *New Politics* XIV (4). https://newpol.org/issue_post/ecosocialism-putting-brakes-going-over-cliff/. Accessed 10 April 2024.

Löwy, Michael. 2015. *Ecosocialism. A Radical Alternative to Capitalist Catastrophe*. Chicago: Haymarket Books.

Löwy, Michael, and Daniel Tanuro. 2021. "Introduction. Le XXIe Siècle Sera le Siècle de la Question Écosociale." In *Luttes Écologiques et Sociales dans le Monde. Allier le Vert et le Rouge*, edited by Michael Löwy and Daniel Tanuro, 7–35. Paris: Les Éditions Textuel.

Löwy, Michael, and Daniel Tanuro. 2022. "The Fourth International's Contribution to Ecosocialism." In *The Routledge Handbook on Ecosocialism*, edited by Leigh Brownhill, Salvatore Engel-Di Mauro, and Terran Giacomini, Ana Isla, Michael Löwy, and Turner Terisa, 346–356. Oxon and New York: Routledge.

Magdoff, Fred, and Chris Williams. 2017. *Creating an Ecological Society: Toward a Revolutionary Transformation*. New York: Monthly Review Press.

Magdoff, Harry, and Fred Magdoff. 2005. "Approaching Socialism." *Monthly Review* 57 (3): 19–61.

Malm, Andreas. 2021. *How to Blow Up a Pipeline*. New York: Verso Books.

Marquina, Cira Pascual, and Chris Gilbert. 2020. *Venezuela, the Present as Struggle: Voices from the Bolivarian Revolution*. New York: Monthly Review Press.

Marx, Karl. 1992 [1867]. *Capital: A Critical Analysis of Capitalist Production, Volume 1*. Translated by S. Moore and E. Aveling. New York: International Publishers.

Mendoza Suárez, Jesús. 2023. "Desarrollo de la economía turística comunal ecosocialismo en el Consejo Comunal la Concepción del "B"." *Miradas* 18 (1): 207–214.

Nelson, Anitra. 2022a. "Ecosocialism from a Post-Development Perspective." In *Post-Capitalist Futures: Paradigms, Politics, and Prospects*, 31–40. Singapore: Springer Singapore.

Nelson, Anitra. 2022b. "Money-Free Economies and Ecosocialism." In *The Routledge Handbook on Ecosocialism*, edited by Leigh Brownhill, Salvatore Engel-Di Mauro, Terran Giacomini, Ana Isla, Michal Löwy, and Terissa Turner, 262–270. New York: Routledge.

Núñez, Miguel Ángel. 2021. *La Ciencia del Konuco y Su Visión Integral*. Caracas: Fonacit.

Núñez, Miguel Ángel. 2022a. "Contributions of Peasant Farmer Communities: The Case of Venezuela." *Science for the People* 25 (2). https://magazine.scienceforthepeople.org/vol25-1-the-soil-and-worker/contributions-of-peasant-farmer-communities-the-case-of-venezuela/.

Núñez, Miguel Ángel. 2022b. "Model Learning Spaces Towards Transitioning to Ecosocialism." In *The Routledge Handbook on Ecosocialism*, edited by Leigh

Brownhill, Salvatore Engel-Di Mauro, Terran Giacomini, Ana Isla, Michal Löwy, and Terissa Turner, 301–309. New York: Routledge.

O'Connor, James R. 1998. *Natural Causes: Essays in Ecological Marxism*. New York: Guilford Press.

Pepper, David 1984. *The Roots of Modern Environmentalism*. London: Croom Helm.

Pepper, David. 1993. *Eco-Socialism. From Deep Ecology to Social Justice*. London: Routledge.

Pepper, David. 2010. "On Contemporary Eco-Socialism." In *Eco-Socialism as Politics: Rebuilding the Basis of Our Modern Civilisation*, edited by Qingzhi Huan, 33–44. Dordrecht: Springer.

Raymundo, Sarah. 2022. "International Solidarity Against US Counterinsurgency." In *Sanctions as War. Anti-Imperialist Perspectives on American Geo-Economic Strategy*, edited by Stuart Davis and Immanuel Ness, 333–344. Leiden: Brill.

Ritchie, Hannah, Max Roser, and Pablo Rosado. 2020. "CO_2 and Greenhouse Gas Emissions." OurWorldInData.org, https://ourworldindata.org/co2-and-other-greenhouse-gas-emissions. Accessed 16 April 2024.

Ritchie, Hannah Max Roser, and Pablo Rosado. 2023. "CO_2 and Greenhouse Gas Emissions". Data adapted from Global Carbon Project, Various sources, https://ourworldindata.org/grapher/co-emissions-per-capita. Accessed 17 April 2024.

Rodríguez Cota, Fernán Enrique. 2018. *El Ecosocialismo. Teoría y práctica transformadora para promover una cultura ambientalista en la Universidad*. Caracas: El Perro y La Rana.

Ryle, Martin. 1988. *Ecology and Socialism*. London: Radius.

Saito, Kohei. 2022. *Marx in the Anthropocene: Towards the Idea of Degrowth Communism*. Cambridge: Cambridge University Press.

Satgar, Vishwas. 2018. "The Climate Crisis and Systemic Alternatives." In *The Climate Crisis: South African and Global Democratic Eco-Socialist Alternatives*, edited by Vishwas Satgar, 10–16. Johannesburg: Wits University Press.

Satgar, Vishwas, and Jacklyn Cock. 2022. "Ecosocialist Activism and Movements in South Africa." In *The Routledge Handbook on Ecosocialism*, edited by Leigh Brownhill, Salvatore Engel-Di Mauro, and Terran Giacomini, Ana Isla, Michael Löwy, and Turner Terisa, 179–188. Oxon and New York: Routledge.

Saul, Quincy. 2023. "Red, Black, and Green Destiny Weapon: Cooperation Jackson the Ecosocialist International." In *Jackson Rising Redux. Lessons on Building the Future in the Present*, edited by Kali Akuno and Matt Meyer, 464–479. Oakland: PM Press.

Schwartzman, David. 2011. "Green New Deal: An Ecosocialist Perspective." *Capitalism Nature Socialism* 22 (3): 49–56.

Schwartzman, David. 2013. "4 Scenarios for 2050." *Capitalism Nature Socialism* 24 (1): 49–53.

Schwartzman, David. 2016. "How Much and What Kind of Energy Does Humanity Need?" *Socialism and Democracy* 30 (2): 97–120.

Schwartzman, David. 2022. "Solar Communism." In *The Routledge Handbook on Ecosocialism*, edited by Leigh Brownhill, Salvatore Engel-Di Mauro, and Terran Giacomini, Ana Isla, Michael Löwy, and Turner Terisa, 280–289. Oxon and New York: Routledge.

Schwartzman, David, and Salvatore Engel-Di Mauro. 2022. "Prefiguration and the Emergence of the Global Subject." *Science & Society* 86 (4): 564–583.

Schwartzman, David, and Quincy Saul. 2015. "An Ecosocialist Horizon for Venezuela: A Solar Communist Horizon for the World." *Capitalism Nature Socialism* 26 (3): 14–30.

Schwartzman, David, and Peter Schwartzman. 2018. *The Earth is Not for Sale: A Path Out of Fossil Capitalism to the Other World That Is Still Possible.* Singapore: World Scientific Publishing.

SERA [The Socialist Environment and Resources Association]. 1980. *Eco-Socialism in a Nutshell.* London: Writers and Readers Publishing Cooperative, also available at https://climateandcapitalism.com/wp-content/uploads/2013/11/Ecosocialism-in-a-Nutshell.pdf. Last accessed 9 February 2024.

Sica, Carlo E. 2020. "For a Radical Green New Deal: Energy, the Means of Production, and the Capitalist State." *Capitalism Nature Socialism* 31 (4): 34–51.

Singh, Pritam. 2022. "Green Reforms and Individual Interventionsin the Green New Deal Transition to Ecosocialism." In *The Routledge Handbook on Ecosocialism*, edited by Leigh Brownhill, Salvatore Engel-Di Mauro, and Terran Giacomini, Ana Isla, Michael Löwy, and Turner Terisa, 235–241. Oxon and New York: Routledge.

Smith, Richard. 2021. "Six Theses on Saving the Planet." In *The New Systems Reader: Alternatives to a Failed Economy*, edited by James Gustav Speth, and Kathleen Courrier, 173–192. New York and London: Routledge.

Somerville, Peter. 2022. "Problems with Ecosocialism." *The Ecologist* December 5. https://theecologist.org/2022/dec/05/problems-ecosocialism. Accessed 7 January 2022.

Suing, Guillaume. 2018. *L'Écologie Réelle. Une Histoire Soviétique et Cubaine.* Paris: Éditions Delga.

Tanuro, Daniel. 2020. *Trop Tard our Être Pessimiste! Écosocialisme ou Effondrement*. Paris: Textuel.

Turner, Terisa E. 2022. "Ethnicized, Gendered Class Analysis: A Theoretical-Methodological Framework for Analyzing Ecofeminist, Ecosocialist Praxis."

In *The Routledge Handbook on Ecosocialism*, edited by Leigh Brownhill, Salvatore Engel-Di Mauro, and Terran Giacomini, Ana Isla, Michael Löwy, and Turner Terisa, 39–48. Oxon and New York: Routledge.

Vettese, Troy, and Drew Pendergrass. 2022. *Half-earth Socialism: A Plan to Save the Future from Extinction, Climate Change and Pandemics*. New York: Verso Books.

Wackernagel, Mathis and Bert Beyers, 2019. *Ecological Footprint: Managing the Biocapacity Budget,* Gabriola Island: New Society Publishers.

Wall, Derek. 2010. *The Rise of the Green Left: Inside the Worldwide Ecosocialist Movement*. London: Pluto

Wallis, Victor. 2001. "Toward Ecological Socialism." *Capitalism, Nature, Socialism* 12 (1): 127–145.

Wallis, Victor. 2018. *Red-Green Revolution: The Politics and Technology of Ecosocialism*, .ePUB edition. Toronto: Political Animal Press.

Wallis, Victor. 2020. *Socialist Practice: Histories and Theories*. London: Palgrave Macmillan.

Wallis, Victor. 2022. "Technology and Ecosocialism." In *The Routledge Handbook on Ecosocialism*, edited by Leigh Brownhill, Salvatore Engel-Di Mauro, and Terran Giacomini, Ana Isla, Michael Löwy, and Turner Terisa, 290–300. Oxon and New York: Routledge.

White, Damian. 2020. "Just Transitions/Design for Transitions: Preliminary Notes on a Design Politics for a Green New Deal." *Capitalism Nature Socialism* 31 (2): 20–39.

Wilpert, Gregory. 2022. "The US War on Venezuela." In *Sanctions as War. Anti-Imperialist Perspectives on American Geo-Economic Strategy*, edited by Stuart Davis and Immanuel Ness, 273–289. Leiden: Brill.

Witherspoon, Alex, Yu Zhou, and Alle Fang. 2022. "The Korean Miracle's Rural Legacy." *Cosmonaut Magazine*. https://cosmonautmag.com/2022/11/the-korean-miracles-rural-legacy/. Accessed 19 February 2024.

Wolf, Frieder O., and Michael Löwy. 2007. "Party-Building for Eco-Socialists: Lessons from the Failed Project of the German Greens." In *Coming to Terms with Nature: Socialist Register*, edited by Leo Panitch and Colin Leys, 310–336. London: The Merlin.

CHAPTER 10

Ecosocialist Prospects

Endless profit-making (capital accumulation) has led to devastating planetary effects ecologically and socially and is the principal impediment to address such effects seriously, much less prevent them. Simply put, compared to implementing preventative measures and ecologically sustainable living, destructive environmental impact is more lucrative, including by way of remediation programmes (when it is even feasible) and, more insidiously, renewable energy systems. This, among other realisations about how destructive capitalist relations are, is what has animated the development of ecosocialist perspectives and organising. These are rooted in socialist histories dating back to the early 1800s at least.

Ecosocialism is a movement that combines the struggles for socialism with those for ecological sustainability (Brownhill et al. 2022). It is a perspective that arose in the 1980s and traceable to social movements originating in the 1960s within liberal democracies, where it is still mainly confined. It grew from within a wider context of leftist opposition to the specifically neoliberal variants of capitalism prevailing by the late 1970s. More specifically, ecosocialism is a response to the often anti-socialist character of environmental movements and to existing socialist institutions or states, perceived as environmentally and socially destructive and/or not socialist at all. In other words, ecosocialism arose not just out of a repudiation capitalism, but also a rejection of other forms of socialism that, according to most ecosocialists, have lacked ecological concerns or have

S. Engel-Di Mauro, *Ecosocialism*, Marx, Engels, and Marxisms,
https://doi.org/10.1007/978-3-031-74288-0_10

contributed to the socio-ecological crisis. This, as argued in this book, vitiates ecosocialist prospects.

Several variants of ecosocialism have developed and exist in different parts of the world, but predominantly in the Americas, Europe, South Africa, and Australia. Their content may differ in relation to varied backgrounds and social contexts, but also because of the historical diversity of socialist movements. Much of the differentiation that has occurred is traceable to ecosocialists' relative commitment to more centralistic or more de-centralistic positions or principles. At times, the difference may be tacit, but is reflected in how Bolivarian Venezuela and socialist states are judged.

Relative to ecosocialist achievements so far, only state institutions in Bolivarian Venezuela feature ecosocialism as part of a country's policies and constitutional framework. Otherwise, ecosocialists are present on the fringes of Green, Leftist anti-capitalist, or social democratic parties. Rarely, as in the case of the Trotskyist Fourth International, ecosocialism is expressed as a platform common to a consortium of adhering political parties spread over many countries Otherwise, ecosocialists have created or take part of loose, often international affinity groups, or localised projects that have limited social influence and longevity. None of these relatively centralist or decentralist strategies have yet to demonstrate the level of effectiveness in terms of wider social and institutional reach that ecosocialism has in Venezuela. Some of this is in part related to the philosophical foundations underpinning ecosocialism (and that may be in need of some tweaking).

The principal philosophical current informing ecosocialism is a variant of Marxism called ecological Marxism or eco-Marxism. It is an essentially defensive effort to demonstrate the ecological content of mainly Marx's thought. Though largely drawing from the writings of Marx, ecosocialists sometimes take inspiration from Engels, and more sparsely from other socialist thinkers, including Utopian socialists. Another main philosophical strand is Ecofeminism. Ecofeminist Ecosocialists (also known as Ecosocialist Ecofeminists) build on the works of ecologically minded socialist feminists like Mies and indirectly D'Eaubonne. Ecofeminist Ecosocialists have drawn attention to the importance of defeating patriarchy to arrive at a just and ecologically sustainable society. Recently, there has been an increasing affinity with leftist degrowth perspectives so that an additional current of thought could become just as influential within ecosocialism.

Eco-Marxism is comprised of different schools of thought, differing over interpretations of how the works of Marx and Engels relate to environmental concerns. Some view Marx's original writings, including at times his unpublished notes, as providing the basis for a socialist ecological critique of capitalism and existing socialism. Much of this variant of eco-Marxism rests on the assumption that a correct reading of classical texts leads to better ecological policies and outcomes for socialist projects. It is also founded on alleging that ecological thought was snuffed in the USSR of the 1930s purges and that all socialist societies that followed went through a similar anti-ecological or ecologically insensitive trajectory. Other eco-Marxists, while agreeing with that position, see a need to elaborate further on Marx's and Engels' insights, considered necessary but insufficient for the ecosocialist cause. Hence, they draw from other kinds of adjacent political philosophies, such as socialist feminist, and neo-Marxist varieties. A few ecosocialists, especially within the last decade, have started recognising the importance of Marxist thinkers' ideas to the development of ecosocialism, beyond those of Marx and Engels. This is a line of research that awaits further exploration and may reveal a much greater ecological content to Marxism than has so far been appreciated.

It may thereby start to become clear to more ecosocialists that it would be more useful to expand one's horizons beyond Marx's and Engels' canonical texts and read later Marxist works addressing the biophysical sciences (e.g., Lenin's *Materialism and Empirio-Critcism*) as well as scientific ecology to create an ecologically informed form of socialism. This is a major challenge for eco-Marxism and ecosocialists generally. There are two manifestations to this challenge.

First, as some eco-Marxists have finally discovered for themselves, ecological thought was not extinguished in the USSR of the 1930s, nor thereafter.[1] One might consider, among numerous examples, the ecological preserves or *zapovedniki*, expanded under the Stalin government and based on ecological science. There are anyway other cases of Marxists elsewhere and in later periods, including those in power,

[1] Claiming the existence of a "late Soviet ecology" in the 1970s, furthermore, leaves unexplained how it arose, since it is asserted that ecology was suppressed. Moreover, the evidence marshalled is of ecologists or ecological thinkers purged or murdered by the Stalin administration. Yet most ecologists lived through that terrible period, continued with their scientific work in ecology, and contributed to environmental policy. This already would explain "late Soviet ecology", but as continuous with what was achieved throughout the previous periods, not some novel development.

who expressed deep ecological thinking and put it into practice as much as feasible, like Amílcar Cabral. This occurred without any eco-Marxist theory, but by applying Marxist method (dialectical and historical materialism). When eco-Marxists will look into socialist countries' policies and accomplishments, they will have plenty of illustrations with which to prove anti-Marxist environmentalists wrong about the ecologically crucial importance of a Marxist outlook.

Second, environmentally destructive cases under existing socialism have been decontextualised and used cynically for anti-socialist ends. The environmental records of socialist countries are at worst mixed, especially as a result of the need for (re)building infrastructure and rapid industrialisation to raise workers' living standards often in the aftermath of wars inflicted by liberal democratic or fascist powers. Despite all sorts of horrors, overall, the impacts of socialist countries have been more constructive than those of capitalist countries, including especially the imperialist liberal democracies, the most globally destructive entities in human history both socially and environmentally. In Cuba, there is a particularly impressive record of combining material well-being with ecological sustainability while under relentless US government siege. Ecosocialists who praise the achievements made in Cuba should at the very least reconcile that praise with their rejection of "first epoch" socialism.

The implication of the above is that finding ecological principles in Marx or Engels is not as significant as learning from the achievements and faults of post-revolutionary socialist countries. Uncovering allegedly suppressed or unpublished insights from Marx's writings provides no guidance for political strategy. Praxis[2] does, and only in positions of responsibility at the scale of countries, if the objective is changing the world. The further development of ecosocialist thought and politics is impeded by insisting on textual reinterpretation and/or borrowing from socialism-adjacent frameworks to the detriment of deepening knowledge about and building on Marxist theories, debates, policies, and practices in socialist countries and communist movements broadly.

Ecofeminism, the other main influential theoretical framework running through much of ecosocialism, tends to be poorly understood in wider

[2] The self-aware engagement with the mutually transforming relation of theory and action.

ecosocialism. Ecosocialist Ecofeminist thought emerged from the materialist current of Ecofeminism and further refined by a class-based understanding of society relative to multiple forms of oppression. Several theoretical frameworks have been forwarded to inform ecosocialist analysis and political action, including embodied materialism, meta-industrial labour theory, and the gendered ethnicised class analysis framework. These contributions and critiques of an often masculinist understandings imbuing eco-Marxism have been much less influential among ecosocialists compared to the metabolic rift school of eco-Marxism.

A major issue that remains to be clarified is the relationship between the objectives of replacing capitalism with ecosocialism and of patriarchy with Ecosocialist Ecofeminism. At times, the matter has even been presented in terms of overcoming socialism, as it is regarded as patriarchal as well or not to be necessarily gender egalitarian. In other words, a challenge may be to reconcile Ecosocialist Ecofeminism, based on the notion that capitalism is a recent form of patriarchy, with an understanding of capitalism as a social totality, rather than a derivative of an allegedly deeper kind of social structuring rooted in biologically related (but not determined) gendered differentiation.

This underlying theory of capitalist patriarchy, which is often not made explicit, seems to be a substantive and politically consequential divergence between Ecofeminist and other currents of ecosocialism that has yet to be fully addressed. The assumption of a gender-egalitarian society being necessarily class-undifferentiated may be as tenable as the notion of a class-free society being necessarily devoid of patriarchal relations. But this quandary can be multiplied several-fold. One can similarly incorporate the problem of the racialised mode of oppression as deeper and beyond capitalism as well as other sorts of relations of domination like heterosexuality and ablism. Yet social relations evolve as a unified whole, rather than as sets of independent axes that may or may not cross paths.

There are also some issues of explanation and political strategy that ensue from the theory of capitalist patriarchy. If patriarchal relations are viewed as arising millennia ago and undergoing mutations (the capitalist one being the latest), more attention should be given to determining how such mutations have occurred if the ultimate point is to overcome patriarchy, not just capitalism. In terms of political action, it could be useful to explain the time frame expected to be necessary to overturn thousands of years of patriarchy. Regardless, the ultimate aims proposed by Ecosocialist Ecofeminists need to be clearly articulated and reconciled with respect to ecosocialism generally if there is to be at least a coordination of ecosocialist efforts or projects. This is in addition to Ecofeminist Ecosocialism suffering from the same blind spots as eco-Marxism when it comes to socialist countries' achievements, including in the struggles against patriarchy.

Thus, the philosophical frameworks underlying ecosocialism, Eco-Marxism and Ecofeminism, could find room for improvement. More generally, ecosocialists could also draw much more from living examples of ecologically sustainable and socially egalitarian lifeways, but with due care. Many ecosocialists often point to First Peoples' perspectives and environmental practices as an inspiration and as showcase of what is already feasible, especially those of state-free peoples. However, there is an astonishingly large discrepancy between the importance ascribed to First Peoples and the scant details about First Peoples' perspective and struggles in the ecosocialist literature. An example of how to address and support First People's struggles is shown by The Red Nation, Yet few ecosocialists engage with that perspective. In some respects, unless ecosocialists devote most of their energies to overcoming settler colonialism, ecosocialism is largely irrelevant to First Peoples. The converse, however, is not the case; that is, for ecosocialism to develop, settler colonial systems must first be dismantled. The question of settler colonialism is important and regrettably it is not at the forefront of ecosocialist thought and therefore does not figure in discussions of strategy. However, for ecosocialists to be of use to First Peoples' or any people's self-determination struggles and at the same time build ecosocialism, there need to be, among other issues of political strategy and long-term goals, knowledge about First Peoples' histories and cultures, a thorough understanding of contradictions within and between First People communities, clarity about the relationship between ecosocialist objectives relative to the aims of national self-determination, knowledge of the context of the anti-colonial struggle and ecosocialists' place in it, and awareness about capitalist relations within and between First Peoples' communities. These are a few of the issues that to date have not been considered in ecosocialist writings, but should be crucial to developing an ecosocialist movement capable of involving First Peoples in a common struggle.

Another major challenge for ecosocialists is with respect to the People's Republic of China (PRC) and the recent instituting of policies aiming to establish an Ecological Civilisation. Though environmental and social injustices have occurred under various administrations in the PRC over the decades, there have also been lasting ecosocial advances that tend to be little appreciated by many ecosocialists living in liberal democracies. An under-appreciated issue is the fate of Ecological Civilisation efforts, which are having and will have global consequences. If successfully implemented, Ecological Civilisation policies would raise major political questions regarding the relevance of building ecosocialism compared to supporting the Communist Party of China's project for an ecologically sustainable socialist future. An important feature of this process for ecosocialists is to be discerning of the various currents that comprise the PRC's politics, including within the Communist Party of China, instead of painting the entire country or the governing political party as hopelessly capitalist or an internally undifferentiated political mass. Furthermore, given liberal democratic governments' military pressures and attempts at economic sabotage against the PRC, ecosocialists living in liberal democracies should consider whether their writings or political actions support the imperialist aims of the governments in their respective countries. The matter is similar to that of ecosocialists' often self-contradictory position regarding Cuba, where a lot of progress has been made towards ecological sustainability but under horrific circumstances imposed by US imperialist forces and their NATO satellite governments. The question to be posed regarding prospects for an ecosocialist world is how to position oneself politically with respect to the PRC's Ecological Civilisation project, but also with all socialist countries generally, and to think carefully about the consequences for ecosocialist objectives entailed in any actions that favour the US and other satellite liberal democracies' imperialistic designs.

A main problem with most existing ecosocialist thought is asserting that existing socialist countries deviate from Marxist or Ecofeminist principles and are therefore false or debased forms of socialism or are inadequate in the fight to overcome capitalist patriarchy. In all this, they differ little from Utopian socialism in that, by avoiding or distancing themselves from the contradictory and often terrible processes involved in actually gaining substantial social influence and responsibilities, they can only offer abstract principles, imagined future outcomes, or pontificate on the wrongs in existing socialist projects. A peremptory rejection of environmental practices and policies in socialist states, along with the historical experiences they represent, denies the possibility of building on (or, currently, being actively engaged with) historical processes and experiences involving tens of millions of people. Thus, a question for most ecosocialists becomes how to reconcile a quest for global social change, deemed necessary to confront the socio-ecological crisis, with an exclusive insistence on phenomena of globally marginal influence, often involving miniscule groupings.

Even if one were to insist on repudiating "first epoch" socialism, Bolivarian Venezuela should be of immense importance to all ecosocialists, since it is the highest achievement so far of any ecosocialist project. It is helpful in providing lessons for political strategy and for the development of efforts to join disparate ecosocialist efforts that value local autonomy with the benefits of coordination at national and international scales. The Bolivarian process can be a model on which to build and the springboard for viable political strategies to be implemented elsewhere, combining, as has been done in Venezuela, pre-existing collectivist or communal forms in a society with party structures capable of channelling popular discontent towards ecosocialism while organising within and taking over the military establishment to ensure the protection of political gains. This is so far the only model of political action that has brought ecosocialism into a country's state institutions and constitutional framework. It is perplexing that most ecosocialists still ignore or dismiss this major achievement and the political strategies that have rendered it possible. The biggest success so far in spreading ecosocialism and implementing ecosocialist policies at a national scale, in Bolivarian Venezuela, besieged by the US state and their NATO cronies, is arguably the most poorly supported by ecosocialists outside of Venezuela. Yet the Bolivarian process entails an attempt to bridge the divide between grassroots organising and state institutions that ecosocialists should find of decisive interest and seek to emulate.

When put under scrutiny, ecosocialist strategy and accomplishments, topics that tend to be avoided in ecosocialist writings, are few and far between and almost entirely inconsequential compared to the lofty goals trumpeted by ecosocialists themselves. The few ecosocialists who confront the issue of strategy default to parliamentary and/or pacifist roads and issue proposals remiss on historical experiences of communist or other revolutionary parties, including those involved in the making of the Bolivarian Revolution. The scarcity of attention ecosocialists give to inventorying their own accomplishments is troubling, as it is difficult to develop praxis without analysing what has been done well and what has gone wrong. Another aspect that should be more pressing to ecosocialists is addressing the ruling class violence that ensues whenever any gains are made, even potential ones. Typically, model cases of small geographical significance are indicated that often may bear similarities with aspects of ecosocialism but are not ecosocialist projects in themselves. It appears that, excepting Bolivarian Venezuela, very little has been achieved generally and of lasting consequence in terms of building ecosocialism at any scale.

There are, then, some deep challenges to building ecosocialism. There is a discrepancy between ecosocialist advocacy for, among other policies, a global redistribution of wealth and the analysis of concrete and discomforting linkages between and within imperialist countries and the rest of the world. Given the histories and mixed results of socialist projects, one could and should add that even organising to struggle for an ecosocialist transition courts contradictory developments, with potential gains and setbacks, even horrors. Somerville (2022) may surmise that Albert "exaggerates the technical and financial difficulties" and loses what is anyway, according to Somervile, largely a non—if not anti-socialist working class. Albert (2023) may fear the coming of an authoritarian turn in an imagined ecosocialist government facing emergencies—a government that would be elected, no less. Camfield (2022) reacting to Malm's direct action dreams, may harbour quaint preoccupations with the intensification of state repression brought on by the naughty actions of small unaccountable radical groups. These are political processes certainly worth addressing but may in most places be the least of ecosocialist concerns. The primacy of place given by such ecosocialists to the above-described challenges largely reflects the anxieties of those mainly living under liberal democratic regimes.

Since warfare is intrinsic to capitalism and capitalists and their institutional allies treat many people in unspeakably brutal ways as part of business as usual, it is strange that it does not dawn on most ecosocialists that not a few members of the ruling classes will have little compunction about having ecosocialists executed or murdered en masse if even so much as they attempt to realise but a small fraction of the changes they have in mind. The experiences of socialist struggles, those that came close to or succeeded in bringing about revolutions, have always been scarred by capitalist terror and war. This is why "exaggerating the difficulties", as Somerville (2022) puts it, is much more prudent, even if regrettably other ecosocialists fail to recognise even the most rudimentary challenges like military intervention or police repression, than pretending away the ends to which capitalist ruling classes have proven to be willing to go to defend their political power and depredation system. Appealing, as some do, to some likelihood of such "exaggerations" discouraging people from joining ecosocialist movements, especially in the light of repressive measures against environmentalists and of hundreds of murders each year of environmental activists, including in liberal democracies, seems out of touch with many, possibly most people's everyday life and with the histories of socialist movements. When dealing with a form of society with a magnitude of deadliness that the world has never before witnessed, it is unconscionable not to prepare for the worst.

A general underlying problem with much of ecosocialism, excepting in Venezuela, is often the absence of concrete practices and of contextual analysis of what is feasible in the present and of what was feasible in the past. Much of ecosocialist activity is concentrated on publishing critiques of capitalism, sometimes also of former and existing socialist states, and of other socialist strategies that allegedly downplay or omit environmental questions. When more engaged in delineating aims, ecosocialist thinkers and formations tend to be more involved in drawing up future possibilities and delineating desirable ends and political programmes than building movements or occupying institutional posts that could enable the realisation of ecosocialist objectives. This imbalance in favour of theory and imagination (or vision) hinders the development of praxis. This situation could be reversed if all ecosocialists focused their energies on overcoming their existing fragmentation (which entails setting aside differences for the sake of increasing political reach and effectiveness), learning from and supporting Bolivarian Venezuela, and appreciating achievements and contributing constructively to the development of policies in socialist countries. These could be ways for ecosocialists to put together and carry out projects worthy of the globally transforming aims to which they aspire.

References

Albert, Michael J. 2023. "Ecosocialism for Realists: Transitions, Trade-Offs, and Authoritarian Dangers." *Capitalism Nature Socialism* 34 (1): 11–30.

Brownhill, Leigh, Salvatore Engel-Di Mauro, Terran Giacomini, Ana Isla, Michael Löwy, and Terisa E. Turner. 2022. "Preface: An Introduction to Ecosocialism." In *The Routledge Handbook on Ecosocialism*, edited by Leigh Brownhill, Salvatore Engel-Di Mauro, Terran Giacomini, Ana Isla, Michael Löwy, and Terisa E. Turner, 1–12. New York: Routledge.

Camfield, David. 2022. *Future on Fire: Capitalism and the Politics of Climate Change*. Oakland: PM Press.

Somerville, Peter. 2022. "Problems with Ecosocialism.' *The Ecologist*, December 5. https://theecologist.org/2022/dec/05/problems-ecosocialism. Accessed 7 January 2022.

Index

S. Engel-Di Mauro, *Ecosocialism*, Marx, Engels, and Marxisms,
https://doi.org/10.1007/978-3-031-74288-0

SPRINGER NATURE

GPSR Compliance

The European Union's (EU) General Product Safety Regulation (GPSR) is a set of rules that requires consumer products to be safe and our obligations to ensure this.

If you have any concerns about our products, you can contact us on ProductSafety@springernature.com

In case Publisher is established outside the EU, the EU authorized representative is:

Springer Nature Customer Service Center GmbH
Europaplatz 3
69115 Heidelberg, Germany

The manufacturer's authorised representative in the EU is Springer Nature Customer Service Centre GmbH, Europaplatz 3, 69115 Heidelberg, Germany. If you have any concerns regarding our products, please contact ProductSafety@springernature.com

Printed and bound by CPI Group (UK) Ltd, Croydon, CR0 4YY
30/09/2026
02225595-0011